A Review of Science and Technology During the 1971 School Year

Science Year
The World Book Science Annual

1972

Field Enterprises Educational Corporation
Chicago   London   Rome   Sydney   Tokyo   Toronto

The publishers of *Science Year* gratefully
acknowledge the following for permission to use
copyrighted illustrations. A full listing of illustration
acknowledgments appears on pages 442 and 443

22 © Paul Peck
23 © Copyright 1969 CIBA Pharmaceutical
    Company, Division of CIBA Corporation
    Reproduced with permission.
    From the CIBA COLLECTION OF MEDICAL
    ILLUSTRATIONS by Frank H. Netter, M.D.
24 © RASSEGNA MEDICA LEPETIT,
    Milan, Italy, 1970, *Rassegna
    Medica E Culturale* 1970.
    Plates by Giorgio Gordoni,
    drawings by Elisa Paternagni.
71 © California Institute of Technology and Carnegie
    Institution of Washington. From Hale Observatories
78 © California Institute of Technology and Carnegie
    Institution of Washington. From Hale Observatories
233 © Peter Clayton
257 James Stevenson © 1971 The New Yorker Magazine, Inc.
260 © 1971 Alexander Marshack
276 Chas. Addams © 1971 The New Yorker Magazine, Inc.
283 Booth © 1970 The New Yorker Magazine, Inc.
301 Lorenz © 1970 The New Yorker Magazine, Inc.;
    Shirvanian © 1970 The New Yorker Magazine, Inc.;
    Richter © 1971 The New Yorker Magazine, Inc.;
    © *Punch,* London; © *Punch,* London
322 © 1970 by the New York Times Company. Reprinted by
    permission
323 Ed Fisher © 1970 Saturday Review, Inc.
324 © 1970 by the New York Times Company. Reprinted by
    permission
338 W. Miller © 1971 The New Yorker Magazine, Inc.
374 Ed Fisher © 1970 The New Yorker Magazine, Inc.

The Cover: Detail from the painting by Frank Armitage for the
Special Report "Cell Fusion and the New Genetics."
See page 190.

Preface

Each year the staff of *Science Year* labors to bring forth a publication that is interesting, informative, and attractive. It is never an easy birth, but it is a successful one. Much of the success is due to the obstetrical teamwork between editors skilled in delivering clear and concise text and artists skilled in design and layout.

The artist's role in *Science Year* is to conceive of the best use of visual aids—paintings, photographs, diagrams, charts, and maps—that will complement and supplement the text. In science, with its difficult concepts, a creative illustration can greatly aid understanding.

The use of illustration to explain science has its roots in anatomy and medicine. The 1972 edition of *Science Year* explores this with a picture essay on the rich history of medical illustration. The essay traces man's first crude attempts to draw bodily organs as religious symbols, with examples from the ancient Chinese and Etruscans, through the beginning of the Renaissance, when restrictions on dissecting were relaxed and accurate drawings could be made. It was this freedom that permitted Leonardo daVinci and his successors to create the first accurate sketches of the human body.

Over the next several centuries, dissection techniques and methods of reproduction continued to improve, and medical illustration grew more accurate, more detailed, and more valuable to the medical student. Today, the medical illustrator must be more than just an accurate anatomist, copying organs and tissues. He must deal with the submicroscopic world of the cell, where he must combine his training with his imagination. As an example, one of these modern artists, Frank Armitage, created three outstanding paintings for the Special Report; "Cell Fusion and the New Genetics." They depict, as no camera could, the intricate mechanisms by which the basic units of life act, react, and interact.

This is not to demean the camera, for it is an invaluable aid in creating scientific illustration. Photography can capture the breadth of science; from the lark on the prairie to the dolphin in the sea; from the mysteries of the universe through the telescope to the mysteries of the cell through the microscope.

The speed and versatility of a camera in the hands of an expert photographer can provide a fascinating record of science in progress. For example, photographer Ed Hoppe took his camera into the operating room and recorded an eight-hour heart operation. As a result, readers of "New Life Line for the Heart" can look over the surgeons' shoulders and watch the step-by-step process in which their skilled hands bring new life to a failing heart.

One of the most rewarding parts of watching the birth of a *Science Year* is to see its pages come alive with the imaginative use of illustration, the blending of color and design. There is much beauty in science. And one of the aims of *Science Year* is to share that beauty with its readers.   ARTHUR G. TRESSLER

Contents

Dr. Walsh McDermott is Livingston Farrand Professor of Public Health and Chairman of the Department of Public Health at the Cornell University Medical College. He received his B.A. degree from Princeton University in 1930 and his M.D. degree from Columbia University's College of Physicians and Surgeons in 1934. In 1967, Dr. McDermott was named chairman of the Board of Medicine of the National Academy of Sciences. He is editor of the *American Review of Respiratory Diseases*.

Roger Revelle is Richard Saltonstall Professor of Population Policy and director of the Center for Population Studies at Harvard University. He received a B.A. degree from Pomona College in 1929 and a Ph.D. degree from the University of California in 1936. He is former chairman of the U.S. National Committee for the International Biological Program. Much of his current work involves studies on the crisis in world hunger and malnutrition.

Allan R. Sandage is astronomer and a member of the observatory committee of the Mount Wilson and Palomar Observatories. He received a B.A. degree at the University of Illinois in 1948 and a Ph.D. degree in astronomy at the California Institute of Technology in 1953. Dr. Sandage is a member of the Royal Astronomical Society and the National Academy of Sciences. He is perhaps most noted for his discovery of quasi-stellar objects in the far reaches of space. He found the first "quasar" in 1960, and the first X-ray star in 1966.

Alvin M. Weinberg is director of the AEC's Oak Ridge National Laboratory in Tennessee. He is a graduate of the University of Chicago, 1935, and earned his doctorate in physics there in 1939. After three years of researching and teaching mathematical biophysics at Chicago, he joined the Metallurgical Laboratory (part of the Manhattan Project) in 1941. He moved to Oak Ridge in 1945, becoming director of the Physics Division in 1947, research director in 1948, and director of the laboratory in 1955.

Contributors

Auerbach, Stanley I., Ph.D.
Director, Ecological Sciences Division
Oak Ridge National Laboratory
Ecology

Barbour, John, B.A.
Science Writer
Associated Press
Probing the Pawnee Grassland

Bardach, John E., Ph.D.
Director, Hawaii Institute of
Marine Biology
University of Hawaii
Livestock from the Sea

Bass, Allan D., M.D.
Chairman, Pharmacology Department
Vanderbilt University
Drugs

Belton, Michael J. S., Ph.D.
Associate Astronomer
Kitt Peak National Observatory
Astronomy, Planetary

Bell, Joseph N., B.A.
Free-Lance Writer
Eugene Shoemaker

Boffey, Philip M., B.A.
Staff Writer
Science Magazine
The World Series of Science;
Science Support

Bromley, D. Allan, Ph.D.
Professor and Chairman, Department
of Physics, and Director, Wright
Nuclear Structure Laboratory
Yale University
Physics, Nuclear

Bryson, Reid A., Ph.D.
Director of the Institute for
Environmental Studies
University of Wisconsin
Is Man Changing His Climate?

Budnick, Joseph, I., Ph.D.
Professor of Physics
Fordham University
Physics, Solid State

Cowen, Robert C., M.S.
Natural Sciences Editor,
Christian Science Monitor
A New Look at Prehistoric Europe

Cromie, William J., B.S.
Senior Vice-President
Universal Science News, Inc.
The Moon's Unfolding History

Deason, Hilary, J., Ph.D.
Director of Libraries
American Association for the
Advancement of Science
Books of Science

DeBakey, Michael E., M.D.
President and Chairman of the
Department of Surgery
Baylor College of Medicine
Medicine, Surgery

Drake, Charles, Ph.D.
Professor of Geology
Dartmouth College
Geophysics

Edgar, N. Terence, Ph.D.
Chief Scientist
Deep Sea Drilling Project
Scripps Institution of Oceanography
Drilling into the Ocean Floor

Edgington, David N., D.Phil.
Associate Chemist
Argonne National Laboratory
Close-Up, Ecology

Edson, Lee, B.S.
Free-Lance Writer
Attempting to Understand the Brain

Ensign, Jerald C., Ph.D.
Associate Professor of Bacteriology
University of Wisconsin
Microbiology

Evans, Earl A., Jr., Ph.D.
Professor and Chairman,
Department of Biochemistry
University of Chicago
Biochemistry

Goss, Richard J., Ph.D.
Professor of Biological and
Medical Science
Brown University
Zoology

Gray, Ernest P., Ph.D.
Chief, Theoretical Plasma Physics
Staff, Applied Physics Laboratory
Johns Hopkins University
Physics, Plasma

Griffin, James B., Ph.D.
Director, Museum of Anthropology
University of Michigan
Archaeology, New World

Hawthorne, M. Frederick, Ph.D.
Professor of Chemistry
University of California, Los Angeles
Chemistry, Synthesis

Henahan, John F., B.S.
Senior Editor,
Chemical and Engineering News
Close-Up, Chemical Technology

Hilton, Robert J., D.D.S.
Associate Professor
Operative Dentistry Department
Northwestern University
Medicine, Dentistry

Hines, William
Science Correspondent
Chicago Sun-Times
Space Exploration

Isaacson, Robert L., Ph.D.
Professor of Psychology and
Neurosciences
University of Florida
Psychology

Jahns, Richard H., Ph.D.
Professor of Geology and Dean,
School of Earth Sciences
Stanford University
Geology

Kessler, Karl G., Ph.D.
Chief, Optical Physics Division
National Bureau of Standards
Physics, Atomic and Molecular

Lapp, Ralph E., Ph.D.
Secretary, Quadri-Science Incorporated
The Cultivation of Technology

Lederberg, Joshua, Ph.D.
Professor and Chairman,
Department of Genetics and
Director of the Kennedy Laboratories
for Molecular Medicine
Stanford University
Cell Fusion and the New Genetics

Lo, Arthur W., Ph.D.
Professor of Electrical Engineering
Princeton University
Computers

Lockwood, Linda Gail, Ph.D.
Assistant Professor of Natural Sciences
Teachers College, Columbia University
Education

Maran, Stephen P., Ph.D.
Head, Advanced Systems and Ground
Observations Branch
Goddard Space Flight Center
Astronomy, Stellar

March, Robert H., Ph.D.
Associate Professor of Physics
University of Wisconsin
Physics, Elementary Particles

Marsh, Richard E., Ph.D.
Senior Research Fellow (Chemistry)
California Institute of Technology
Chemistry, Structural

Melloni, Biagio John, Ph.D.
Director, Department of Medical-Dental
Communication
Schools of Medicine and Dentistry
Georgetown University
Anatomy and the Artist

Merbs, Charles F., Ph.D.
Associate Professor of Anthropology
University of Chicago
Anthropology

O'Neill, Eugene F., M.S.E.E.
Executive Director
Transmission Systems Development
Division
Bell Telephone Laboratories
Communications

Price, Frederick C., B.S.
Managing Editor,
Chemical Engineering Magazine
Chemical Technology

Randal, Judith, B.A.
Science Writer
The Washington Star
Albert Szent-Györgyi

Reese, Kenneth M., Ch.E.
Free-Lance Writer
Close-Up, Oceanography

Rodden, Judith, M.Litt.
Research Archaeologist
Archaeology, Old World

Romualdi, James P., Ph.D.
Professor of Civil Engineering and
Director of Transportation Research
Institute, Carnegie-Mellon University
Transportation

Ross, John E., Ph.D.
Associate Director, Institute
for Environmental Studies
University of Wisconsin
Is Man Changing His Climate?

Schücking, E. L., Dr.rer.nat.
Professor of Physics
New York University
Astronomy, High Energy; Cosmology

Snider, Arthur J., M.S.
Science Writer
Chicago Daily News
New Life Line for the Heart

Spar, Jerome, Ph.D.
Professor of Meteorology
New York University
Meteorology

Spinrad, Hyron, Ph.D.
Professor of Astronomy
University of California, Berkeley
Close-Up, Astronomy, Stellar

Steere, William C., Ph.D.
President, New York Botanical Garden
Professor, Columbia University
Botany

Stumpf, John H., B.A.
Special Projects Editor,
Nuclear Industry
Energy

Sulzberger, Marion B., Ph.D.
Professor Emeritus of New York
University and Technical Director
of Research, Letterman General
Hospital, San Francisco
The Unwelcome Badge of Adolescence

Sutton, H. Eldon, Ph.D.
Professor and Chairman,
Department of Zoology
University of Texas, Austin
Genetics

Temin, Howard M., Ph.D.
Professor of Oncology
University of Wisconsin
Close-Up, Biochemistry

Thaddeus, Patrick, Ph.D.
Research Physicist
Goddard Institute for Space Studies
Adjunct Professor of Physics
Columbia University
Molecules in Space

Tilton, George R., Ph.D.
Professor of Geochemistry
University of California,
Santa Barbara
Geochemistry

Treuting, Theodore F., M.D.
Professor of Medicine
Tulane University School of Medicine
Medicine, Internal

Vetter, Richard C., M.A.
Executive Secretary, Ocean Affairs
Board, National Academy of Sciences
Oceanography

Weber, Joseph, Ph.D.
Professor of Physics
University of Maryland
Pulses of Gravity

Weber, Samuel, B.S.E.E.
Executive Editor,
Electronics Magazine
Electronics

Weinberg, Alvin, Ph.D.
Director, Oak Ridge National Laboratory
Curbing the Energy Crisis

Wittwer, Sylvan H., Ph.D.
Director, Michigan State University
Agricultural Experiment Station
Agriculture

Zare, Richard N., Ph.D.
Professor of Chemistry
Columbia University
Chemistry, Dynamics

Contributors not listed on these pages
are members of the *Science Year*
editorial staff.

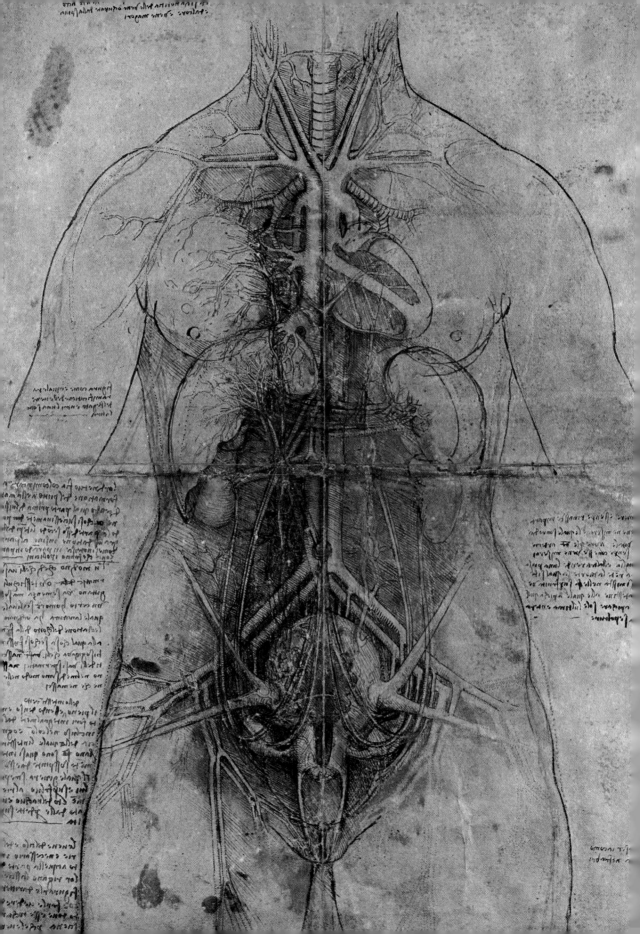

Anatomy and The Artist

By Biagio John Melloni

Illustrating the human body has progressed from crude symbolism to sophisticated reproductions requiring both medical knowledge and artistic skill

While Christopher Columbus was braving the unknown Atlantic Ocean to find a new route to the Indies, another Italian was charting new courses in the human body. In 1492, at the age of 40, Leonardo da Vinci was studying and systematically recording the body's structures. During his lifetime, he made nearly 800 anatomical sketches with detailed, written descriptions based upon his dissection of some 30 cadavers. The originality and comprehensiveness of his work earned him the title Father of Medical Illustration.

Prior to Leonardo's work, medical illustration was generally artistic, social, or religious, but not intended to teach medical facts. Something approaching authentic medical illustration appeared in Chinese diagrams detailing *acupuncture* (the piercing of skin or tissues) around 2700 B.C., in Greek vase paintings of the time of Hippocrates, the 400s B.C., and in Arabic manuscripts of the Alexandrian Age, about 300 B.C. However, the best of them are elementary and inaccurate, for those times were dominated by mysticism and religious dogma.

Accurate drawings were possible to do only after the restrictions of church and state on human dissection began to relax. This happened at the beginning of the Renaissance period, which was the cradle of true medical illustration. In 1543, 24 years after the death of Leonardo, the most complete and best-illustrated treatise on anatomy yet produced was published in Basel, Switzerland. This textbook, *Concerning the Fabric of the Human Body*, by Andreas Vesalius, contained beautifully executed woodcuts that set a standard of excellence which would endure for centuries. In some of the plates, the human figures were posed in lifelike attitudes, with landscape backgrounds.

The simple 16th century drawing materials—this sketch done on dirty brown paper—did not hamper the skills of Leonardo da Vinci.

Bronze casting of an Etruscan liver shows divisions representing the regions of heaven. Most early medical art served the purpose of teaching religion, rather than anatomy.

The author:

Biagio John Melloni is Director, Department of Medical-Dental Communication at Georgetown University, and Director, Archive of Medical Visual Resources at the Francis A. Countway Library of Medicine of Harvard University.

The book revolutionized the teaching of anatomy. The illustrations are by some of the foremost artists of the time. But Vesalius was a petulant and jealous man, and he refused to acknowledge these artists. Nevertheless, historians credit some of the pupils of Titian, especially Jan Stephan van Calcar, with contributions to the book.

During the 1700s and 1800s, a number of physician-scientists, such as Charles Bell and Joseph Lister, created illustrations for their own books. Some of the artists who helped develop the visual language of medicine are better known for their eminence in other fields. An example is Sir Christopher Wren, the architect, whose drawings of the circle of Willis, at the base of the brain, long stood as a classic work.

During the late 1800s, there was a new surge in medical illustration spearheaded by the detailed work of Bruno Héroux and H. Unger who illustrated Werner Spalteholz' *Hand Atlas of Anatomy* published in three volumes from 1895 to 1903. These textbooks made use of the new technique of lithography. The illustrations represented a giant step forward in detailing the human body as the dissecting techniques improved and better methods of preserving cadavers were found. K. Hajek provided such accurate paintings in Johannes Sobotta's *Atlas of Human Anatomy*, published in 1904, that they are still widely used in medical and dental schools. During this period, Franz Froshe produced superb anatomical wall charts, which also are still in use.

Until the middle of the 1800s, the United States had little need to develop medical illustration because of the material available from Europe. Then came the Civil War and the establishment of the U.S. Army's pathology laboratory. The large numbers of wounded soldiers provided both a purpose and models needed to train a new group of medical artists. Medical literature of this period contains many of the drawings of Herman Faber.

In 1894, 23-year-old Max Brödel was invited to the United States from Germany. He spent the next 48 years at Johns Hopkins Medical Center in Baltimore, Md., where he raised medical illustration to a new level of accuracy and detail. Brödel realized that independence of judgment and originality of conception were only to be gained by original study. He systematically and repeatedly dissected and studied many regions of the body. He succeeded because of his self-discipline and endless patience, and because he was in the center of a huge medical complex where he had access to the most brilliant minds of the time.

Although Brödel's contributions were significant in teaching students anatomy, physiology, and surgery, nothing gave him more satisfaction than training a "new generation of artists to illustrate the medical journals and books of the future and to spare them the years of trials and disappointments of their self-taught predecessors." Many great artists around the world flocked to Johns Hopkins to study under the undisputed master.

Another important American medical illustrator was Tom Jones of Chicago. He developed medical illustration in coordination with other media—two- and three-dimensional material for motion picture films, filmstrips, videotapes, scientific exhibits, and lantern slides as an aid in the teaching of medicine. Because of the "quick-view" nature of these illustrations, Jones designed them so that their important points could be quickly grasped.

In 1928, while Brödel and Jones were enjoying international fame, a young artist, Frank H. Netter, was enrolled at New York University medical school in New York City. He found he could study his subjects better by making sketches. After receiving his M.D. degree, he continued to draw. Netter exemplifies the modern medical illustrator who must contend not only with the structure of tissues, which he can see, but with the microscopic and submicroscopic aspects of the body which he cannot see. Moreover, he must convey these complexities in simple, graphic language.

With the constant development of better art and reproduction techniques, outstanding medical illustrations are being made throughout the world today. In the spirit of Leonardo da Vinci, medical illustrators continue to create new ways to chart man's body in its infinite architecture.

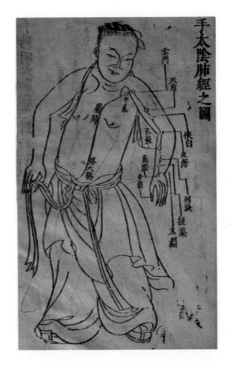

Chinese medicine of the 1000s featured acupuncture—piercing the tissues to clear the blood and air passages, *above*. Crude European drawing from the 1200s, *below*, shows bone, muscles, and circulatory system.

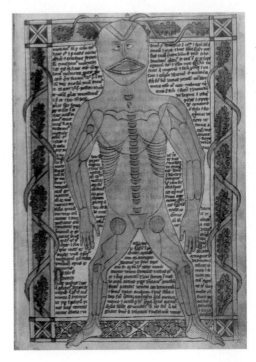

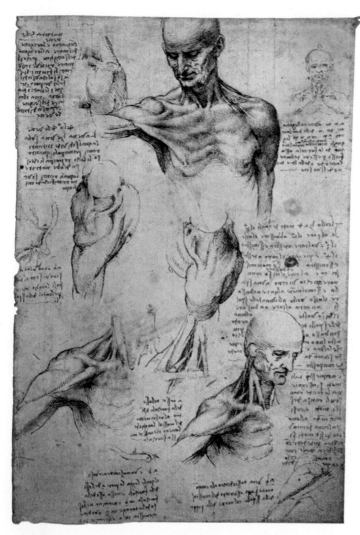

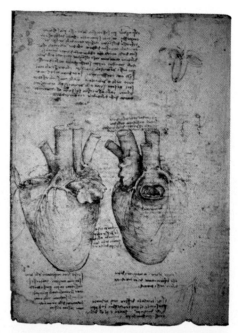

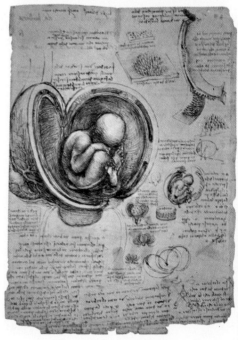

Leonardo da Vinci's amazing skill in depicting the details of musculature, *above,* the heart, *above right,* and a fetus, *right,* seem even more remarkable considering the handicaps he worked under—too few daylight hours, unembalmed cadavers, and no prior knowledge of anatomy. The ambidextrous Leonardo wrote copious notes on his sketches with his left hand—but backward, so that they could not be readily plagiarized.

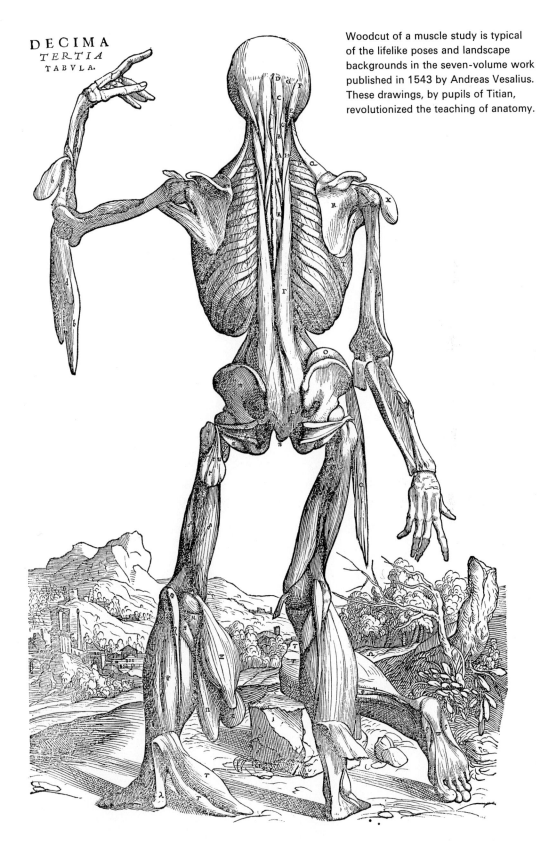

Woodcut of a muscle study is typical of the lifelike poses and landscape backgrounds in the seven-volume work published in 1543 by Andreas Vesalius. These drawings, by pupils of Titian, revolutionized the teaching of anatomy.

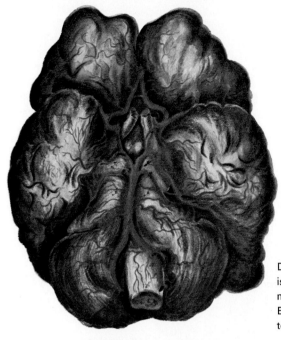

Drawing of the divisions of the brain is an illustration for the 19th century manuscript on anatomy by Sir Charles Bell. This surgeon, anatomist, and teacher had no formal art training.

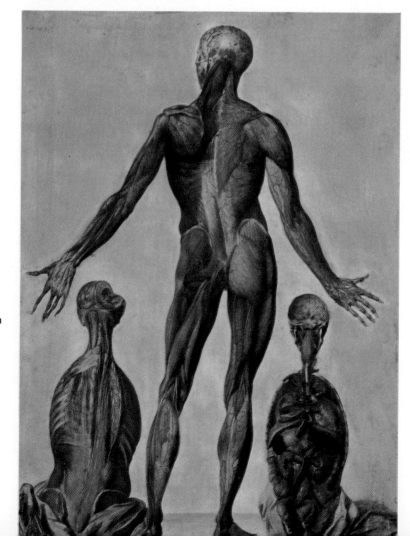

Crayon drawing by Dutch artist Jan Van Riemsdyk is one of a series eventually given to the Pennsylvania Hospital in Philadelphia in 1775. The drawings were used in the earliest efforts at medical education in the new United States.

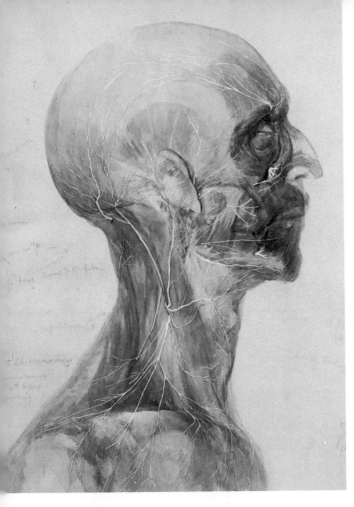

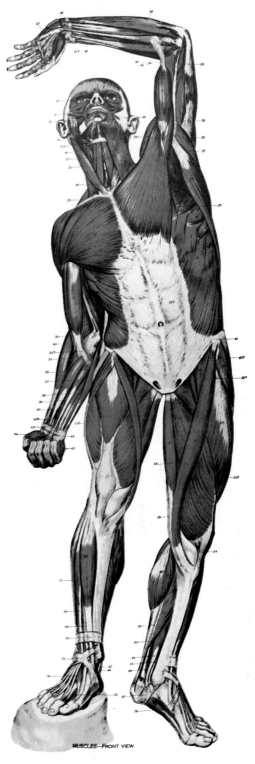

Pencil and water-color drawing of the nerves of the head is by Herman Faber, one of America's first medical illustrators. Faber sketched wounds and treatments of soldiers during the Civil War.

Lateral view of the brain is by Bruno Héroux, a principal illustrator of Werner Spalteholtz' *Hand Atlas of Anatomy*. Héroux introduced fine art to anatomy.

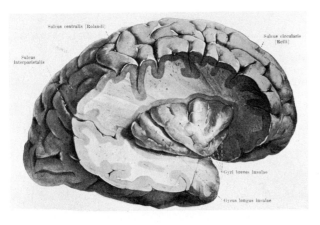

MUSCLES—FRONT VIEW

Anatomical wall charts were the chief work of German artist Franz Froshe. Later retouched by Brödel, these charts are widely used in classrooms today.

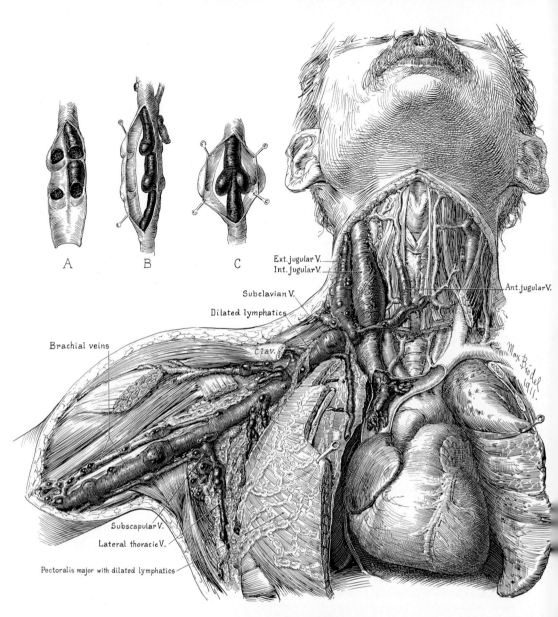

A   B   C

Ext. jugular V.
Int. jugular V.
Ant. jugular V.
Subclavian V.
Dilated lymphatics
Brachial veins
Clav.
Subscapular V.
Lateral thoracic V.
Pectoralis major with dilated lymphatics

Whether in the details of the cause of death from a blood clot in a vein, *above,* or in the reconstruction of a liver, *right,* the genius of Max Brödel displays an exceptional amount of patience and self-discipline. This German-born artist not only brought medical art to a new degree of accuracy, but also left a priceless legacy of pupils trained to illustrate the growing medical literature.

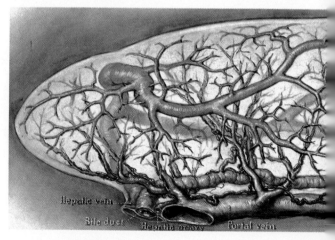

Hepatic vein
Bile duct   Hepatic artery   Portal vein

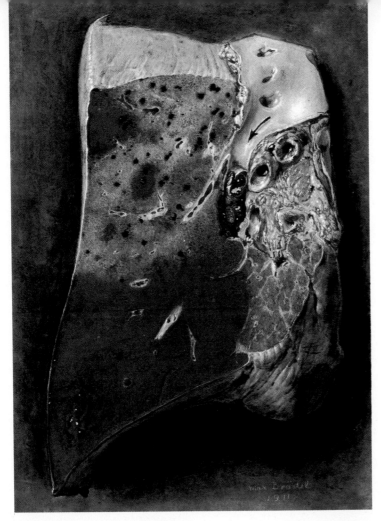

With his wide-ranging skills, Brödel
could deliver a water-color portrait
of the flesh that was astonishingly
lifelike, *above,* or a clinically
detailed representation of a disease
and its surgical correction, *right.*
He reproduced his subjects faithfully,
because he studied them endlessly.

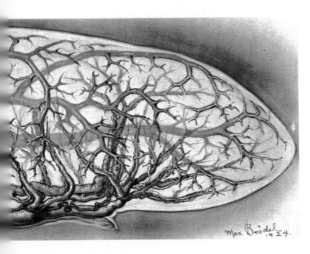

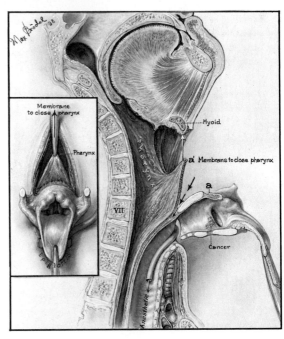

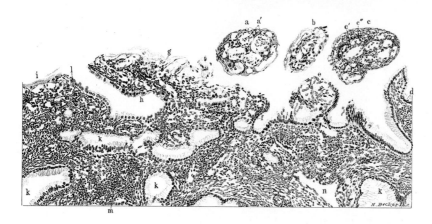

A schoolmate of Brödel's at Leipzig Academy was Hermann Becker who joined him at Johns Hopkins. Becker became a master of microscopic drawings such as this ink-and-wash portion of the cervix.

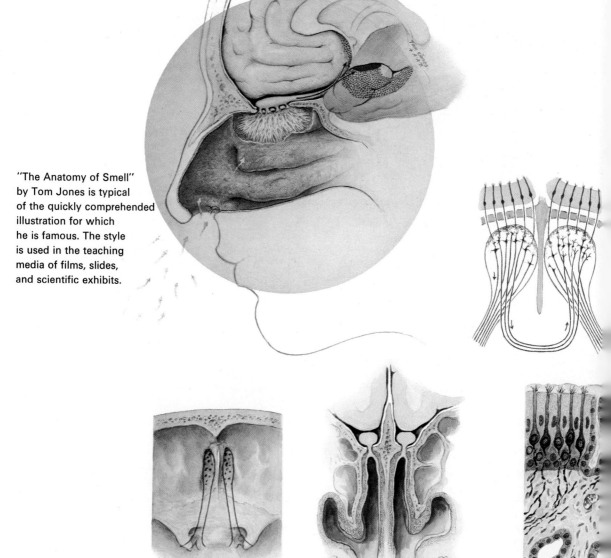

"The Anatomy of Smell" by Tom Jones is typical of the quickly comprehended illustration for which he is famous. The style is used in the teaching media of films, slides, and scientific exhibits.

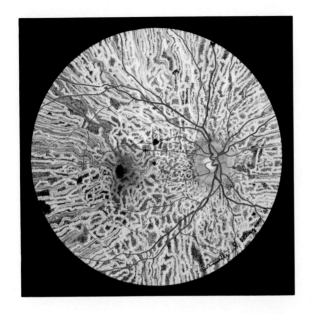

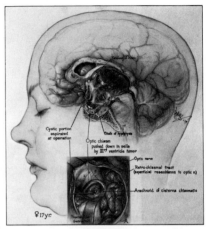

Successful students of
Brödel included Annette
Burgess, an illustrator
of ophthalmology, *above
left,* and Dorcus Padget,
whose early drawings
were concentrated in
neurosurgery, *above.*

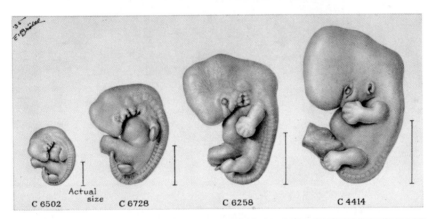

C 6502   Actual size   C 6728   C 6258   C 4414

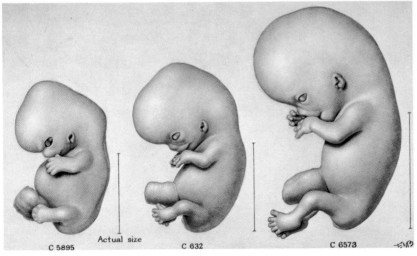

C 5895   Actual size   C 632   C 6573

Stages in development
of the human fetus by
Elizabeth Brödel shows
that she learned well
from her famous
father the value of
attention to detail.

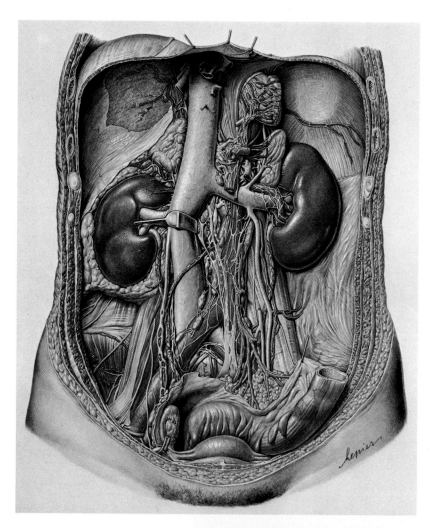

Detailed water-color rendering
of blood vessels, nerves, and
lymphatics is by contemporary
Austrian illustrator Erich Lepier.

This microscopic view of the
weblike bone structure is one
way Paul Peck visualizes the
delicacy and beauty that is
inherent in the human body.

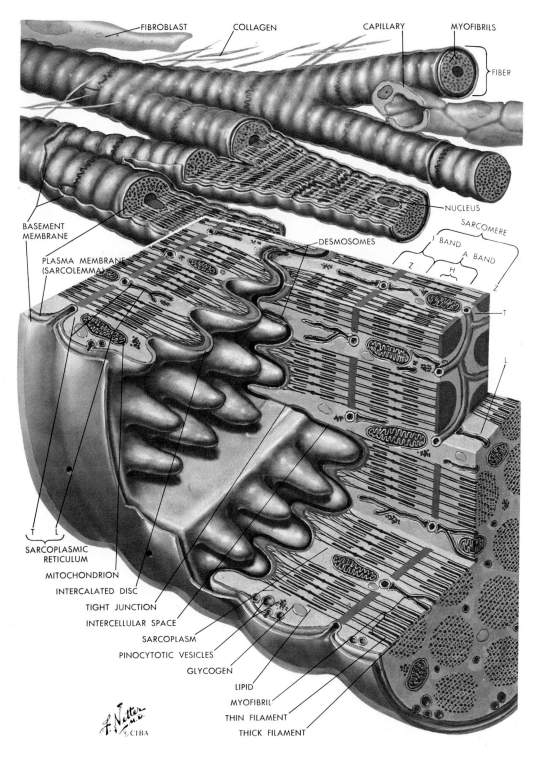

FIBROBLAST   COLLAGEN   CAPILLARY   MYOFIBRILS   FIBER   NUCLEUS

BASEMENT MEMBRANE

PLASMA MEMBRANE (SARCOLEMMA)

DESMOSOMES

SARCOMERE
I BAND   A BAND
Z   H   Z
T
L

T   L
SARCOPLASMIC RETICULUM
MITOCHONDRION
INTERCALATED DISC
TIGHT JUNCTION
INTERCELLULAR SPACE
SARCOPLASM
PINOCYTOTIC VESICLES
GLYCOGEN
LIPID
MYOFIBRIL
THIN FILAMENT
THICK FILAMENT

Details of a portion of the cardiac muscle is in the style of a leading
modern illustrator, Dr. Frank Netter. He believes that medical illustration
not only teaches surgery, but has also contributed to its development.

Illustration of the inner ear by the author, left, depicts trend in modern medical illustration to strip away extraneous elements and concentrate on the detailed portrayal of pertinent parts.

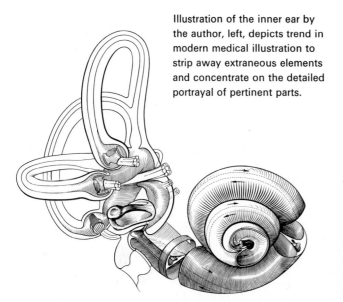

The explosion of knowledge in cell biology is reflected in work of artists such as Italy's Elisa Patergnagni. Her drawings, right, attempt to visualize the miniature cellular world.

This extraordinary view of the human eye with a bacterial infection is the interpretative work of another of today's modern illustrators, Frank Armitage. For more examples of his technique see "Cell Fusion and the New Genetics."

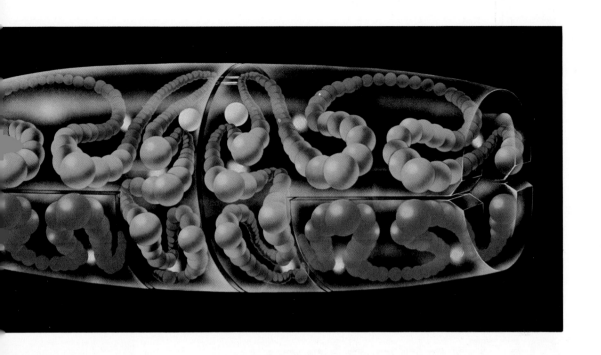

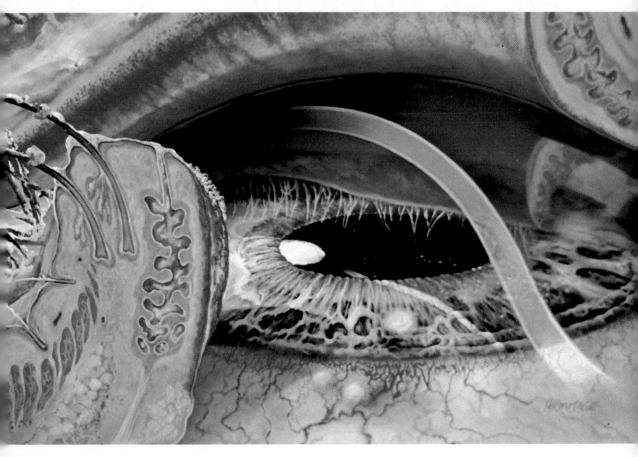

The Cultivation Of Technology

By Ralph E. Lapp

**Technology can be guided to benefit mankind,
but only if scientists, legislators, and the public
work toward understanding it and each other**

Our senses are trying to tell us that something is radically wrong
with the world around us. Our eyes smart from the fumes exhausted
by millions of automobile engines and industrial smokestacks. Our
noses wrinkle from the noxious reek of polluted streams, rivers, and
lakes. Our eardrums reverberate with the rumbling of subways or the
piercing shriek of powerful jet engines.

More subtle hazards escape our sensory perceptions. We cannot see,
smell, or hear the dangers of pesticides or of other chemicals, such as
mercury, that are insinuated into our bodies. We discover, often too
late, that drugs meant to ease pain and combat disease can produce
adverse, sometimes deadly, side effects.

On top of this, conveniences we take for granted have begun to
desert us. Our transportation systems are slowing down and becom-

"The problem is that the fruits of technology have been allowed to grow untended."

The author:
Ralph E. Lapp is a
Washington-based
observer of science
and author of numerous
books and magazine
articles. A nuclear
physicist, he is a
member of the board
of Quadri-Science, Inc.

ing more unwieldy and undependable. In some parts of the United States, entire regional power distribution systems break down during certain periods of increased electrical demand. And on recent occasions, that unparalleled example of American technological efficiency —the telephone system—has refused to work.

What is happening? Many people blame science for many, if not most, of our present woes. Actually they mean technology, but laymen rarely distinguish between science and technology—between scientist and engineer. And, as is usually the case with complicated ills, these people seek simplistic solutions.

For example, some want to turn back the technological clock and return to the uncomplicated ways of the last century, uncluttered by the complexity of modern life. These "primitives" troop off to pastoral communes—but, ironically, their electric guitars, transistor radios, and expensive cameras accompany them. Another group wants to stop the clock. They say our problems stem from sheer growth—the increase in population and in the manufacture of things—and they advocate no further growth. I call these people "zeroists." Then there are the modern Luddites—the revolutionary radicals who want to destroy the clock. They throw bombs at computers, and would like to rip society apart and rebuild it from some unclear blueprint in their minds.

If these groups could look back to life in America a few generations ago they would see that few people had time to really enjoy life. Most men and many women worked long hours at backbreaking jobs. Much of the nation's economy was keyed to the farm. Many homes had no running water, no indoor plumbing. Few had electricity.

As industrialization in the United States brought the shorter work-week, low-priced consumer goods became available and a wide variety of choice in life style opened up for many Americans. People today take technology for granted, not realizing the immense changes that were wrought earlier in the 20th century. I remember the first battery-powered crystal radio set in my home in the early 1920s, the thrill of the first family car, the first talking pictures, frozen orange juice, and even a gas stove in the family kitchen. I was a teen-ager during the Great Depression of the 1930s and I remember what zero growth—the stilling of technology—meant and the misery that it inflicted on millions of Americans.

There is no denying, however, that technology has become a massive boomerang. Man has used it to produce damaging, as well as beneficial, changes. The problem is that the fruits of technology have been allowed to grow untended. Consider the example of the automobile. At the turn of the century there were 8,000 cars in the United States. Automotive transportation caught on so quickly that by 1912 motor-car registrations had passed the 1-million mark. Shortly after World War II there were 10 million vehicles on U.S. roads and in 1971, there are nearly 100 million.

The automobile's impact on society 70 years after its introduction is overwhelming. During rush hours in metropolitan areas the traffic barely moves. Commuters daily turn superhighways into parking lots as they struggle through traffic jams. In the process the millions of tail pipes wage chemical warfare, laying down gas-fume attacks in almost every major city in the United States.

The automobile has turned out to be one of the most lethal weapons ever invented. In 1970, traffic fatalities in the United States totaled 56,000, and over 4 million people were injured. Simply as a disposal item, the automobile is a great nuisance. In New York City alone, over 250,000 cars were abandoned on city streets in the past 10 years.

The American preoccupation with the automobile, and the private and public investment in its behalf, deny growth to other modes of transportation. For example, government roadbuilding programs use funds that might otherwise go to help the ailing railroads, still the most efficient form of transportation. More important, some of these funds might be better used for research and development of new modes of transportation, such as the monorail.

The sad fact is that the major facets of modern technology—transportation, communication, energy—continue to develop with little thought for the problems they may cause. But you do not solve the problems of society by turning off or by blowing up its technology. Rather, you try to control it. We must not simply control technology, however, but try to cultivate it, to direct it to man's advantage and to a condition of ecological harmony. I use the word "cultivate" because I think it connotes a compassionate control, one that involves careful planning to see that the seeds are planted at the right time so the fruits will mature on schedule.

If we are to cultivate technology we need both understanding and planning. And if complex issues are to be understood, we will need more science, not less. For example, the development of an ecologically good detergent requires that chemists study a wide variety of possible compounds and determine how they degrade when interacting with the environment. In this way, they can assure that decomposition products that could cause a great deal of harm do not find their way into our food and water supplies.

". . . we are going to have to make long-range plans about our technological future."

Planning is basically a matter of making the right choices. This means laying out alternative programs leading to a given goal and rationally assessing the risks and benefits associated with each alternative. Planning also includes the choice not to do something. The choice not to build an American supersonic transport (SST) was a step in the direction toward controlling unfettered technology. Unfortunately, however, both as individuals and as a nation, we do not always make wise choices. Too often we think only in terms of benefits or short-term gains and do not evaluate the risks.

If our society is to avoid collapse, we are going to have to make long-range plans about our technological future. We are going to have to make them with scientific guidance and leadership. And we are going to have to start now.

The scope of the problem can be appreciated by looking at a major aspect of technology—the growth of energy in the United States. At the turn of the century the nation's power plants turned out about 60 kilowatt-hours of electrical energy per person per year. In 1971, our electrical energy consumption amounts to more than 8,000 kilowatt-hours per person. This Gargantuan growth shows no sign of letup. Indeed, the public utilities are building and planning to build plants that will increase their power capacity nearly 7 times by the year 2000. Then every U.S. citizen will have at his disposal some 30,000 kilowatt-hours of electrical energy per year.

It is not at all clear how this energy will be best developed. Atomic energy eventually will be the dominant power source, but not immediately. In 1971, the U.S. economy is powered primarily by fossil-fuel energy. A coast-to-coast system depends on a continuous flow of fuel. The energy of coal is converted into electricity and fed into a vast labyrinth of tower-suspended wires. Oil is shipped through pipelines to its point of use. Natural gas is distributed by a nationwide system of underground pipelines. At present, we use these fossil fuels at an annual rate of about 550 million tons of coal, 5 billion barrels of oil and 22 trillion cubic feet (tcf) of natural gas.

If we convert the energy of oil and gas to heat units equivalent to tons of coal, it turns out that the United States today burns the equivalent of 2.5 billion tons of coal per year. That is 12 tons for every person. But the U.S. energy machine is not running at constant speed. It is accelerating—to supply more power for a growing population and to provide more energy for each person. A recent government estimate puts the total U.S. energy consumption in the year 2000 at the equivalent of 7 billion tons of coal. Between now and the end of the century the United States will demand energy equivalent to that obtained from burning over 130 billion tons of coal.

Now if all the energy were to come from coal, I think we would probably end up by coughing our way into the 21st century and groping our way through the fume-darkened streets. But here the uranium atom is cast in a rescue role. The pattern of nuclear power indicates that by the year 2000 it will supply two-thirds of utility-generated energy. Yet fossil fuels will still be very much in the energy picture, amounting to 4.5 billion tons of coal equivalent.

Obviously, it will be impossible to fill this annual fuel requirement by digging coal from the ground. Even if it could be done, no one knows now of an economical way to remove the sulfur from the coal that is found near the big-city markets of the northeastern United States. Let us assume, however, that about 1 billion tons of coal is mined in the year 2000; this means gas and oil will have to make up for the energy equivalent of 3.5 billion tons of coal.

The energy future is very clouded for gas and oil. Natural gas is an ideal fuel—cheap at the wellhead, easy to transport, and clean to burn. But it is in short supply. I estimate that even with tapering off, the consumption during the remaining 30 years will use up 1,110 tcf. If one adopts the prudent policy of not pumping more than one-tenth the reserve capacity from the wells each year, we will still have to find 1,300 tcf of gas before the century ends.

Finding more gas is increasingly difficult; the U.S. continental land mass is already perforated with drill holes. This is where technology can offer an alternative—the gasification of coal. In this process, hydrogen is added to the carbon of coal to form an artificial hydrocarbon equivalent of pipeline gas. This marriage of carbon and hydrogen can be accomplished in a variety of ways, although as yet no large plants have been built to demonstrate the economic feasibility.

In 1971, 5 billion barrels of oil supplied 43 per cent of the fossil fuel energy. By the year 2000, an equal percentage would require about 15-billion barrels per year. This threefold jump in oil burning would simply overwhelm U.S. domestic production and force the nation into dependence on foreign oil.

We must conclude then that the United States will have reached a peak production of gas and oil, allowable in terms of reserves, and will be sliding down the other side of the production curve well before the turn of the century. Fortunately, the United States has some 1,600-billion tons of coal that could be mined, which constitutes an immense energy reserve of fossil fuel.

"... technology can offer help to prevent thermal pollution or minimize its impact ..."

It is clear that to make the best use of this reserve, some sort of plan must be set up. A system of priorities is urgently needed for research and development in many aspects of energy—in gasification of coal, in desulfurizing coal, in tapping the oil potential of oil shale, in developing efficient means of transmitting power over long distances, and in replacing the internal-combustion engine with a more efficient, and less polluting, means of vehicular propulsion.

Simultaneously, we must plan for the age of nuclear energy. There has been great public opposition to construction of new nuclear-powered electric plants. Apart from those who argue against growth per se, opponents cite the dangers of thermal pollution and low-level radioactive discharges from nuclear plants.

Thermal pollution is not uniquely characteristic of a nuclear plant. Any steam generator produces heat that must be disposed of. In a coal-fired boiler, some heat goes up the smokestack into the atmosphere, but most of it is released (other than that converted into electric energy) in the condenser. The latter is usually cooled by water from a river, a lake, reservoirs, or the ocean.

The discharge of water heated 10 to 30° F. by passing through a condenser can harm the local environment if the volume of water is great enough. Heating bodies of water can damage or destroy marine life and upset the ecology.

Here too, technology can offer help to prevent thermal pollution or to minimize its impact on the environment. Electrical utilities are now building cooling towers to chill condenser discharges before returning the water to the river or waterway. One such tower now being designed—the hyperbolic, or natural-draft, chimney—may rise 450 feet above ground level and cover an area the size of a football field. Such towers may become the hallmark of the nuclear industry.

The hazard posed by the emission of radioactivity from nuclear power plants has caused great debate. Much of the argument over this hazard focused on the adequacy of Atomic Energy Commission radiation standards rather than on the actual level of emissions from operating nuclear plants. It must be recognized, however, that a nuclear power plant is a unique hazard. It contains an immense amount of radioactivity that could, in the event of an accident, contaminate the surrounding community. Admittedly there is little chance that such accidents will occur, but the consequences of a single accident could be extremely serious if the plant is located close to a populated area.

At the moment, I feel we do not know enough about the safety inherent in the design of present nuclear plants, especially those exceeding 1,000 megawatts. I have accordingly publicly urged a policy of restricting the power rating of new nuclear units and of building them far from regions of high population density.

Since most problems of modern technology are studded with possible projects, each of which could cost a billion dollars, it is obvious that only in Congress can funds be found to finance programs of such magnitude. Thus technological planning involves politics. But it also must involve the scientific community. Scientists, being closest to the intricate technology of the future, bear a special responsibility to society. Unfortunately, there is no government of science to act as a coherent force for relating the potentials of science to society.

The basic dilemma of a democracy attempting to arrive at good decisions in an age of galloping technology is that the vast proportion of the citizenry is ignorant of science. Scientists find it very difficult to bridge the communications gap between themselves and society. This difficulty is compounded because the scientist is often looked upon with some suspicion.

Why should this be? Presumably some of it may be traced to the ancient distrust of the expert-priest—the user and withholder of secret information. Does the public fear a conspiracy of scientists—a take-over by technicians? Such conspiracy requires great coordination and thinking alike, conditions hard to achieve in any group of scientists I know. I suspect much distrust of scientists is a revulsion against technology itself, against a lack of humanism in applying the force of science and against an inexorability that leaves the average man helpless to determine his fate. He fears domination by a computer mentality that will leave him little sense of individuality in the future. Also, he is afraid that technological blunders will wreck the environment. And in recent years he has seen good evidence for this fear.

Scientists certainly do not reign supreme in our national decision-making. In the case of the Safeguard antiballistic missile (ABM) system for example, the majority of scientists opposed the defense project but Congress voted for it anyway. The SST, similarly opposed, was defeated. But here politics, rather than the technical arguments, was the driving force. In each case scientists were compelled to act negatively. How can they become a positive force?

One way would be through an extension of science advisory posts in federal agencies. The White House already has various advisory groups, such as the President's Science Advisory Committee, to give the Executive Office advice on scientific and technical issues. The infusion of competent scientific advice into the federal hierarchy is a

useful mechanism, but not sufficient to ensure that the public interest
is served. Too often the advice-givers act behind closed doors. Admittedly, decisions made in public are not always wise ones. But in a
democracy, unless the give-and-take of argument occurs in public, the
wisdom of any decision is in doubt.

Recently, both scientist and layman were given a new instrument
for public discussion of issues where federal action poses a threat to
the environment. The National Environmental Policy Act signed into
law on Jan. 1, 1970, set up the Council on Environmental Quality and
required that each agency of government submit to the council an
"environmental impact statement" detailing the probable detrimental
effects a proposed project may have on the environment. The Trans-
Alaska Pipeline is an example of how this can bring problems to
public view. This pipeline is planned to carry oil from Prudhoe Bay,
Alaska, to Valdes, Alaska. In response to the Department of Interior's
environmental impact statement, independent scientists publicly challenged estimates of the environmental changes that the hot oil pipeline
might produce on the Alaskan tundra. Thus the American public has
had a chance to hear two sides of the story.

The federal government has also begun to provide positive correc-
tive agents, even though this involves placing restrictions on local
communities and, to some extent, the free enterprise system. For example, the Environmental Protection Agency, created in December,
1970, set air-quality standards in 1971 for six principal pollutants of
urban areas. These were sulfur oxides, carbon monoxide, particulates
(soot and smoke), hydrocarbons, nitrogen oxides, and photochemicals.
Similarly, Congress has specified strict regulation of automotive exhaust emissions that the automobile industry must meet in 1975-1976.
Some people denounce these restrictions. But in the long run, they
may be the only thing that will salvage the democratic system in the
future world of complex technology.

The new environmental requirements, coupled with an aggressive
attitude on the part of scientists and engineers can form the beginning
of the cultivation of technology. Scientists and legislators must work
together to form a prudent national policy that conserves resources,
protects the environment, and at the same time ensures the quality of
life for future generations. Furthermore, this policy must be conceived in public and written in a language that all people can understand. The bridge to the future is not just for scientists to plan and
engineers to build—it is a cooperative venture for every member of
society to see that it is well structured and safe to travel.

Special Reports

The special reports and the exclusive *Science Year* Trans-Vision®
give in-depth treatment to the major advances in science. The
subjects were chosen for their current importance and lasting interest.

The Moon's Unfolding History

By William J. Cromie

Scientists sift the moon's dust and rocks for clues to the mystery of its past

Geochemist John A. Wood examined the table-spoon of coarse moon soil under a microscope in his Cambridge, Mass., laboratory in the fall of 1969. Allotted to the Smithsonian Astrophysical Observatory by the National Aeronautics and Space Administration (NASA), this small sample was part of about 48 pounds of rocks and soil brought back from man's first landing on the moon.

Among the thousand or so rock fragments and chips were some white particles. They seemed completely alien to the dark plain on which the Apollo 11 astronauts had landed. Wood studied these carefully and became convinced they were chips from the mountains of the moon, many miles from where the astronauts had landed. He was to conclude later that he had seen rocks one step removed from primordial (original) material.

In the same year, halfway across the country at the University of Chicago, chemist Anthony L. Turkevich was studying sheets of computer output. They represented the results of an important experiment performed on the moon nearly two years earlier by the robot Surveyor 7 spacecraft. Turkevich had already startled the scientific world with a remote chemical analysis of moon soil made be-

fore the first man landed there. Many doubted the reliability of his technique, but the first rocks brought back from the lunar surface had proven him right. Now, Turkevich was about to pinpoint the chemical composition of the alien white chips that were to become so important in solving the mystery of the moon's origin and history.

At the same time, in Pasadena, Calif., a group of 10 scientists at the California Institute of Technology (Caltech) who call themselves the "Lunatic Asylum" examined a small chip they call "Luny Rock 1." The Lunatic Asylum (the term refers to the ancient belief in the moon's influence on the mind) is part of Caltech's Division of Geological and Planetary Sciences. The head "inmate" is Gerald J. Wasserburg, 44-year-old professor of geology and geophysics. Luny Rock 1 was highly radioactive relative to the other rocks. The Caltech scientists set out to measure the extent of radioactive decay to determine its age. It turned out to be 4.4 billion years old. Except for some meteorites, it was the oldest rock ever discovered by man—and one that dates back almost to the dawn of the solar system.

Wood, Turkevich, and Wasserburg are part of a worldwide detective team that is trying to solve the mysteries of the moon by studying the rocks and soil that the astronauts bring back and data relayed by the geophysical instruments left on the lunar surface. To these scientists, each rock chip or bit of dust, each recorded track of invisible cosmic rays, each tremor of the lunar surface is a clue to the physical character or history of the 81-billion-billion-ton moon. Like those seeking to solve a complex crime, the lunar detectives do not always agree on what the clues reveal. But argument, persistence, logic, and further investigation eventually yield enough information to write another chapter of the cosmic detective story. It is a story that stretches millions of miles into space and billions of years back in time to the birth of the moon, the earth, and the other planets. And it is a story that is forcing geologists to re-examine previously accepted ideas about the early history of the earth as well as the moon.

Man's curiosity about the moon undoubtedly began as soon as he developed enough of a brain to look up at the sky and wonder. As early as 2200 B.C., the Mesopotamians recorded lunar eclipses. Galileo became the first lunar investigator about 1609 when he looked at the moon through a telescope for the first time. He saw two distinct features—rough, light-colored mountain regions and smooth, dark, low-lying areas. Galileo and other early investigators thought the latter were seas; the first moon map called the light areas *terra* (Latin for *land*) and the dark ones *maria* (for *seas*).

When better telescopes were developed, it became obvious that these seas held no water. By the 1800s, most astronomers had concluded that the maria were gigantic craters filled with black or dark-gray lava that had cooled and hardened. As Wood explains: "If you don't believe the moon had seas and you ask yourself what it could be that produced such smooth, regular surfaces, lava springs immedi-

The Author:
William J. Cromie is senior vice-president of Universal Science News, Incorporated. He wrote "Navigating in Space" for the 1969 edition of *Science Year*.

ately to mind." By checking similar formations on the earth, the lunar detectives deduced that this surface was basalt, a fine-grain rock that results when lava containing dark-colored minerals cools quickly. In the early 1960s, notes Wood, most knowledgeable people assumed that the maria were basalt flows. But definite evidence was lacking.

Until the space age, all lunar investigations were earthbound; features less than 800 feet in diameter were hardly distinguishable through a telescope. But space technology extended man's reach far beyond the earth. In October, 1959, the Soviet spacecraft Lunik 3 looped the moon and sent back the first photographs of its far side, which we never see from the earth. The pictures revealed a curious fact that lunar detectives are still unable to explain. The hidden side has fewer maria, but more mountains and craters, than the near side.

In July, 1964, lunar investigators obtained their first close look at the craters on the moon's near side. The unmanned Ranger 7 space-

Two key experiments in lunar geology were performed during the Apollo 14 mission. The lunar passive seismic experiment package in the foreground, *left*, detected evidence of quakes 400 miles deep. The seismic thumper, *above*, showed that the layer of lunar "dirt" was as much as 50 feet deep.

craft snapped photographs of the lunar surface from as close as 1,000 feet before crash landing. Photographs from Rangers 8 and 9 followed in 1965. These pictures showed moon craters that varied from miles in diameter down to pits only 3 feet across.

The Ranger program established once and for all that the lunar craters were caused by meteorite impacts. "Until that time," Wood explained, "there had been a fairly strong school of thought that craters were some kind of volcanic features. But the smooth distribution of different-sized craters in the Ranger pictures fits more closely with the idea that they were produced by impacts. All the pieces of rubbish out in space that had been banging against each other and breaking up would produce the kind of crater distribution we see."

In 1967, two unmanned Surveyor spacecraft made soft landings on lunar maria. They both carried the equipment Anthony Turkevich had designed to make the first analyses of the surface rock. Alpha particles, the nuclei of helium atoms, were released to bombard the soil and rocks and rebound to special detectors on the spacecraft. By measuring the rebound energy, scientists determined which elements were present and how abundant they were in the soil. The experiment showed that both maria were covered with ground-up rock similar to basaltic lava found on the Columbia Plateau in the Pacific Northwest.

In Chicago, Turkevich analyzed data sent over a quarter-million miles from the Surveyors. He was surprised to find one marked difference between earth basalts and the rock in the Sea of Tranquility. The latter, he calculated, are rich in titanium. Other lunar detectives felt Turkevich was wrong, that the titanium-rich composition he suggested was most unlikely.

In mid-1969, the Apollo 11 astronauts landed only about 15 miles from Surveyor 5 in the Sea of Tranquility. The 48.4 pounds of rocks and soil they brought back were given a preliminary analysis in the Lunar Receiving Laboratory at NASA's Manned Spacecraft Center near Houston. One of the first things discovered was that the rocks, were, indeed, rich in titanium. Turkevich was proved right in a triumph that Wood calls "one of the grand moments of science."

The reason for the high titanium content remains a mystery, but the basalts established one major fact about the moon. The fact that they are igneous rocks, solidified from a molten mixture, left no doubt that the moon was once a hot, active body just as the earth is today.

Surveyor 7, the last in the series, landed in the lunar mountains on the rim of the huge crater Tycho in January, 1968. It provided the first good clues to the character of the lighter-colored highlands. Turkevich used the same technique he pioneered on Surveyors 5 and 6 with his remote chemical analyses of the maria. Signals received on the earth revealed that the chemical composition of the Tycho highlands was distinctly different from that of the maria.

At the same time that the Surveyors were landing on the moon's surface, a series of five unmanned lunar orbiters circled the moon and

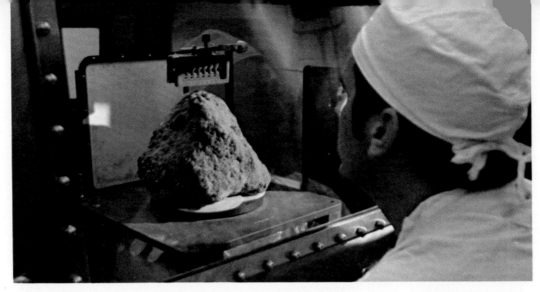

photographed virtually all of its surface. These pictures provided the clearest, most detailed look at the moon that man had ever had.

This, then, was the picture of the moon until the first men walked on its surface. The main shortcoming in the reconstruction of its geological history was a time scale. Lunar investigators had traced out the sequence of many events, but they did not know when these events had taken place. They tried to use craters as a means of obtaining relative dates, reasoning that the older a surface area was, the more meteorite impacts would pock its surface. Comparing the numbers and sizes of meteorites that enter the earth's atmosphere with the patterns of craters photographed in the Sea of Tranquility by lunar orbiters, scientists calculated its age as 200 million years. This is young on the time scale of the solar system and would mean that the moon had been hot and active until relatively recent times. Indeed, some lunar detectives are convinced the moon, like the earth, is still volcanically active.

When the astronauts returned from the Sea of Tranquility, scientists set to work dating their lunar samples by measuring the degree to which their radioactive components had decayed to form other elements. One approach was the rubidium-strontium method used by Wasserburg. Rubidium 87 in the rock is radioactive and decays into stable strontium 87. The Caltech team established the amount of strontium that had resulted from the decay. Then they measured the amount of undecayed rubidium 87. Since the decay rate is known, the proportion that had decayed indicated the age of the rock. The investigators were stunned. The basalt was about 3.7 billion years old, 18 times the earlier estimate. The outpourings of lava had occurred less than a billion years after the moon originated.

The detectives were just getting over this shock when they received another. The dust or soil picked up by the astronauts appeared, from the amounts of radioactive elements and their decay products, to be almost a billion years older than the basalt rock it covered. Since the soil was largely made up of broken rocks similar to the 3.7-billion-year-old fragments, its age should be similar.

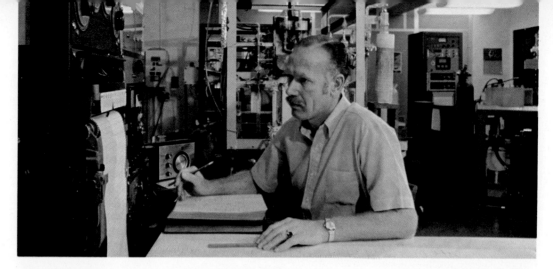

Thousands of laboratory tests, such as on this mass spectrometer at Caltech, provide the chemical and physical clues that geologists need to understand rock samples gathered by the astronauts.

Regarding each particle of the lunar soil as a tiny rock that could provide new clues to this and other mysteries, Wood and his Cambridge group set to work. So did Wasserburg and his co-workers. Wood discovered that under the microscope the light-colored chips of alien material were rich in plagioclase, a lightweight mineral containing much calcium, silicon, and aluminum.

At about this time, Turkevich had finished his analysis of the highland site where Surveyor 7 had landed. He concluded that the rocks that scattered the alpha particles were composed mainly of elements found in plagioclase feldspar. Thus, Wood's observations confirmed Turkevich's remote-controlled analysis. And Wood reasoned that his chips were from the lunar highlands, apparently knocked far onto the maria by the impact of large meteorites.

Meanwhile, Wasserburg's group had picked out a promising-looking fragment from their Apollo 11 sample. This was Luny Rock 1, the specimen that turned out to be 4.4 billion years old. Small amounts of dust of this older material had made the lunar soil from the Sea of Tranquility appear to be much older than the basalts lying beneath it. It is so much more radioactive than the basalt that it greatly influenced the age determination. Luny Rock 1 was obviously quite different from both Wood's alien chips and the rocks Turkevich analyzed.

Rocks like Luny Rock 1 were scarce in Apollo 11 material. Wood found only one piece in his sample, and did not notice it until after the Caltech group published its findings. But such material made up about 30 per cent of the soil brought back in November, 1969, by the Apollo 12 astronauts. They had landed in the large mare called the Ocean of Storms about 600 feet from the silent Surveyor 3. Wood labeled the material "norite" after a similar earth rock. Investigators at the Manned Spacecraft Center gave it the chemical name KREEP. (K is the chemical symbol for potassium, R-E-E stands for rare earth elements, and P is the symbol for phosphorus.)

When the Apollo 14 crew brought back some samples in February, 1971, that were chemically and mineralogically similar to Luny Rock 1, lunar detectives felt still more confident that this rock type plays a major role in the basic crust of the moon. "It is probably the material that is under the basalt of the maria," Wood speculates.

Apollo 14 landed on the Fra Mauro Formation, a deposit of broken rock formed from material thrown out when an asteroid, comet, or other huge mass created the 700-mile-wide Imbrium Basin a few hundred miles away. Since filled with lava, the basin area is now called the Sea of Rains. The Fra Mauro Formation is covered with a blanket of soil and debris that formed during the billions of years since that impact. The covering was penetrated, however, by the projectile that created Cone Crater. For that reason, astronauts collected rocks from the crater's flanks. Scientists believe that these rocks were first blasted out of the moon by the Imbrium impact, then by the smaller impact that produced Cone Crater.

Robin Brett, chief of the geochemistry branch at the Manned Spacecraft Center, says most of the 95 pounds of Apollo 14 rocks are "complex, mixed, and broken up. Most of them are breccias—rocks made up of the fragments of several different rock types welded together by impact energy."

The light-colored rock, rich in plagioclase, was relatively rare in the Apollo 14 samples. Wood and others expect, however, that when Apollo 15 and 16 missions explore other parts of the lunar highlands, they will find much of the highland crust composed of light-colored, anorthosite-like igneous rock. It will be similar to the material Wood found, to Luny Rock 1, and to the rocks Turkevich analyzed.

Samples from the first two landings showed that the lunar interior was hot enough to melt rocks early in its history. As a working hypothesis, the scientists assume that the anorthosite solidified from this melt, probably floating to the top because it is lighter in weight.

The lunar detectives use complicated methods to determine the thickness of this crust, the material that lies underneath it, and the deep interior structure of the moon. Each successful Apollo landing

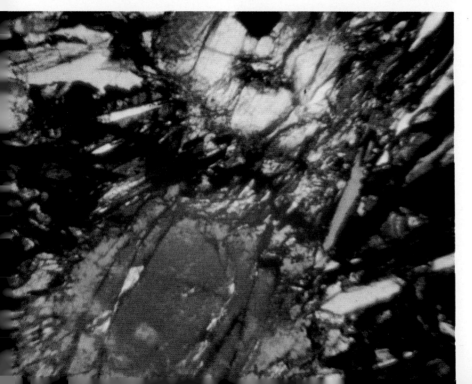

Apparently of volcanic origin, this Apollo 12 sample reveals the differing growth rates of its crystals under the microscope. Such rocks were most common at the maria sites of Apollo 11 and 12 missions.

The moon's first surface, formed about 4.5 billion years ago, was a rock "scum." Bombarding material had heated and melted the surface several hundred miles deep. Meteorites continued to pound the hardening crust, shattering it, cratering it, and altering its minerals. Magma streamed up, *left*, into the cooler, more viscous region near the surface. Rock 12013, *below*, dates from this fiery epoch when the newly formed earth was close by. The rock may be a relic of the moon's primal crust.

mission leaves behind an array of instruments to detect moonquakes, measure the moon's magnetic field, and record the influx of cosmic rays and particles of the solar wind. Scientists have learned much about the interior of the earth from the way that earthquake shocks move through it. The speeds with which different types of sound waves travel from the quake's origin to the seismometers tell geophysicists much about the rock types the signals passed through. Such activity is infrequent on the moon, so investigators measure the impact of spent spacecraft, such as the lunar module, plus explosions set off by the astronauts or by remote control from the earth. Data from such experiments, transmitted to earth from seismometers left by the astronauts, provide important clues to the mysteries of the moon, but their interpretation can be difficult and is rarely unanimous.

Seismologists support the existence of a regolith layer on the lunar surface. The regolith, or blanket of rubble, created by meteorite impacts was first described by Eugene M. Shoemaker, chairman of Caltech's Division of Geological and Planetary Sciences, and his colleagues. Its depth varies, reaching 50 feet at Fra Mauro, where it was measured by setting off small "thumper" charges equivalent to firing a .22-caliber bullet. In the words of seismologist Gary Latham, one of the principals in seismic experiments on Apollo 11, 12, and 14, the layer under the regolith "scatters the seismic energy intensely." In the area where Apollo 12 and 14 landed, impact waves made when

In the thickened crust of about 4 billion years ago, a
giant meteorite gouged a crater many hundred miles wide.
Fingers of lava oozed up through cracks in the crust, *left*, partly
filling the basin with basaltlike rock several miles thick.
Meteorites continued to rain on the layered surface, dug
deep holes, and created new layers of debris. The
crystalline rock, *below*, found at the Apollo 14 site,
was probably ejected from Mare Imbrium by a meteorite.

the abandoned lunar module crashed back to the surface showed this lower layer is from 30 to 60 miles thick. While details about this layer are missing, the best interpretation holds that it is made up of large broken blocks. This can be explained in one of two ways, according to Latham, who works at the Lamont-Doherty Geological Observatory. He believes the lower layer may correspond to the final stages of accreting (clumping together) of the moon material. Or, it could be the remnant of a smooth, homogeneous layer that was shattered by the rain of meteors over billions of years.

Below the fragmented layer, the moon behaves like a solid mass of homogeneous rock. The lunar detectives have not yet determined what this rock is. Instruments left at Fra Mauro by Apollo 14, however, have detected the first evidence of quakes deep within the moon. "Up to this point we thought all moonquakes were shallow," said Latham. "But combining information from the Apollo 12 and 14 seismometers reveals they could be as deep as 400 miles. The disturbances seem related to volcanic activity—that is, they seem to be caused by the movement of small pockets of *magma* (molten rock). It turns out that the moon is not as dead as everyone expected."

The sun contributed more clues to the interior structure. Charged particles streaming from the sun, the solar wind, carry a magnetic field. In the body of the moon, the field generates a current that sets up a small field of its own. Surface magnetometers measure the latter field

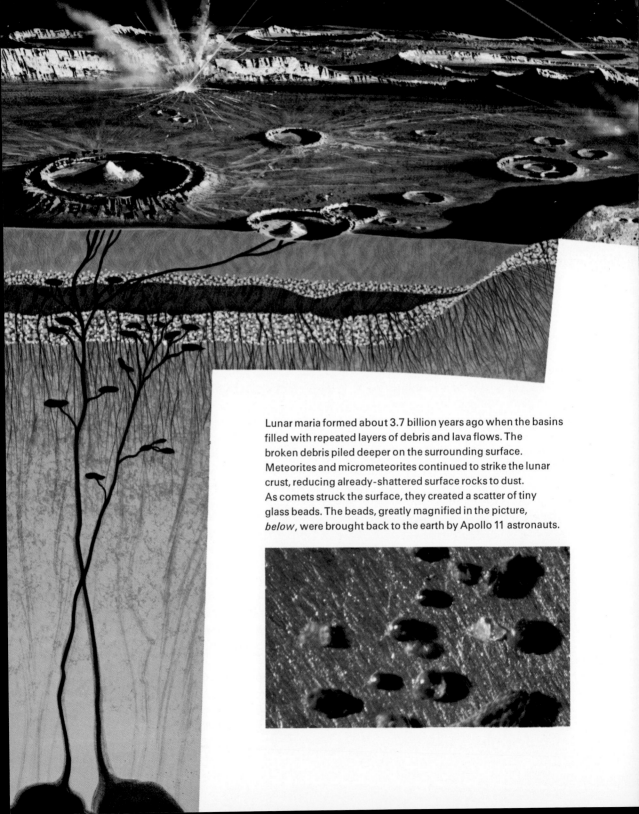

Lunar maria formed about 3.7 billion years ago when the basins filled with repeated layers of debris and lava flows. The broken debris piled deeper on the surrounding surface. Meteorites and micrometeorites continued to strike the lunar crust, reducing already-shattered surface rocks to dust. As comets struck the surface, they created a scatter of tiny glass beads. The beads, greatly magnified in the picture, *below*, were brought back to the earth by Apollo 11 astronauts.

—and changes in it—for comparison with variations in the solar wind. The comparison yields an estimate of electrical conductivity, from which internal lunar temperatures can be inferred.

From the fact that the moon has no strong permanent magnetic field, lunar scientists conclude that it probably has no large, partly molten, nickel-iron core at its center, as does the earth. Currents of molten material moving in the earth's 2,200-mile-diameter core are thought to generate the magnetic field that causes iron-rich minerals in rocks to become magnetized and compass needles to line up in a north-south direction. "Since this kind of field is lacking on the moon," Latham pointed out, "you expect no fluid core."

Evidence pointing to the absence of such a core poses a cosmic puzzle that bears strongly on the moon's origin. Lunar investigators offer three possible explanations. First, the moon could have formed in another part of the solar system where materials available for building a planet differed chemically. Then it was captured by the earth's gravitational attraction. "Hardly anybody believes this anymore," Wood pointed out. However, that does not mean the theory is invalid, nor that it will never come back into prominence.

Another theory holds that the moon was thrown out of the earth's surface after heavy elements had already settled. This pictures the early earth as molten or nearly molten. Catastrophic separation could have been due to tides building up in the earth's body. Rotational instability (with the earth spinning more rapidly as heavier material moved to the center) or the passage nearby of some massive body, are other proposed reasons. This, too, is now considered unlikely.

The third theory holds that the earth and the rest of the solar system condensed out of a cloud of gas and rock particles. The first materials to condense were iron and other heavy elements with a high melting point. These heavy substances came together to form the earth's core. Then, as temperatures dropped, lighter material condensed and accreted in layers around the core. According to this explanation, the moon did not start to form until after most of the cloud's available iron had been used up. But physicist Harold Urey's comment, that "all theories of the origin of the moon are improbable," still has wide appeal among scientists.

This part of the lunar detective story is pure speculation. But one fact relating to the moon's origin and early history was well established from samples brought back by the Apollo 11 and 12 astronauts. It is clear that the outer part of the moon was very hot during the first billion years of its existence. "This is a point of view that many did not hold," said Wood. "The theory of the moon accreting as a cold planet was fashionable for 20 years." Proponents of this theory still insist that the rocks so far prove only that parts near the surface were melted, not the deeper interior. The bulk of the moon was always cold, they say. But most scientists favor the hot accretion theory, and evidence continues to accumulate in its favor.

By mid-1971, lunar detectives had put together all the clues and evidence and reconstructed the events that produced the moon we know today. The planet was probably formed by large chunks of matter that fell together perhaps as much as 4.6 billion years ago. This accretion took place at high temperatures, but where the heat came from remains a mystery. If the moon formed rapidly, the accretion itself would generate enough heat to melt rocks. Or decay of radioactive materials could have contributed the required heat.

The melted part of the early moon, Wood estimates, could have extended 250 miles below the surface. As the layer cooled, various compounds separated from the melt at different times and crystallized into different rock types. According to Wood: "The mineral plagioclase, crystallizing in this melt, floated up to the surface and formed a scum which became anorthosite when it hardened."

Thus, when the moon first solidified 4.6 or more billion years ago, it would have been covered by light-colored material with no dark maria. Under the surface, heat was building up from the decay of such radioactive elements as potassium, thorium, and uranium. About 4 billion or more years ago, according to this theory, a series of large meteorites struck the moon, blasting out the maria basins.

The earth remains thermally active today because of radioactivity. Heat and the erosive effects of wind and water on the surface have constantly reworked its rocks. But the moon, except for the diminishing meteorite bombardment, generally settled down after its first billion years. Some of these later impacts were spectacular, creating huge craters like Tycho and Copernicus with their bright rays of material extending for hundreds of miles. This bombardment left the record of the moon's first billion years battered, but not destroyed. The original rocks remain, keys to the origin and early history of the solar system. "They have been just sitting there, waiting for us to come and collect them," says Wood.

What investigators have learned about the moon has caused them to reappraise their ideas about the early history of the earth. The two bodies are the same age, which suggests that their origins are closely linked. The earth and the moon were much closer together in their early history. Since the moon was hot in the beginning, so must the earth have been partly molten. The same chemical processes have occurred on both bodies, and Brett speculates that the far side of the moon may look much like the earth before the ocean basins formed. "Earth's original crust may have been anorthosite," Wood added.

One of the Apollo 12 samples is a piece of fractured norite-like rock with veins and pockets of granite in it. Called Rock 12013, it is almost as old as the solar system—4.5 billion years. "Up to that point, we had always believed it takes many cycles of melting and rock evolution to produce granite," Wood noted. "Now it appears that the moon could make granite very easily because it occurs in the oldest rock found so far. This is a major breakthrough in geology. Apparently, the planets can make granite with great ease. This has caused us to re-examine our thinking about the early history of continents on earth."

"The important thing to understand about lunar science," Brett has pointed out, "is that it isn't an academic exercise on a dead planet by a group of small-thinking scientists. The moon allows us to look back beyond any data that we have on earth, to the sort of processes that first occurred on and within Earth, Mars, Venus, and the other planets. By learning about the moon, we will learn much more about the earth at present, putting its past in context. We will also understand more about the formation of the crust, which to most of us is the earth itself."

For further reading:
Cooper, Henry, Jr., *Moon Rocks*, Dial Press, 1970.
Cromie, William J., "Exploring the Moon by Proxy," *Science Year*. 1968.
National Aeronautics and Space Administration, *Science at Fra Mauro*
    U.S. Government Printing Office, 1971.
Wood, John A., "The Lunar Soil," *Scientific American*. August, 1970.

Pulses of Gravity

By Joseph Weber

A physicist's patient search for gravitational waves yields results that may revolutionize astronomy

In my laboratory at the University of Maryland a large aluminum cylinder hangs suspended in a vacuum, isolated and protected from all outside disturbances. An identical cylinder is at the Argonne National Laboratory near Chicago. Both cylinders are monitored day and night by electronic instruments of unusual sensitivity. Two recording pens jiggle steadily, tracing on charts the infinitesimal vibrations of each cylinder. Once each day, on the average, an unexpected and sudden vibration is recorded simultaneously at both installations.

Over the past 10 years I have examined half a million feet of these chart records, and have found thousands of simultaneous vibrations. They indicate that an unfamiliar force coming from far beyond the earth penetrates the thick iron vacuum chambers and shakes the cylinders. I believe the force is that of gravitational waves—gravitational energy moving at the speed of light!

We all know gravitation as stationary and unchanging. We struggle against its constant force when we rise from bed each morning or climb the stairs each night. That gravitation could also travel through space as a wave was a remarkable idea that Albert Einstein had in 1916. He calculated that a spinning rod (such as a drum majorette's twirling baton) generates such waves. But the gravitational waves are incredibly weak. It would take more than a thousand billion billion billion years for the rod to lose just 1 per cent of its rotational energy in this way. How could such feeble radiation ever be detected?

Progress in science usually results from theory and experiment stimulating each other. But by 1960, when Einstein's theory was more than 40 years old, many critical experiments were still waiting for the technology and resources needed to perform them. Most theoretical

The author solders a connection to one of the crystals that emit signals whenever waves of gravitational energy shake the cylindrical antenna.

physicists thought that experiments on gravitational waves were impossible. A few believed that only through a national effort, comparable to the exploration of the moon, could we build a source of gravitational radiation suitable for study.

My theoretical research on gravitational waves from 1957 to 1959 convinced me that experiments could be done to test Einstein's prediction. It occurred to me that large masses of matter in the stars might be spinning rapidly. Such cosmic events could give off gravitational waves strong enough to be observed on earth once we solved the problem of how to detect them.

My first task, then, was to invent an antenna that could detect gravitational radiation. And, through a long mathematical analysis using Einstein's equations, I discovered that a certain type of solid object would serve as an antenna.

In my calculations, I used the geometry of space to describe the physics of gravity. Geometry treats both flat and curved spaces. In a flat space, the shortest paths between points are straight lines and these lines can be used to construct triangles whose angles always add up to 180°. This is not true in a curved space. For example, on the surface of a sphere, the shortest path between two points is along a great circle, and the sum of the angles of any triangle on the surface will be more than 180°.

Is the three-dimensional space around us flat or curved? The straightest lines we know are rays of light, and constructing triangles with them should tell us. Light has weight and falls in the gravitational field of the earth. This means that a light ray is curved, much like the "straightest" lines on the surface of a sphere are curved. A triangle constructed of these light rays will have an angle sum that is greater than 180°. The difference from 180°, from flat space, shows that space itself is indeed curved in a gravitational field. Einstein, in his General Theory of Relativity, described gravitation in terms of this space-time curvature.

Gravitational waves are curvatures of space that are always changing. As the waves move across a triangle made of light rays, the sum of the angles in the triangle will change, alternating between values greater and less than 180°.

The changing curvature of gravitational waves will create changing forces. To detect these forces, I calculated exactly what motions should result when a moving wave encounters matter. Two opposite forces would squeeze an object from the sides, compressing it. At the same time, a second pair of forces at right angles to the first pair would stretch the object. For example, a passing gravitational wave will squeeze and stretch a circular object, such as a wheel, into an ellipse. Then, both pairs of forces will reverse, and distort the object in the opposite direction. These deformations are greatest if the circle directly faces the approaching waves. The in-and-out distortion continues as long as the gravitational waves pass through it.

The author:
Joseph Weber, professor of physics at the University of Maryland, has unusual talents in being both a theorist and an experimentalist in general relativity.

Undulating Space

Gravitational waves, traveling distortions of three-dimensional space (dashed white lines), carry energy away from a deformed collapsing star. This curved space creates gravitational forces (white arrows). Objects in normal flat space, such as the black circle and square, will be physically distorted one way, then another, when they encounter gravitational waves.

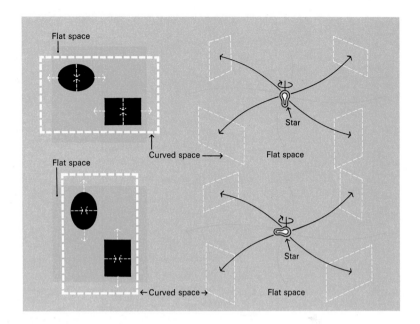

A gravitational wave, according to my calculations, distorts any solid in much the same way. The waves will move the ends of the solid in and out in relation to the center of mass. This distortion should create strains that might be detected using sensitive strain gauges.

The frequency of the gravitational waves that distort the solid depends on how fast their distant source rotated or vibrated. The antenna should be designed to be most sensitive at that frequency. Then, the oscillations of the antenna will stay in step with the waves, gaining energy until the power transferred from the waves is matched by the power consumed by internal friction in the antenna plus the power transferred to the strain gauges. By choosing an elastic material —one that has little internal friction—for the antenna, the greatest strains would be produced. My mathematical analysis showed that an elastic-solid antenna would measure the oscillating curvature of gravitational radiation precisely.

In 1960, after two years of theoretical work, I began the first laboratory experiment. My guess that the most abundant and powerful sources of gravitational radiation would be stars determined the size of the laboratory antenna I used. Most stars rotate slowly until their nuclear fuel is consumed. Then they collapse. As a star collapses, however, it spins faster, becomes appreciably deformed, and, like an enormous spinning rod, should radiate away much of its rotational energy as one burst of intense gravitational waves. Most of the radiation is emitted at frequencies of more than a thousand cycles per second. To detect radiation at these frequencies, an elastic-solid antenna need be only a few feet long.

For my antenna, I chose a solid cylinder of aluminum because this metal has a high sound velocity and can be forged economically. The cylinder was 5 feet long, 2 feet in diameter, and weighed about 3,000 pounds. At this length, it is most sensitive to radiation having a frequency of 1,661 cycles per second.

To observe its vibrations, I bonded piezoelectric crystals to the cylinder near its center. These crystals produce a small voltage in response to strains. I then attached wires from the crystals to carefully designed electronic amplifiers, and from there to a pen-and-ink chart recorder. A 1,661 cycle-per-second wave striking the cylinder from the side would cause in-and-out motions of the end faces of the cylinder. This system permitted me to detect exceedingly small motions—as small as 10^{-15} inch, only 1/100 the diameter of an atomic nucleus.

There was no way for me to predict how large a motion gravitational waves would cause. Their energy near the earth was unknown. Motions of the cylinder might also be caused by seismic, electromagnetic, or cosmic-ray disturbances. Such nongravitational responses could not be completely eliminated, but I hoped that isolating and shielding the detector would reduce them to a level well below the response produced by gravitational radiation.

The heavy cylinder was balanced on a single cable that passed like a sling under its middle. This isolated it from earth vibrations and minimized its energy loss to the supporting structure. Acoustic filters also reduced the transfer of earth vibrations. A large iron vacuum chamber surrounding the cylinder insulated it from outside temperature changes and from sound waves in the air. And electromagnetic shields protected the electronic equipment. These precautions isolated the system so well that it could not detect average environmental effects. The occasional large environmental effects it did detect were easily identified as such by a seismometer array and other instruments.

The detector's ability to detect gravitational radiation is ultimately limited, however, by thermal "noise." At room temperature, which is about 500°F. hotter than absolute zero, heat agitates the atomic nuclei and electrons in both the cylinder and the electronic equipment. The recording pen traces this noise as a steady series of random pulses of varying heights. I could only detect gravitational waves that produced pulses large enough to stand out against this background noise. Using probability theory, I could predict just how often I might expect to see large pulses of thermal noise. If large pulses occurred more frequently, the extra pulses might be caused by an external source. However, in my careful study of the charts of this first gravitational-radiation detector, I seemed to observe only random noise. Gravitational radiation was either too weak or arriving too infrequently.

I could increase the chances of identifying gravitational pulses by using a second detector that responded to the same frequency. This would allow me to search for simultaneous pulses. Simultaneous large increases should result from random noise occasionally, of course, but these coincidences would always be expected from an energetic gravitational wave.

To further minimize the influence of nongravitational effects, the detectors should be at different locations. A disturbance originating on earth is less likely to arrive simultaneously with equal amplitude at both sites. In 1964, a cylinder 24 inches in diameter and an 8-inch cylinder were placed about a mile apart. After two years of continuous operation, I analyzed the results. When the coincidences due to noise were subtracted, there remained an unexplained coincident increase from both antennas about every six weeks, on the average.

I decided to modify the experiment, using two larger antennas of 26-inch diameter, lower noise electronics, and increased separation of the antennas. If gravitational waves had indeed caused the coincidences, increasing the separation should have no effect, and the increased sensitivity should result in more coincidences.

One detector was placed in my laboratory at the University of Maryland, near Washington, D.C. The other detector was placed 600 miles away at the Argonne National Laboratory, near Chicago. Argonne was chosen because of the deep interest of Argonne physicist G. Roy Ringo and his colleagues in this research, and because of the excellent facilities there.

An Antenna for Gravitational Waves

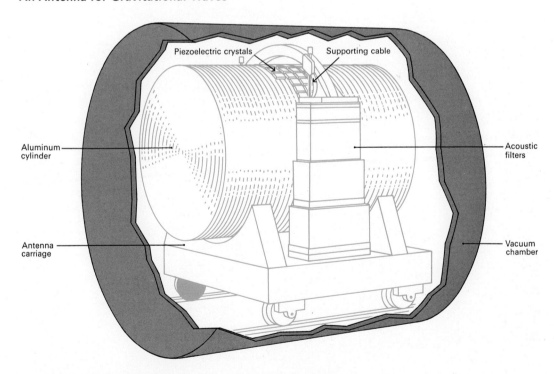

The 1½-ton cylinder is carefully shielded so that other forces cannot obscure its weak response to gravitational signals. It is girdled with a cable that is supported by vibration-damping acoustic filters. The antenna assembly is wheeled inside a thick-walled vacuum chamber.

Sensitive, low-noise amplifiers that boost the exceedingly weak crystal voltages are checked out by the author, *above*. The crystals, *left*, are attached by epoxy resin.

Chart recorder, *right*, produces a history of an antenna's vibrations. Just as a bell rings when struck, so the large cylinder generates the sequence of infinitesimally weak rings due to its thermal noise.

Antennas at the Maryland laboratory are the 7-foot-diameter aluminum disk, *above*, and three cylinders, one of which is temporarily removed from its vacuum chamber, *below*.

The ideal way to have performed this experiment would have been to have clock-synchronized tape recorders at each location, and have a computer later analyze the tapes and count the coincidences. A very modest budget forced us to choose a more economical way. We used a telephone line to transmit the Argonne detector's signals to Maryland. A small on-line computer was built to monitor the two signals at all times and to record simultaneous increases in their strength by marking both charts. In the new experiment, unexplained coincidences increased from roughly one every six weeks to about one per day, presumably because of the more sensitive system.

I have examined the charts regularly since 1968, and have found almost 2,000 coincidences. For each coincidence, I measure the heights of the peaks and sort the coincidences into groups. Each group consists of coincidences having voltages within a certain range. I then compute the number of chance coincidences for each group from the laws of probability. The results show that five out of six of the large voltage coincidences were not caused by thermal noise.

As a check, in September, 1969, I began a time-delay experiment. A delay in the signal from one of the cylinders should not change the number of coincidences caused by random noise, but it should eliminate every coincidence caused by a gravitational wave. I carried out two experiments at the same time, one with a two-second delay and the other without a time delay, for about two months. The results were decisive. The five out of six coincidences not due to chance disappeared in the experiment with the time delay. We repeated these experiments twice, at time delays of three and five seconds, and got the same result. There was little doubt that the two widely separated cylinders were responding to the same external signal.

Was it possible, however, that I had overlooked something, and that the force causing the coincidences was not due to gravitational radiation? If seismic effects were responsible, they could only be unusual events not recording on the seismometers, eluding the acoustic filters, and originating in a narrow zone of the earth with equal travel time to the Argonne and Maryland antennas. But even this unlikely possibility must be ruled out. Similar seismic events at other places would increase the noise at each site, and this was not observed.

I most suspected electromagnetic effects because they usually propagate at the same velocity as gravitational waves. Electromagnetic disturbances could be caused, for example, by electrical storms or local power-line fluctuations. The wide separation of the cylinders should help protect the experiment from simultaneous signals of this type. However, some electromagnetic disturbances travel worldwide by reflection from the ionosphere, and there might also be nationwide power-system fluctuations.

Such electromagnetic signals can enter the gravitational radiation detectors in a number of ways, so I decided to test for them. In 1969, I performed an experiment to see if signals were leaking into the

Counting Coincidences on the Rotating Earth

As the earth turns in the Milky Way once each 24 hours, the eastern U.S. faces the galactic center at 18 hours sidereal time. Gravitational waves from there are "seen" identically through the earth at 6 hours.

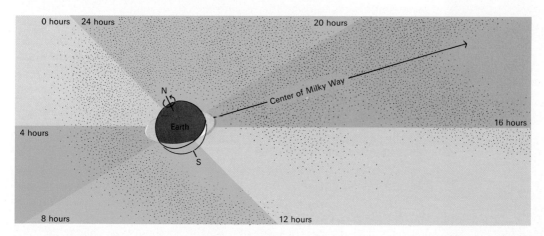

Counts of coincident signals from antennas in Maryland and Illinois, *right,* over many months gave the results, *far right.* Two cylinders produced peaks at both 6 and 18 hours.

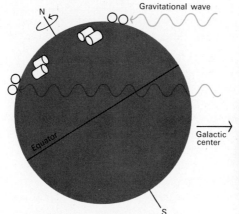

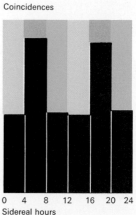

A disk and a cylinder, *right,* produced a peak at around 6 hours and a dip at 18 hours, *far right.* Results of both tests indicate that a gravitational radiation source may be at the galactic centers.

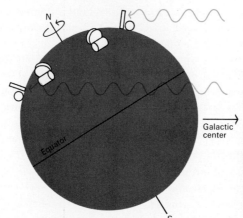

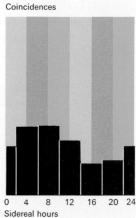

sensitive electronic amplifiers. I used a detector modified to cause an 11-second time delay for signals originating in the cylinders, but no appreciable time delay for any signals that might originate in the amplifiers. The results indicated that the coincidences were caused only by the cylinders' vibrations.

Electromagnetic signals could also induce electric currents in the aluminum cylinder. These currents might interact with the earth's magnetic field or other fields and cause oscillations of the cylinder. Electromagnetic signals at 1,661 cycles per second, especially, could produce oscillations in this way. These signals might result from combinations of signals from radio and television stations. In January, 1970, I set up a monitor at the Maryland site, using an amplifier especially designed to respond to any combination of electromagnetic signals giving rise to one signal of 1,661 cycles per second. This monitor showed no significant correlations between the signals and the cylinders' coincidences.

Cosmic rays can introduce highly energetic particles into the cylinder as well as electric charges into the electronics. But no cosmic-ray effects are known that extend over the 600 miles separating the two detectors. I had previously calculated that cosmic rays would not influence the detectors because the instrumentation had been designed to minimize their effect. Nevertheless, cosmic-ray counters had been placed around one of the Maryland detectors by University of Maryland physicists Nathan S. Wall, Gaurang B. Yodh, and David H. Ezrow in November, 1969. The counters verified my calculations; cosmic-ray effects were too small for the gravitational radiation detectors to observe.

It now appeared very likely that the cylinders were not reacting to any previously observed forces. Therefore, the next step was most important: to identify the direction of the source.

Using Einstein's equations, I computed the directivity pattern of the antennas exactly. The cylinders are suspended horizontally with their ends facing east and west. Thus, as the earth rotates, they view different parts of the sky. Gravitational radiation from one region of the sky is viewed every 12 hours, however, rather than every 24 hours, because the radiation passes easily through the earth.

The sun is the closest object having enough energy to produce strong gravitational radiation. If it were the source, the coincidences would occur mainly around noon and midnight. When the thousands of unexplained coincidences are counted in solar time, however, they are not significantly higher at either 12 o'clock.

For a source outside the solar system, a different kind of time must be employed. The earth rotates one more time per year relative to the stars than relative to the sun because it revolves about the sun. The time of rotation relative to the stars is called sidereal time, and the length of a sidereal day is about 23 hours 56 minutes. When the coincidences are counted in sidereal time, two groups of high counts

The Disk's Response

A circular disk antenna, *top,* will be distorted into an ellipse when it faces a strong source of gravitational radiation. But the area of its faces does not change. Turned edge-on to the source, *bottom,* the disk changes its area as the vertical forces stretch it. The horizontal forces now change its thickness.

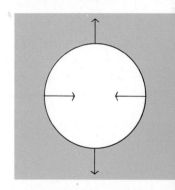

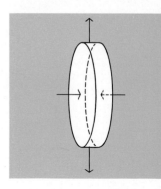

A new type of antenna was being built at the Argonne Laboratory in 1971. The author discusses its assembly, *left.* The cylindrical antenna rests in half of its stainless steel casing, *above.* Crystals that ring the cylinder on both sides of the groove for the supporting cable, *right,* must remain bonded even after the antenna is cooled more than 500°F., to only 7°F. above absolute zero.

appear, half a day apart. The signals must come to, and through, the earth from the same region of the sky. That region contains the center of the Milky Way, our Galaxy, in the constellation Sagittarius.

Only two known things could pass through the earth so easily—neutrinos and gravitational radiation. Neutrinos are subatomic particles that also travel at the speed of light. The antennas' response to neutrinos, however, should depend only on the number of the antenna's atomic nuclei in the path of the radiation. Since this number is constant, and does not change with antenna orientation, neutrinos would not cause twin peaks half a day apart.

Our knowledge of the sun, the planets, and the rest of the universe has been based almost entirely on the light, radio emission, and cosmic rays we receive. Now, there seemed to be a completely new channel of information—gravitational waves. More observations are needed, of course, to confirm it, but I felt certain that the detectors had observed gravitational radiation.

Einstein's theory predicts that a sphere will be deformed into an ellipsoid by a gravitational wave, but that its volume will not change. A number of physicists have explored the possibility that the wave might change the volume of the sphere in addition to its shape. Einstein himself and physicist Peter G. Bergmann theoretically investigated some such modifications of general relativity theory in

Entombed in its steel casing, *left*, the cylinder
will be further encased in a bright copper vessel
that contains cooling coils, *above left*. When this
vessel for liquid helium and a surrounding vessel
for liquid nitrogen are in place, the entire assembly
will be sealed in an iron vacuum chamber, *above*.

the late 1930s at the Institute for Advanced Study in Princeton, N.J.
In 1955, physicist Pascual Jordan of the University of Hamburg,
Germany, and in 1961, physicist Robert H. Dicke of Princeton
University suggested similar modifications.

There are some difficult experimental problems if a sphere is employed. I found that I could avoid them, however, by using a section
of a sphere—a disk. When the disk faces the source of the radiation, its
area, and thus the volume, should not be changed by a gravitational
wave that obeys Einstein's equations. The two pairs of forces that the
wave produces are at right angles to each other and are oppositely
directed. One pair will compress the surface, and the other pair will
stretch it, and their effects on the area should cancel each other.

When the disk is turned, so that the same wave strikes the edge of
the disk, one pair of forces now acts to increase or decrease its thickness.
The other pair of forces still acts on the face of the disk, however, and
is no longer canceled. Thus, according to Einstein's theory, the area
of the face of the disk only changes when the edge is toward the source,
but the modifications of the theory say the area will also change when
the face is toward the source.

With the help of engineer Darrell Gretz, I constructed an aluminum
disk antenna in 1970. The disk, 7 feet 1 inch in diameter, oscillates
best at the same frequency as the cylinders, 1,661 cycles per second.

These oscillations increase and decrease the area of the faces of the disk. I bonded piezoelectric crystals to the disk so that they would detect strains caused by a change in the area. The disk is suspended in a vacuum tank. As the earth rotates, a face of the disk directly views the galactic center once each sidereal day. Half a day later, the disk views the galactic center nearly edge-on. The cylinders, of course, view the center identically at both times.

If Einstein's 1916 theory is correct, and if the center of our Galaxy is the source of the radiation, there should be relatively few coincidences between the disk and a cylinder during the time that the face of the disk views the galactic center. Many more coincidences should occur during the time when the disk sees the galactic center with its edge. If the proposed modifications of the theory are correct, there should be coincidences at both times.

Two six-month experiments were carried out simultaneously. In one experiment, the two cylinders at Argonne and Maryland operated in coincidence, as before, and produced the same twin peaks in the region containing the galactic center. The second experiment used the Argonne cylinder and the Maryland disk in coincidence.

Relatively few coincidences appeared during the time that the disk faced the galactic center. Almost twice as many coincidences occurred during the time, half a day later, when the galactic center was nearly edge-on to the disk. The minimum seen when the disk faced the galactic center implies that Einstein's Theory of General Relativity needs little or no modification.

The galactic center is far away—nearly 25,000 light-years. To account for the number of coincidences observed, a gravitational radiation source at that distance must, each year, radiate energy equivalent to between 1/10 and 1/100 of the total mass of the sun at frequencies around 1,661 cycles per second. But the radiation probably has a much wider frequency range. In addition, the detectors may actually detect fewer than 10 per cent of the signals, because they cannot identify signals much weaker than the average thermal noise. These assumptions suggest an even more enormous loss of gravitational energy by the galactic center—at least a thousand suns, and possibly as many as 100,000, converted entirely to gravitational waves each year.

This is an incredible transformation, so large that it casts doubt on my experiments. I have always realized this energy problem existed and noted the large energy loss in my earliest published papers. I fully expected that a storm of criticism would follow the publication of my results in 1969 and 1970, and that few physicists would find such energy values acceptable.

I was surprised and pleased when three astrophysicists, Dennis W. Sciama and Martin J. Rees of Cambridge University in England and George B. Field of the University of California, Berkeley, published a paper in 1969 which pointed out that a large galactic energy loss had long been suspected. Stars seem to be leaving the Milky Way. This

A Chilly Confinement

Each successive wall of the Argonne supercooled antenna further isolates the cylinder from the environment. The entire assembly is supported by acoustic filters.

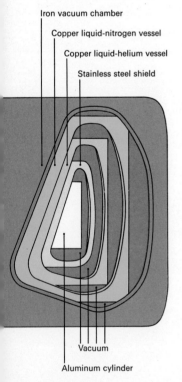

Iron vacuum chamber

Copper liquid-nitrogen vessel

Copper liquid-helium vessel

Stainless steel shield

Vacuum

Aluminum cylinder

suggests that the Galaxy is undergoing an expansion that is fueled by the total conversion into energy of a few hundred solar masses per year in the galactic center. While this energy is very large compared with other kinds of radiated energy, it is still much smaller than the energy implied by my observations.

It is also conceivable that the observed radiation originated beyond the galactic center. Only the direction of the source was observed, not its distance. Gravitational radiation is not appreciably absorbed by matter, and it may have been accumulating in the universe since the beginning of time. It could not escape if the universe is finite and unbounded, as Einstein thought. The strong gravitational field at the center of the Galaxy might focus this gravitational radiation, and make it appear stronger in that direction.

Today, there is a widespread impression among physicists, many of whom have visited my laboratory, that these gravitational-radiation experiments are sound and that the primary conclusions based on the results are correct. Still, the very large energy release that is implied is the strongest reason to suspect that the results of the experiments are not completely understood.

I am now recording all detector outputs on magnetic tape. At the University of Maryland, computer programmer Brian K. Reid is developing a program that will process the coincidence data. In late 1971, it will replace all human observations. More sensitive detectors operating at very low temperatures are also being developed – one at the Argonne National Laboratory by cryogenic engineer John R. Purcell, his associates, and myself, and another at Stanford University by physicists William M. Fairbank and William O. Hamilton. By going to low temperatures – first to $-451\,^{\circ}$F., which is attainable using liquid helium, then down to thousandths or millionths of a degree above absolute zero by special cryogenic techniques – we will greatly reduce the thermal noise. We can then study finer details of the signals and should detect fainter signals arriving from much greater distances, even from other galaxies.

The huge energies that now make the experiment suspect give it, at the same time, enormously greater potential importance than earlier discoveries of new kinds of radiation. Using improved detectors, we may confirm that we are witnessing the release of gravitational energy so enormous, so unimaginable, that all other forms of energy are dwarfed in comparison.

A sensitive gravimeter, scheduled to be placed on the lunar surface by Apollo 17 astronauts, may detect gravitational waves. Because it has little seismic noise, the moon is an ideal low-frequency antenna.

For further reading:

Bergmann, Peter G., *The Riddle of Gravitation: From Newton to Einstein to Today's Exciting Theories,* Scribner, 1969.
Gamow, George, "Gravity," *Scientific American,* March, 1961.
Gamow, George, *Gravity,* Doubleday, 1962.
Thorne, Kip S., "The Death of a Star," *Science Year,* 1968.
Weber, Joseph, "The Detection of Gravitational Waves, *Scientific American,* May, 1971.

Molecules
In Space

By Patrick Thaddeus

A bonanza between the stars is the complex chemical lode
from which stars, planets, and perhaps living things evolve

Faint radio signals from distant interstellar molecules converge on the microwave detector at the focal point of a giant radio telescope.

In November, 1968, a team of scientists from the University of California, Berkeley, pointed a radio telescope at the distant, dense clouds of gas and dust that lie in the direction of the center of the Milky Way, our Galaxy. The telescope picked up two faint microwave signals that could be emitted only by molecules of ammonia (NH_3).

A few weeks later, the same group—headed by electrical engineer William J. Welch and physicist Charles H. Townes—detected a much stronger microwave signal coming from other high-density clouds, this time emitted by water vapor (H_2O). Then, in March, 1969, radio astronomers Lewis E. Snyder and David Buhl of the National Radio Astronomy Observatory (NRAO), Benjamin Zuckerman of the University of Maryland, and Patrick Palmer of the University of Chicago detected a number of far-off regions rich in formaldehyde (H_2CO). By mid-1971, water and ammonia molecules had been found in only a few directions, but the more fragile formaldehyde—an

The 140-foot telescope of the National Radio Astronomy Observatory scans the Milky Way day and night from a quiet valley near Green Bank, W.Va. It was used to discover five different molecules.

The author:

Patrick Thaddeus is engaged in the search for molecules in space. He is a physicist at NASA's Goddard Institute for Space Studies and adjunct professor of physics at Columbia University.

organic molecule—had been found virtually all over the dusty areas of the Milky Way.

These discoveries marked the beginning of a new and exciting phase of astronomy that is advancing rapidly—the study of molecules in interstellar space. Prior to 1968, astronomers had discovered only four molecular fragments, or free radicals, CH, CH^+, CN, and OH, in the spaces between the stars. Ammonia, water, and formaldehyde were the first complete and chemically stable molecules to be found there. By June, 1971, however, a total of 21 species of interstellar molecules had been discovered, and the list was increasing at the rate of better than one new molecule a month. Their discovery has radically changed our ideas about the chemistry of space.

Much of the excitement is because formaldehyde and nine of the other molecules contain both carbon and hydrogen atoms. These are organic molecules, members of the enormous class of compounds on which all life on earth is based. There is no reason to suppose that living organisms or any life process produced these cosmic organic molecules. Yet, their existence has aroused the keenest interest among scientists who ponder how life on earth began.

Our solar system condensed by gravitational attraction out of the interstellar clouds of dust and gas about $4\frac{1}{2}$ billion years ago. Today, stars, and presumably planets, are being formed in contracting regions in the dense interstellar clouds. The discoveries of complex organic molecules suggest that, at their creation, planets already have an extensive set of the basic chemical building blocks from which living things may in time be synthesized.

Astronomers believe that the original matter of the universe, and thus of the Milky Way, was entirely hydrogen and helium, and that the first interstellar clouds and the first stars, were "clean." But stars derive energy by thermonuclear burning, which changes hydrogen into helium, and helium into heavier elements. Therefore, with the passage of time, the first stars became increasingly "dirty" as the ashes of their thermonuclear infernos accumulated. An important property of stars appreciably heavier than the sun is that they ultimately shed much—maybe most—of their gases back into space. They may do this gently, when they puff up into red giants as much as 100-million miles in diameter after their hydrogren is used up. Or they may expel matter catastrophically in gigantic supernova explosions that mark the star's death. In either case, gas richer in heavy elements than the gas from which the star was formed is injected back into cold space and mixes into the interstellar clouds. In this way, the clouds gradually acquired their present chemical composition over the 10 billion years the Galaxy has existed.

The tens of thousands of gas and dust clouds in the Milky Way vary greatly in size and density. Many of them, apparently, are rich in molecules. Hydrogen and helium are still by far the dominant elements in the clouds, however. Everything else—all the heavier elements, such as carbon, oxygen, nitrogen, silicon, and iron—probably account for only 1 per cent of the cloud's mass.

The dust grains, tiny flecks of cosmic dirt that make the clouds visible in photographs, are aggregates of the heavier elements. The clouds all seem to have about the same chemical composition and proportion of gas to dust. There is roughly one dust grain per trillion hydrogen atoms. Studies of starlight scattered by and passing through the clouds reveal that the dust grains are microscopic—about 1/250,-000 inch in diameter—comparable in size to the smoke particles from a cigarette. But little is known about their chemical and physical structure and how and where they were formed. They probably contain most of the metals and rocklike material of which the earth is made, and perhaps some of the carbon, oxygen, and nitrogen, as well. Largely from ignorance, astronomers have long pictured the dust grains as being simple in composition—flecks of sand, graphite, or ice, for example. But it is conceivable that they actually are complex, and consist, at least in part, of a snarl of molecular chains and rings, like a tangled ball of yarn.

It takes such small dust grains a long time to build up, atom by atom, in the near vacuum of interstellar space. This process can scarcely account for the vast number and size of the grains. It seems more likely that the grains are formed comparatively rapidly in the denser atmospheres of red giant stars as gas is being lost, in somewhat the same way that soot particles condense out of a candle flame as the hot gas cools down.

Molecules detected by mid-1971 at various points in the Milky Way include many organic molecules, containing both carbon and hydrogen.

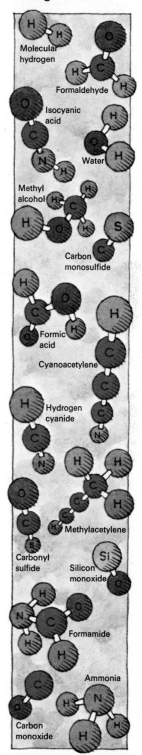

Molecules Among the Stars

Molecular hydrogen

Formaldehyde

Isocyanic acid

Water

Methyl alcohol

Carbon monosulfide

Formic acid

Cyanoacetylene

Hydrogen cyanide

Methylacetylene

Carbonyl sulfide

Silicon monoxide

Formamide

Ammonia

Carbon monoxide

Where there is much dust, there also are many molecules. Undoubtedly one reason molecules exist in regions where dust is abnormally dense is that the dust protects them. The molecules are split apart by starlight—especially by the intense ultraviolet light produced by hot, young stars. Cosmic dust, however, absorbs ultraviolet light and is therefore extremely effective in shielding the molecule-filled interiors of the dense interstellar clouds against these energetic and destructive rays.

Because the dust in the clouds is so opaque, however, only one of the recently found molecules has ever been found by astronomers using optical telescopes. It is the most plentiful molecule in the entire universe, molecular hydrogen (H_2). Virtually all the hydrogen in the interiors of the dark clouds is probably in this form. Molecular hydrogen is elusive. It does not absorb or emit signals that can be detected by ground-based telescopes peering through earth's absorbing atmosphere. H_2 was discovered by physicist George Carruthers of the Naval Research Laboratory in 1970 by means of an ultraviolet spectrometer lifted miles above the earth's absorbing atmosphere by rocket. The spectrometer observed absorption of light coming from a bright star by a cloud of hydrogen molecules.

Radio signals that reveal nearly two dozen other molecules in the black clouds result from a quite different sort of atomic motion than that which produces the more familiar optical signals. Optical signals result when the electrons of an atom or molecule gain or lose energy. This usually changes the actual size and shape of the atom or molecule. On the other hand, the radio signals coming from within the very cold, low-energy interstellar clouds are most often produced by a change in the rotation of a molecule as it tumbles end over end, and this change does not usually change the molecule's structure. In both cases, however, the wave length of emitted or absorbed radiation is inversely proportional to the energy that the molecule loses or gains. Optical signals have very short wave lengths, about 1/20,000 centimeter. The wave length of the radio signals are very much longer, about 1 centimeter, and are thus in the middle of the microwave band of radio wave lengths.

Not all microwave signals originate from changes in molecular rotation. Sometimes they, like the shorter wave length visible radiation, result from changes in the molecule's structure. The ammonia molecule, for example, is a three-sided pyramid with the three hydrogen atoms at its base points and the nitrogen atom at its apex. Ammonia's 1¼-centimeter wave length signal results from the inversion, or turning inside out, of the molecule. This occurs about 24 billion times each second, the nitrogen shuttling back-and-forth between the equilibrium positions on either side of the base. In methyl alcohol (CH_3OH), discovered in September, 1970, microwave signals are produced by the hydroxyl group (-OH) twisting back-and-forth with respect to the methyl group (CH_3-).

The molecules in space are found inside the Milky Way's dense dust clouds. There, safe from the destructive ultraviolet rays pouring from hot stars, they can form and survive.

How Radio Waves Detect the Molecules

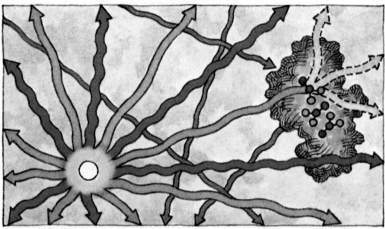

Radio waves pour from the ionized hydrogen gas surrounding hot stars, *top.* Waves at 21-cm wave length (pink arrows), also come from clouds of hydrogen atoms. These signals penetrate the dust clouds. Each type of molecule within can absorb signals (green arrows) only at specific wave lengths. We can detect the molecules by their absorbed signal, *bottom left,* or by the signal they emit in our direction, *bottom right.*

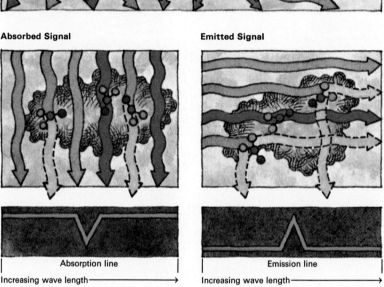

Absorbed Signal

Emitted Signal

Absorption line

Emission line

Increasing wave length⟶

Increasing wave length⟶

Each of these transitions, or changes in the molecules, are accompanied by either emission or absorption of radiation over a very narrow wave-length interval. Whether the microwave astronomer detects a specific molecule by absorption or emission of radiation depends on the physical conditions, such as temperature and density, to which the molecule is subjected, and whether a source of background radiation exists. The precise wave lengths of transitions—called the spectral lines—are characteristic of a given molecule and are often very precisely known. Because spectral lines are very narrow, it is unlikely that two molecules will have a line at the very same wave length. In fact, a molecule can often be identified on the basis of a single spectral line. Two or more lines establish the identity beyond question.

Except for molecular hydrogen, ammonia, and water vapor, all of the recently discovered molecules have been found with two radio telescopes, both constructed and operated by the NRAO. One is a large antenna, 140 feet in diameter, located near Green Bank, W.Va., in a sparsely populated valley to avoid radio interference. This instrument is best suited for observations at moderately long wave lengths – from about 1 meter down to 1 centimeter.

The other telescope is at the site of Kitt Peak National Observatory in southern Arizona. It has a 36-foot antenna with an exquisitely precise surface that permits observations from 1 centimeter or longer wave lengths down to about 1 millimeter, the shortest wave length at which water vapor in the earth's atmosphere permits the transmission of radio waves. Rain, clouds, and high humidity, which offer little hindrance to longer wave-length observations with the 140-foot telescope, can appreciably absorb short radio waves from space. So the 36-foot telescope is located at a dry, clear site.

Neither of these instruments is unique from the standpoint of size alone. Much of their success in the discovery of molecules in space must be credited to the sensitive electronic receivers and data-handling systems with which they are equipped.

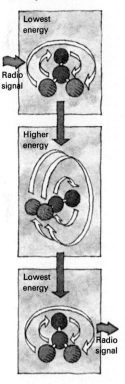

A Microwave Flips The Spin Axis

Lowest energy

Radio signal

Higher energy

Lowest energy

Radio signal

Formaldehyde molecules absorb and also emit 6.21-cm radio signals by changing rotation. In the strange quantum world of molecules, the two spins having the least energy are around different axes.

All the molecules discovered with the 140-foot telescope have been organic. In July, 1970, astronomer Barry E. Turner of NRAO found cyanoacetylene (HC_3N), a straight chain of atoms arranged in the order listed in the formula. Scientists from Harvard University, the Smithsonian Astrophysical Observatory, and the University of Maryland detected methyl alcohol and formic acid (CH_2O_2) in late 1970. Then, in February, 1971, a team from the University of Illinois found formamide (H_3NCO), which is similar to formaldehyde except that one of the hydrogen atoms is replaced by an amide group (-NH_2).

Of the 11 molecules detected with the 36-foot telescope, 5 are organic, and the structure and composition of one is not known. Snyder and Buhl, members of the team that first observed formaldehyde, detected hydrogen cyanide (HCN) in 1970 at a wave length of about 3 millimeters. At the same time, they accidentally found a radio signal of an unknown substance they call X-Ogen at an adjacent wave length. Physical chemist William Klemperer of Harvard suggested that X-Ogen might be the electrically charged fragment HCO^+. The same two observers also discovered isocyanic acid (HNCO) and methylacetylene (CH_3C_2H) at about the same wave length in 1971.

The very first molecular discoveries using the 36-foot telescope were made by three Bell Telephone Laboratories scientists, Arno A. Penzias, Robert W. Wilson, and Keith B. Jefferts. In 1970, this team detected the cyanogen radical (CN) at a wave length of 2.6 millimeters. This is the only substance so far detected both in the optical and radio regions. They also detected a diatomic molecule, carbon monoxide (CO), that because of its stability and the relatively high abundance of its constituent atoms is likely to prove the most widely distributed radio molecule of all.

The Bell Labs group has continued to specialize in observations at very short wave lengths, from 2.0 to 2.7 millimeters. In February,

Lowering the detector
housing into the
elevated shack, *above,*
allows technicians to
change the electronics,
right, so that the
telescope can detect
signals in the wave
length range of a
suspected new molecule.

1971, in collaboration with Philip M. Solomon of Columbia University, they discovered carbon monosulfide (CS) and, two months later, carbonyl sulfide (OCS) and the organic molecule methyl cyanide (CH_3CN). And with Marc Kutner of Columbia and me in April, 1971, they also found silicon monoxide (SiO) and three new lines of the formaldehyde molecule.

On the basis of the relatively small number of observations already made, there is good reason to believe that the 2.5 millimeter line of the carbon monoxide molecule will turn out to be one of the workhorses of radio astronomy. The line should provide astronomers with much new information about the structure and internal motions of dust clouds all over the galaxy.

It seems certain that the interstellar molecules we have already discovered are beginning to produce a major revolution in our knowledge of the Milky Way. For instance, the molecules' radio lines are marvelous tools for investigating the dense interstellar clouds where stars are being formed. Astronomers have trouble observing these regions in visible light because of the dust grains. Radio astronomy's traditional tool for studying clouds of interstellar gas—the

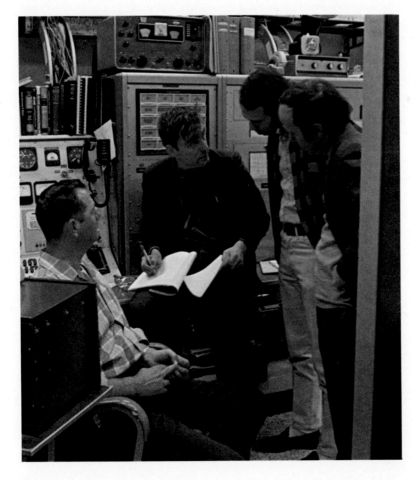

The author, center, reviews plans for a week's hunt for distant molecules with a team of scientists and technicians who man the 140-foot telescope.

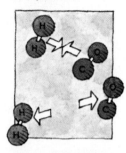

Collisions between
atoms and molecules
usually involve too
much energy to form
a new molecule, *top*.
Collisions while on a
dust grain, which
can absorb the energy,
above, may allow a
new molecule to form.

signal emitted and absorbed by atomic hydrogen at a wave length of
21 centimeters–is of limited value for two reasons. First, the dense
clouds are usually very small as seen from earth, often less than a
tenth of a degree in angular size. These small clouds cannot be resolved
by radio telescopes working at the long 21-centimeter wave lengths.
In addition, hydrogen atoms, at high densities, tend to form hydrogen
molecules, and thus cease to emit signals at 21 centimeters, or at any
other radio wave length.

Many of the new molecular lines, on the other hand, have wave
lengths of only a few millimeters, and can be observed with existing
telescopes even if the source is a cloud no larger than a sixtieth of a
degree in angular size.

Further, both the high density of the gas and the presence of dust
particles favor the formation of molecules. The higher the density, the
more likely two dissimilar atoms are to collide and stick together.
If, for example, a C and an H join, they form a CH fragment, which
might be the first stage in the formation of HCN or CH_3CN. It is
also likely that the surface of the dust grains serves as a catalyst, on
which reactions can occur between absorbed atoms or molecules
that would be impossible in a direct collision. Scientists believe this is
the way in which molecular hydrogen forms. It is never produced
when two H atoms simply collide. Dust grains may be important in
the formation of more complex molecules, as well.

It is just this tendency for molecules to form in high-density clouds
that makes them such valuable tools with which to investigate the
complex chain of events that leads to the formation of stars. The
large, low-density clouds that have been studied intensively at 21-
centimeter wave length contain about 10 hydrogen atoms per cubic
centimeter. Because the gravitational attraction within such clouds is
weak, they are usually disrupted before they can contract enough to
form stars. But when a cloud reaches the density of 1,000 or more
atoms per cubic centimeter that is favorable to the creation of molecules, it is tightly held together by its own gravity. Thus, stars are
almost certain to form in this region in the near future.

The famous nebula in the constellation Orion is a spectacular
example. This great cloud of glowing gas located in Orion's belt is
1,500 light-years away, yet it is so bright that it is readily visible with
a pair of binoculars on a dark night. The cloud is excited by four
young and exceedingly hot stars called the Trapezium, which are
located near its center. The nebula's brilliant core and filaments
consist mainly of ionized hydrogen–free electrons and protons–a
plasma heated to about 18,000°F. Given the susceptibility of molecules
to destruction by high temperatures and ultraviolet light, the Orion
Nebula would be the last place to expect to find molecules. There is
protective dust in the area, however, seen as dark lanes on a photograph. And, it is logical to suppose that where stars have recently
formed, others may be in the process of being born.

Shifts of Molecular Signals

Motions along the line of sight slightly change the wave length of the molecules' signals. Such Doppler shifts of the 6.21-cm formaldehyde signal were apparent when the molecule was discovered in 1969, *below.* Each dip corresponds to a cloud having a different velocity.

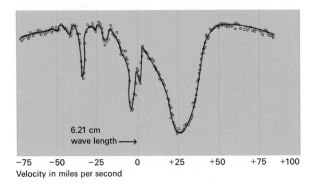

6.21 cm
wave length ⟶

| -75 | -50 | -25 | 0 | +25 | +50 | +75 | +100 |

Velocity in miles per second

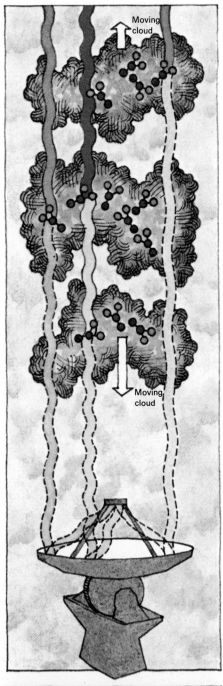

Moving cloud

Moving cloud

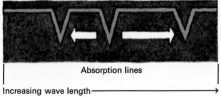

Absorption lines

Increasing wave length ⟶

In fact, observers have found the Orion nebula to be a splendid source of molecular lines, particularly at short wave lengths. Carbon monoxide is spread over the entire area. Formaldehyde, hydrogen cyanide, and several other molecules are found in a small source less than a tenth of a degree in size and slightly displaced from the Trapezium. Analysis of the radio lines indicates that the gas in this small source is extremely dense, and that its total mass is several hundred times that of the sun—about equal to that of the four Trapezium stars combined. At this density, the molecular cloud must be totally opaque to starlight, and therefore cannot lie between the earth and the visible Trapezium stars. Evidently, we are observing molecules in a part of the far side of the Orion nebula that is totally invisible on optical photographs.

From the mass deduced from their molecular observations at millimeter wave lengths, it was calculated that this molecular cloud is rapidly contracting, and will soon form stars. Other observations indicate that some stars have already formed. Near the center of the cloud, intense sources of OH and H_2O radio signals have been observed that probably come from objects not much larger than the solar system. This size has been determined from observations with intercontinental radio interferometers—simultaneous observation with telescopes located on different sides of the earth and synchronized by atomic clocks. Such interferometers

Two Views of Orion's Glow

The Orion Nebula glows almost as brightly in molecular signals as in visible radiation, *below*. Visible light comes from the four brilliant trapezium stars, but the intense formaldehyde radio signals, *right*, come from a dark cloud behind the nebula.

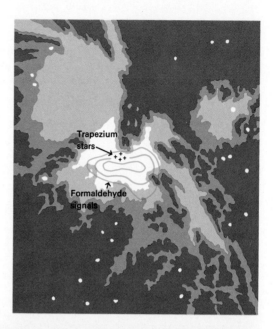

can determine the size of a small source as well as a single antenna as wide as the earth. The observed energy coming from the small sources is much too great to be the energy released by contracting clouds. Evidently, there are internal sources of energy, and these are most reasonably young, hot stars.

Microwave observations made in other parts of the Milky Way show that the hidden molecular cloud behind the Orion nebula is not an isolated case. The numerous appearances of the millimeter wave-length signals of formaldehyde and other molecules is probably a general sign of other high-density regions where stars are being born. It is most likely that astronomers who have no interest in the intriguing chemical questions raised by the interstellar molecules will seize them as observational tools they can use to study the processes of stellar formation.

What of the future? Millimeter wave-length receivers, which have proven so successful in the hunt for interstellar molecules, could theoretically be made even more sensitive. It is likely that this maximum sensitivity can be achieved as commercial communications technology moves into the millimeter wave length region of the electromagnetic spectrum.

The capacity of present systems, such as the telephone and microwave networks, to meet the rising demands of modern communications is strictly limited. The remedy must be to take advantage of still higher-frequency channels. The radio astronomer who uses millimeter wave lengths therefore finds himself in the enviable position of riding the crest of a major technological revolution. It is no coincidence, for example, that scientists from Bell Telephone Laboratories, and instrumentation developed by them, have played a significant part in the molecular discoveries at short wave lengths. There is every prospect that this basic instrumentation on earth will soon be made much more sensitive to the molecules in the vast, active regions of the Milky Way.

A great deal of additional work will be needed to establish the importance of the cosmic molecules to the origin of life. It will perhaps hinge on whether the hectic searches currently underway continue to turn up an increasingly complex sequence of molecules, or whether the vein is largely played out. I will hazard a prediction that the present flood of discoveries will continue to grow during the next few years, and I would not be in the least surprised if many dozens—even hundreds—of molecules were eventually found in interstellar space.

For further reading:

Barrett, Alan H., "Radio Signals from Hydroxyl Radicals," *Scientific American,* December, 1968.

Snyder, Lewis E., and Buhl, David, "Molecules in the Interstellar Medium," *Sky and Telescope,* November and December, 1970.

Wick, Gerald L., "Interstellar Molecules: Chemicals in the Sky," *Science,* Oct. 9, 1970.

Probing the Pawnee Grassland

By John Barbour

**A virgin Colorado plain is the object of intensive
scrutiny by scientists who seek to understand
and ultimately protect its natural balances**

The land at first glance seems to be a high plain, or plateau, falling
away from the eastern edge of the Rocky Mountains. But the view is
misleading. This is not plateau, but a low-lying piedmont, a plain at
the base of mountains. It is the Pawnee National Grassland of northeastern Colorado—15,000 acres of rolling, treeless, windswept terrain.
It is dry and brown most of the year, although heavy summer rains
and the hot summer sun awaken it and briefly make it green.

Today, this Colorado grassland is like a patient on an examination
table, being studied and measured, probed and biopsied in its largest
and in its most minute details. The scrutiny continues through wind
and storm, on quiet, rainless days, and throughout the year.

This unprecedented scientific analysis of an entire biological environment is part of the Grassland Biome Program of the International
Biological Program (IBP). The IBP is a vast, five-year study begun in
1967 as a joint effort by scientists of many disciplines from many
nations. IBP scientists are surveying the environment to learn how
best to maintain its delicate balances as man intrudes. They are closely

Chirping for their food, lark bunting nestlings illustrate the grassland
riddle : What part will they play in the life cycles of the plain?

scrutinizing deserts, forests, the oceans, and other biomes—commu-nities of plants and animals—to try to reduce nature to an understand-able, predictable system.

The Pawnee study involves more than 80 scientists from 20 univer-sities and laboratories working at a 320-acre site in Weld County, Colorado, 40 miles northeast of Fort Collins. These scientists study everything imaginable: the bacteria and fungi in the soil, the digestion of buffaloes and cows, even the lowest struggling blade of grass and the shadow it casts. They examine the mating of the lark bunting, watch the 13-lined ground squirrel rob nests, and scrutinize the golden eagle that hunts game from the blue sky. They measure the decom-posing of vegetation and the ability of the soil to hold water.

The scientists are trying to translate all of these observable acts of nature in the Pawnee grassland into the numerical language of sci-ence. They hope to reduce this information to mathematical relation-ships, then reduce the relationships to larger, interlocking patterns or models, and feed these models into a computer. Finally, they will use the computer's memory and speed to uncover the secrets of the grass-land they have electronically captured.

George M. Van Dyne, an ecologist at Colorado State University, Fort Collins, who heads the grassland study, explains that all data the scientists are gathering are necessary to promote better conservation and land use, and also to understand the balances that are at work within the grassland.

"Nature has designed these balances," he says, "and we're trying to find out what they are and what man can do with them. For instance, consider the variety of creatures that graze on the grassland—antelope, cattle, sheep, buffaloes, and goats, among others. They are all in the same area," Van Dyne points out. "What grasses do they choose? They are all at the same cafeteria. What do they order?" And once

The author:
John Barbour, science
writer for Associated
Press, has contributed
Special Reports to past
editions of *Science Year*.

you know that, he says, you can ask another question: What effect does the grazing have on the many kinds of grasses that grow there? "Five or six plants provide 95 per cent of the food. On the Pawnee site cattle are the most important single plant-eater. Antelope are second. And third, we think, is a mite—so small you can barely see it."

No one would have guessed the enormous importance of this smallest of creatures that feeds on the grass. The buffalo, the cow, and the pronghorn antelope devour the grass in great mouthfuls. How mighty are their appetites. No wonder the small grasses struggle for survival.

But look below the surface. Dig down only 2½ inches and you find a teeming world. In one square meter of soil there are 17,000 mites. These are the smallest of "cattle," each measuring only 1/50 inch

The calm beauty of the plain, sparkling in the clear autumn air, *above,* disappears with the onslaught of winter's harsh blizzards, *below.*

Plants, basis of most life in the grassland, intrigue scientists who seek to learn what each contributes to the life of the plain. The prickly pear cactus, *left*, blue grama grass, *below left*, and silver sage, *below*, all have a different story to tell researchers.

long—barely an ink dot. They are foraging on the grass. And though they are tiny, they are a mighty force. There are so many mites that together they weigh 250 times more than all the grassland's many small mammals combined.

How much do the mites eat? No one knows for sure. But they are certainly a major grassland plant-eater. Are they a greater grazing force than all the other creatures that live off the grassland? Charles Proctor, an entomologist at the University of Georgia, Athens, who has laboriously corralled, counted, and weighed a selection of mites, says, "I wouldn't be a bit surprised." Some project scientists still do not think the mites are quite that important, however.

Indeed, few scientists would have guessed that mites could have such an impact. The mites are important in deciduous forests, where the oribatid species is a prime force in the decomposition of fallen leaves. Some scientists have estimated that the absence of all mites from a deciduous forest would reduce the rate of decomposition 30 to 40 per cent. A Russian report estimates the rate to be even higher. In the grassland, the population of mites is only one-tenth that of the forests. But there is far less for them to eat in the sparse grass, and far more competition for it.

The mite study went further in the summer of 1971. The scientists measured the caloric value of the mites' bodies and determined how much oxygen the animals use. Both these studies should help to determine the overall energy flow in the grassland. To get the caloric content, mite bodies are burned in a special chamber where the heat produced is measured. To measure the oxygen they breathe, live mites floating in liquid are sucked into tiny glass tubes that hold less than a thousandth of an ounce of liquid. The tops of the tubes are sealed. As chemicals in the tubes remove the carbon dioxide that the mites exhale, the liquid level rises. The scientists then measure how much liquid the mites' breathing displaces.

This operation is a good example of the patience displayed by the men and women who sift through nature's warehouse in the grassland and record everything they see, everything that may hint at how creatures, vegetation, and the weather relate to each other.

Birds are one of the forces at work in nature, transforming one kind of energy into another. More than 150 species of birds pass through the vast grassland each year, most of them unseen and unheard by man. For instance, take the lark bunting, a western finch midway in size between a sparrow and a robin. It has been visiting the grassland each spring for uncounted years. Yet little was known of its habits there. It and other nesting birds are now being studied closely, their diets analyzed, their mating success tabulated.

The lark bunting arrives in May. The males come first and a few days later the females follow. The males spread out through the low, clumpy grass, displaying their black-and-white wing patches in short courtship flights and singing sweet, trilling notes to attract the females.

Many birds make their home in the grassland. The burrowing owl, *right,* nests in the ground like a prairie dog. The male and the female lark bunting, *center,* are summertime visitors, and the size of their young, *below left,* is of interest to the scientists. The horned lark, *below right,* is the most common bird in the winter months.

The small, brown females move through the staked-out male provinces, and male and female pair up.

The female immediately begins sampling various plants to choose a nesting site. Finally she selects one and nest-building begins in a shallow depression in the ground at the base of the chosen plant. This is usually red three-awn grass because this plant tends to form a dome over the nest, offering both shade and protection. The birds usually build their nests on the southeast side of the plants for protection from the prevailing northwest winds. Of the lark buntings studied in 1971, almost two-thirds chose the red three-awn grass, and young were hatched in two-thirds of those nests.

To find out what the young nestlings were fed the researchers used a curious technique so they did not have to cut them open to examine their stomachs. They tied a thin thread around a bird's throat to prevent swallowing, and went away while the parents fed the young. The scientists returned 20 minutes later to retrieve food samples with forceps, then removed the thread.

A study of the food of the adult lark bunting, based on its natural choice and its caloric requirements, showed just how much ecological effect this and other birds have during the months of grass growth. Field observation and analysis of the contents of the birds' stomachs showed that the daily food intake of the lark bunting from May to July was enormous. On an average day, a single bird consumed 65 grasshoppers, 103 weevils, 122 ants, 33 scarab beetles, and 352 smaller insects. It also consumed almost 750 seeds. There were even small shreds of leaves and stems found in their digestive tracts, some of it from the grasshoppers the birds had eaten. Another migratory visitor to the grassland, McCown's longspur, a small white-breasted finch the size of a sparrow, was discovered to require more than 700 insects and 400 seeds a day.

The birds are not without enemies, of course, and early in at least one study, the presence of man may unfortunately have encouraged one enemy. It happened this way. The blue eggs of the lark bunting are a source of food for the 13-lined ground squirrel. This small hibernating rodent does not put away food in its dens for the winter. Instead, it stores fat in its body. Consequently, the squirrel is famished and is a heavy eater in spring and summer. When researchers discovered a sudden large loss in lark bunting nests they concluded that the ground squirrels were following man's scent to the otherwise camouflaged and protected nests. In fact, the scientists saw an unusually large number of male and female lark buntings defending their nests against the squirrels. To protect the birds, and the experiment, the scientists spread naphtha flakes around the nests they visited. This seemed to successfully confound the sniffing ground squirrels.

By the end of September, the lark buntings head south again in the order in which they came—males first, then females and finally the young birds. As the grassland fades and begins its winter sleep again,

McCown's longspur leaves too. There is little protection for grassland birds through the winter, and the food supply decreases. Nevertheless, some birds have been seen there every month of the year—notably the horned lark, a small, brownish bird with a white-masked face and two small, feathery horns. Whether this bird spends the entire winter in the grassland is unknown.

There seems to be no end to the vast and apparently unrelated materials being collected in the grassland. Of course the grasses themselves are under close examination. The scientists found, for example, that buffalo grass and blue grama grow best when daytime temperatures are about 90° F. and nighttime temperatures drop to 80°. Western wheatgrass likes it cooler—60° during the day, 50° at night. But when day temperatures rise to nearly 100°, even the warm-season grasses seem to fade quickly.

Direct outdoor study is difficult. Thus it is necessary to use a greenhouse to measure the growth rates of prime grasses—blue grama grass, buffalo grass, and western wheatgrass. And only in a greenhouse have scientists, using radioactive carbon 14, been able to measure the grasses' decomposition rate when they die.

The scientists are especially interested in the uses of water in the nearly arid grassland. Precipitation records have been collected for the past 30 years, but more information is needed. Sensors are being placed in the major types of soil to find out how well the water is received and retained, and how the rainfall is related to the whole flow of energy in the grasslands.

In 1971, the researchers turned to a new measuring technique. They carefully cut out a cylinder-shaped portion of soil 10 feet in diameter and 4 feet deep and installed a delicate weighing device in the hole. Then they returned the soil section to its original position, on top of the device, so they could measure just how much rain water the soil absorbs, and how much evaporates. The scale, read through an access hole 15 feet away, is so sensitive it can measure the gain or loss of one-thousandth of an inch of water from the soil's surface.

So dependent is the grassland on summer rain that a sudden downpour produces an instant growth of green—which disappears in an hour or so. Ninety-nine per cent of the plant roots lie in the top 12 inches of soil, and the rain thus brings a quick transformation to the shallow-rooted grass. What happens during the surge of green? Van Dyne says that the instant greenery changes the entire balance of available food. How do the animals respond? To find out, researchers are irrigating a separate section of land to turn on the instant green artificially so they can carry out carefully controlled experiments.

Two portable stations, about 1,000 feet apart, are measuring almost all other weather phenomena from 4 feet below the ground surface to 6½ feet above. Automatic devices measure wind speed, humidity, soil temperature at various levels, incoming solar radiation, reflected radiation, and barometric pressure.

Grassland mammals like the 13-lined ground squirrel, *below,* black-tailed
jack rabbit, *bottom left,* and Western badger are all being studied by
scientists interested in how these animals affect the ecology of the plain.

The instruments send data directly to a nearby trailer where it is recorded on magnetic computer tape. The tape can store 36 channels of data simultaneously. There are 4 channels for air temperature at various levels above the ground; 14 for soil temperature; 6 for air speed and direction; 4 for humidity; 7 for various radiation inputs; and 1 for barometric pressure. The time and date of each bit of information is coded. With these data, computers will eventually be able to measure the impact of the changing weather and soil conditions on the animals and plants in the grassland.

Even if you know all the factors, however, the problem of understanding the interrelationships is never simple. For instance, take the red three-awn grass, the preferred nesting place of the lark bunting. If the normal grass supply is reduced, the cattle will eat the red three-awn, and then the lark bunting cannot nest. And if the lark bunting does not nest, the grasshoppers it feeds on will increase out of control and eat the other grasses. Without these low grasses the cattle will turn elsewhere for their food.

The interrelationships of a biome are complex—perhaps too complex, although no one knows for sure yet. The interdependence in the grassland of land and water, grass and mite, eagle and jack rabbit affect each other, then together they affect a third relationship, and a fourth, and a fifth. Everything lives together.

American buffaloes still roam here and there on the grassland. Because researchers want to know all about these animals, they strap specimen bags to some of them to collect evidence of what they eat.

Researchers and their colleagues at the Natural Resources Ecology Laboratory at Colorado State University record the tabulated summaries of grassland observations on computer cards in the form of punched holes. Then they carefully construct models to try to make sense of it all. It is a laborious task.

From the beginning in 1967, the scientists found they would have trouble creating the mathematical models—the set of equations representing nature's intricate ways. Often, when they tried to stuff nature

into a computer, it seemed that something would elude them. Nature is something more complicated than it once seemed.

The problem is not like that involved in devising a pattern of a machine or a physical process where all the details are known and can easily be fed into a computer. Those details can be related on a simple one-to-one basis, and the computer can tell you what would happen to all the other components if one component changed.

But in modeling an ecosystem—the network of relationships supporting the organisms—you have too many variables. Each variable affects all the others as they change over time. And each of these changes, in turn, affects all the others, so that you actually have a spider web of changing elements. The system is extremely complex. It defies mathematics, largely because the mathematical theory used to describe these wide-ranging relationships is not fully developed. As a result, attempts to model the complete grassland for a computer have so far not been very successful.

At the University of Georgia, Bernard Patten—a systems designer—has come up with a skeletal model for the Colorado grassland that does fit into a computer. But he had to make some compromises. For instance, rather than include all the creatures and environmental processes he decided to use about 45 central processes—turnover rates, biomass (all life in the area), standing crops, productivity, and populations of major species, for example. Any attempt to duplicate the natural processes completely in the computer would move from the sure world of simple proportional relationships—of linear mathematics—to the unsure world of nonlinear mathematics.

So Patten decided to settle for a model that could be expressed in linear mathematics only, even if nature expressed in this way is less than realistic. The problem with nature, Patten explained, is that "every species is doing something. We decided to look for the processes

The tiny mite, here magnified 1,400 times, may be the major grassland consumer.

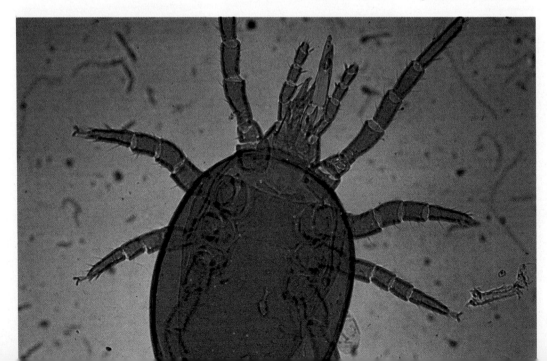

Using a huge "vacuum cleaner," a grassland researcher carefully collects insects for laboratory examination.

An electronic direction finder, *right,* helps researchers keep track of jack rabbits, that have been tagged to help signal their presence. The electronic weather station, *far right,* is one of many collecting around-the-clock data in the Pawnee grassland.

Researchers prepare a huge, but delicate, scalelike device that measures the rate of water evaporation of the soil layer placed on top of it.

that are central, look for the creatures that are central. In the end we had to trade off some realism for abstraction."

Patten and his students developed a model that works. It describes the transfer of biomass from one element to another—that is, from grass to grasshopper to lark bunting, or from fungi and nutrients to grass to mite to the insect that preys on the mite, and ultimately back to the nutrients again.

"If you remove the cows or some other big force, you're going to have to redesign the model, of course, but my interest was in developing a general model to compare one ecosystem with another. To do that, I wanted it all in the same language, using the same mathematics as our other computer projects at the University of Georgia. Now that's different from designing a mathematical theory that will be able to solve specific problems."

Patten's linear model is, in a sense, like a motorboat that barrels along a direct course despite winds or tides. What would be ideal is a sailboat that could make dozens of small correcting maneuvers along its course to accommodate the changes. Someday, perhaps, a more complete model will be fashioned from nonlinear mathematics that can accept more variables and more interrelated changes.

At Fort Collins, in fact, George Innis and George Van Dyne are already working on several nonlinear models. One model has worked well and seems stable, but the final version is far from being able to represent what the researchers understand to be realism in the grassland. More work is needed.

The target date for completion of the Grassland Biome Program is 1972. The problem is so complex, so great mathematically, that it may not be finished by then. One nonlinear model requires about 40 separate differential equations and hundreds of variables within each equation. And even then the researchers admit they still don't know all the details of nature that should go into the computer.

So, as the grassland study goes into its fourth year, nature is still defiant. It has so far hidden many secrets from the prying eyes of scientists, and it has baffled man's mathematics with its complexity. But it is a grand experiment—a nature study on the broadest possible basis, an effort by man to observe all of the myriad details without disturbing nature itself. And for the first time man acknowledges his debt to this lean and hard land and says, in essence, to nature, "I will try to fit into your scheme if you will confide it to me."

For further reading:

Kormondy, Edwin, *The Concepts of Ecology,* Prentice-Hall, paperback (biological science series), 1969.

Smith, Robert L., *Ecology and Field Biology,* Harper, 1966.

Van Dyne, George, editor, *The Ecosystem in Natural Resource Management,* selected papers, Academic, 1969.

Weaver, John E., and Albertson, F. W., *Grasslands of the Great Plains,* Johnsen, 1956.

Is Man Changing His Climate?

By Reid A. Bryson and John E. Ross

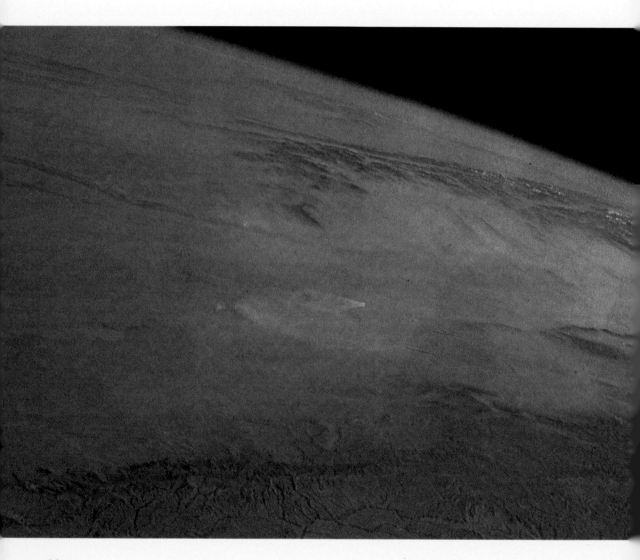

The consequences are not yet known, but man has joined nature as a dust producer in an alliance that may bring about severe climatic shifts

The dust that fills the air over Southeast Asia, as seen in Gemini 5 photo, affects the area's climate.

High in the snowfields of the Caucasus Mountains—some 15,000 feet above sea level—a team of Russian scientists lower themselves deep into a glacial crevasse. Using rope slings, they cautiously edge down hundreds of feet into the icy crack. Carefully, they scoop up columns of *firn* (granular snow) from the sheer walls. The weather is so cold that the snow never melts, and each year a thick new layer covers the old. Altogether, the scientists have found more than 180 distinct layers—a record of the snowfall for a century and a half.

The Russians are members of a team headed by climatologist Fyodor F. Davitaia of the Russian Academy of Sciences. They can read the layers of snow just as other experts read annual tree rings. The snow in each sampled layer is melted and evaporated to find out how much dust fell with that year's snow. In this way, the relationship of atmospheric dust to climatic change may be found.

The Russian study is only one of several dust studies in progress throughout the world. At the University of Wisconsin, we are using computers to try to discover what effect dust may have on worldwide temperature patterns today.

Climatologists have long known that dust particles suspended in the air affect weather and climate, often becoming the nuclei around which raindrops and snowflakes form. Dust can also affect the amount of the sun's radiation that reaches the

The Climatic Engine

The sun's energy is the source of all climatic effects. Some of it is reflected back into space by the atmosphere and land surfaces, while the rest is absorbed by the earth. The mean temperature on earth rises or falls when the absorbed energy increases or decreases, respectively.

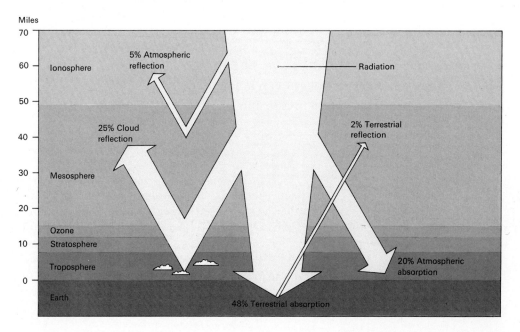

Miles

70

60 — Ionosphere   5% Atmospheric reflection   Radiation

50 —

40 — Mesosphere   25% Cloud reflection   2% Terrestrial reflection

30 — Mesosphere

20 —

Ozone
10 — Stratosphere
Troposphere   20% Atmospheric absorption

0 —

Earth   48% Terrestrial absorption

The authors:
Reid A. Bryson and
John E. Ross are
Director and Associate
Director, respectively,
of the University of
Wisconsin Institute for
Environmental Studies.

earth. However, we still have much to learn about the relationship of dust to climate. These studies have a special urgency now because increased farming and industrial activity are throwing larger amounts of dust and other particulates into the atmosphere.

Climate is the long-term pattern of weather. It includes all the general conditions of heat, moisture, cloudiness, and wind. It is the total average weather over many years, measured in such terms as annual rainfall and mean-temperature patterns. Thus, while climatologists use current weather information provided by meteorologists and weather bureau technicians, they must also study historical weather records of the last few centuries, and the prehistoric records unearthed by geologists, archaeologists, and paleontologists. A climatologist attempts to analyze the climate, discover its patterns, and find out why it changes.

The sun is the source of all climatic effects. About 32 per cent of the solar energy reaching the earth is reflected back into space by the atmosphere, land, and water surfaces. The remaining 68 per cent is absorbed—20 per cent within the atmosphere, and 48 per cent at the earth's surface.

Any change that makes the earth a brighter planet—which causes it to reflect more sunlight—lowers its mean temperature. Likewise, a change that makes the earth a dimmer planet raises its mean temperature. In either case, the change in brightness can be caused by clouds or dust in the atmosphere as well as by changes in the surface cover. Just increasing the normal reflectivity 1 per cent would lower the mean temperature of the earth about 3° F.

The earth, being a sphere, does not receive the sun's energy uniformly. It receives more energy at the equator than at the poles, because the sun's rays hit the earth more directly at the equator. In addition, more energy is reflected at the poles because of the snow and ice cover. Because of these temperature differences, warmer air rises in equatorial regions and flows to cooler regions in a complex pattern that is determined by the energy available, the rotation of the earth, and its changing inclination relative to the sun. This upwelling circulation tends to dominate in the tropics—the low latitudes near the equator. At the higher latitudes, the circulation is dominated by horizontal eddying that moves parallel to the equator rather than by vertical overturning, as in the tropics.

The clouds, water vapor, and carbon dioxide in the atmosphere operate much like a blanket, warming the earth. As the land, the ocean, and the atmosphere absorb solar energy, they radiate it away as infrared (heat) energy, which is absorbed by the blanket of air. Eventually, of course, the heat passes through the atmosphere and is radiated into space. But the radiated energy must equal the solar energy absorbed; the input and output of energy nearly always are in balance.

Many details of these interrelated patterns still cannot be satisfactorily explained. The earth's atmospheric system is highly complex, it operates over a long time span, and it is sensitive to subtle changes. One thing appears clear, however. Climatic patterns are changing and have changed over the centuries. In fact, changes were happening long before man began to walk on the earth.

We know that climate has changed in the past because each change leaves records thousands of years old in peat bogs, ice cores, and seabed clay and silts that scientists can interpret. Records show, for example, that Antarctica had a warm climate many millions of years ago and that glaciers have covered parts of what is now the Sahara. They also show that the mean temperature of the Northern Hemisphere for hundreds of years prior to 1900 was generally lower than it has been during this century.

Fossil pollen and plant spores found in ancient peat bogs and lake sediments can be dated by the radiocarbon method to determine when the fossil plants flourished. In this method, the residue of radioactive carbon atoms in the fossil material determines the time span, because radioactive atoms decay at a known, constant rate. Certain types of plants will grow only in certain types of climate. For example, fossil cypress spores are found only where there was a relatively warm climate, while spruce pollen suggests a cooler one.

Scientists also determine facts about the climate at a specific time by studying the shells of tiny one-celled sea creatures called foramini-

The Earth's Blanket
If the atmosphere did not absorb the infrared energy that is radiated by the earth's surface, *top,* the climate would never be warm enough to support life. However, in a process known as the "greenhouse effect," *bottom,* the atmosphere acts as a blanket to raise the earth's temperature.

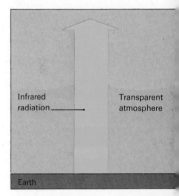

Infrared radiation

Transparent atmosphere

Earth

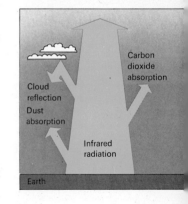

Cloud reflection

Dust absorption

Carbon dioxide absorption

Infrared radiation

Earth

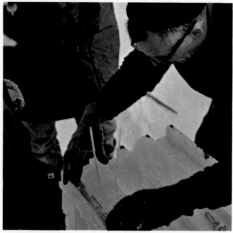

Ice cores are taken from glaciers, *left,* and examined, *above,* for clues to climatic change by measuring dust in layers of snow.

fera that are found in ocean silts. These microfossils also can be dated by the radiocarbon method. Paleontologist John Imbrie of Brown University found in 1967 that the shells of one foraminifera twist in one direction if the water they live in is warm and in the opposite direction if it is cold. Thus, the twist of foraminifera shells taken from a sedimentary layer of limestone is a clue to the climate as far back as 400,000 years. In 1960, Cesari Emiliani of the University of Miami found a way to analyze the ratio of two oxygen isotopes—O^{16} and O^{18}—in the shells of fossil sea creatures. The isotopic ratio is directly related to the temperature of the water at the time these creatures lived. By analyzing the shells from many layers of limestone in this way, the scientists have been able to chart variations in the climate over a period of millions of years.

The actions of man have had wide-ranging effects on the climate. In one sense, man has been trying to change his climate ever since he first threw an animal skin over his shoulders and discovered that it warmed him. Later, when man began to build houses, he extended his control of his personal climate. When he began to plant trees and re-route water for irrigation purposes, he extended his effect on the climate to an ever wider area.

As man built cities, he began to change local climates drastically. This has been particularly true since the development of industry, with its smoke- and heat-producing machinery. Cities generate and trap their own dust and heat. This creates a blanket of pollutants, including dust and soot particles, that hangs over the city. The blanket maintains a man-made city climate that is warmer, wetter, cloudier, and more stagnant than that of the surrounding countryside.

The contaminants that man releases into the air may change the climate on an even wider scale. Carbon dioxide (CO_2) is one of the pollutants most frequently cited. It exists naturally in our atmosphere, and it is used by green plants in photosynthesis. It is also generated by any kind of combustion, including the burning of fuels in homes and factories. At the beginning of the Industrial Revolution in the late 1700s, the atmosphere probably contained about 280 parts per million (ppm) of carbon dioxide. It now contains 321 ppm, and the carbon dioxide content is rising .7 ppm each year. A study conducted in 1970 by Massachusetts Institute of Technology indicates that by the year 2000, the carbon dioxide concentration will probably have risen about 18 per cent to 379 ppm.

Perhaps the most important factor in changing the climate, and the one for which man has the most responsibility, is dust. Man creates dust in farming and in many other ways. Dust is also generated by volcanic eruptions, which throw particles into the stratosphere, the layer of air from 10 to 30 miles above the earth, where they may remain for as long as 10 years. Records show that solar radiation at the earth's surface was sharply reduced following the spectacular eruption of the volcano Krakatoa in Indonesia in 1883. Volcanic dust particles stayed in the atmosphere for five years or more. After the eruption, summers in the Northern Hemisphere were cooler than they had been in the years preceding the Krakatoa eruption.

The record of ancient climates can be found in microscopic shells, called foraminifera, that are taken from seabed sediments, *below.* A study of the direction in which these shells twist, *bottom,* indicates the temperature of water that they once lived in.

Scientists calculate the amount of dust in the air by measuring the scattering of light as it passes through the atmosphere. Studies by Robert McCormick of the National Air Pollution Control Administration show that dust has increased 57 per cent over Washington, D.C., during the past 60 years, and 80 per cent over the weather station at Davos, Switzerland, in the past 30 years. Industrial development around Washington and in western Europe accounts for about two-thirds of the increase, while the rest is due to an increase in dustiness throughout the world.

The study of the snow in the Russian Caucasus Mountains shows that the amount of dust in each layer of snow is about the same for every year from 1790 to 1930. Then, the amount of dust begins to rise dramatically, increasing 19 times by 1963. The Russian scientists found that it leveled off during World War II, perhaps because farming was cut back by the war. The increase was apparently worldwide, because much of this dust is composed of microscopic particles that are carried by the winds around the world.

Some parts of the global atmosphere are actually much dustier than others. The air over northwestern India is probably from 50 to 150 times dustier than the air over Chicago, one of our grimier cities. The dust in India is part of a gigantic cloud that stretches from North Africa to Cambodia. Over northern Africa, the dust cloud is deep and dense. It thins out along the southern coast of Iran and West Pakistan, but it becomes dense and deep again over the Thar Desert and in northwest India's state of Rājasthān, an area of 232,000 square miles. From there the dust cloud diminishes southward and eastward over

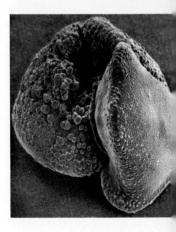

Volcanic Dust Affects Temperature

Volcanic eruptions

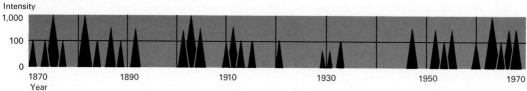

Temperature effect

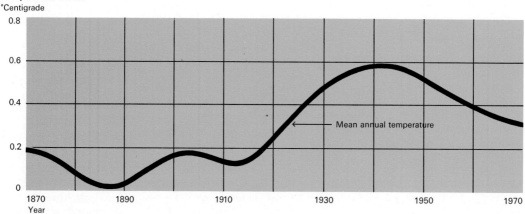

Increase of volcanic activity, which casts great amounts of dust into the atmosphere, appears to affect the average temperature of Northern Hemisphere. Changes occur slowly because of the great thermal mass of oceans.

northeastern India, and can be seen as layers over Burma, Thailand, and Cambodia. The top of this dust cloud is often about 25,000 feet above the Arabian Sea and 30,000 feet above the Thar Desert. Our meteorology team studied it from the air at various times during the past decade, and made extensive field surveys of the quantities of dust on the desert floor.

Meteorological knowledge of the moisture in the air over northwestern India suggests that this Rājasthān area should not be a desert. Indeed, if all the moisture in the atmosphere there came out at once as rain, it would make a layer of water about 1½ inches deep. In the wettest place on earth, such a process at any given time would produce only about 2 inches of water. In other words, the air above the Thar Desert contains about as much moisture as that above Panama, or the Amazon Valley, or the steamy Congo region in Africa, areas that regularly receive heavy rains.

The moisture remains in the air over the Thar, however, because the dust prevents rain from developing. It overseeds the clouds by providing too many condensation nuclei for the water vapor to ever coalesce into raindrops. And it causes the cool upper layers of the dust cloud to fall rapidly, compressing the lower layers of air so that they have a greater capacity to hold moisture. Thus, we have a desert that shouldn't be there, and was not some 3,700 years ago.

This desert was once the home of one of the world's first great civilizations, a vast agricultural empire centered around the cities of

Harappa and Mohenjo-daro. This civilization, sometimes called the Harappan civilization, flourished until about 1700 B.C. The people harvested grain and raised cattle and hogs. As their population increased, they farmed more and more intensively until they had destroyed much of the grass cover on the land. Then, the wind blew the dusty soil into the air, just as it did in the dust bowl area of the Great Plains States of the United States in the 1930s.

As the dust filled the air and the climate became drier, the land produced less. The Harappan farmers, trying to grow enough food to feed the population, plowed up more land and loosened more dust to blow into the air. Finally, they had to abandon the land, and it has never again been as good for farming. There are other explanations for the decline of the Harappan civilization, such as the theory that invading tribes from the north forced the Harappans out of their homeland, but I think this one is equally valid.

Today, the people who live in that area raise goats and sheep that eat what little grass there is. So the soil is exposed, and the wind still blows vast amounts of dust into the air. A group of scientists from the Central Arid Zone Research Institution at Jodhpur, India, demonstrated in 1969 that the vegetation could again cover the land in the area if the goats and sheep were kept away. The scientists built a fence around 40 acres in the desert to keep out the animals. Nothing was planted in the enclosure, but within two years wild grass was growing everywhere in the plot.

Present-day farming methods are changing climate in many parts of the world. Heavy farm machinery throws vast amounts of soil particles into the air in such countries as the United States and

A massive dust cloud spreading westward from Africa was detected in this photograph taken from a weather satellite.

Canada. Primitive tribes in tropical areas of Africa, South America, and the Pacific Islands do much the same thing when they practice slash-and-burn agriculture to produce their food. These people chop down trees and burn away the ground cover of grass and brush so they can plant crops. At the end of the growing season, they move on to clear new areas for the next season's crops. In some tropical countries and in Australia, unharvested stubble of sugar cane and similar crops are burned to clear the land for new plantings. This also increases the amount of smoke particles in the air.

It is not clear whether today's climatic changes are something new or are just a continuation of trends that have existed for a long time. But everywhere we look we can see signs of continuing change. Records show that glaciers on the west coast of Greenland advanced farthest during the 1870s and that they have retreated since. The glacial pattern fits the known global temperatures of the period, which rose about .8° F. between 1900 and 1945. Incidentally, the spread in temperature from an Ice Age climate to a nonglacial world climate is only about 9° F. After 1945, a worldwide cooling trend set in. Many glaciers have advanced since 1945. We are now nearly two-thirds of the way back to the averages of the early 1800s—a colder time than any living person can recall.

Temperature records for the past 45 years in such places as Vienna, Austria; Winnipeg, Canada; Reykjavík, Iceland; Edinburgh, Scotland; and Birmingham, England, show the same pre-1945 warming

Dust from Irazu volcano in Costa
Rica, *above,* may remain in the
stratosphere for years, while
dust-storm particles, *right,*
and those thrown up by farm
machinery, *below,* settle back
much quicker. However, all these
may have an effect on climate.

As Iceland's annual temperatures fall, more of the country's green hayfields, *above,* turn to barren ones, *right.*

spurt and then the return to cooling. Climatologists B. L. Dzeerdzeevskiy of the Russian Academy of Sciences, John E. Kutzbach of the University of Wisconsin, and Hubert H. Lamb of the British Meteorological Office have all published statistics showing that air-circulation patterns have changed in the last 20 years. In turn, local climates have changed, some regions becoming much wetter, some much drier, some much colder, and some much warmer.

African lake levels have risen rapidly in the last decade because the balance between evaporation and precipitation has changed slightly. Also, weather records at the University of Wisconsin show that midsummer frosts in the Great Lakes region of the United States have increased. And Icelandic weather stations report that drift ice from the Arctic has been appearing again along the coast of Iceland.

The people of Iceland are particularly vulnerable to changes in climate because they live on the edge of the highly inhospitable Arctic region. In 1945, meteorologist Lauge Koch used weather information found in ancient Norse sagas, as well as modern records, to determine what the ice conditions of Iceland were from A.D. 800 to 1939.

Koch found that the sailing route from Iceland to Greenland, pioneered by Eric the Red about 982, went directly west from Iceland to Greenland and then followed the coast south around the tip to the Greenland colony, near present-day Julianehåb. The Vikings used this route for about 400 years, and the sagas say nothing about the early Norsemen encountering difficulties with sea ice. This must have been a warm period. By 1350, however, they had shifted to a route farther south, and the route was moved progressively southward as ice conditions apparently became more hazardous.

Since 1959, there has been much more sea ice around Iceland than there was during the 1920s. There is about as much sea ice now as there was in the 1880s, a severe period in Iceland's history. In addition, the mean annual temperature, the mean annual sea-surface temperature, and the number of days with no snow cover have dropped in Iceland. The number of days of frost has sharply increased. Fog, always more prevalent during heavy ice years, has also increased. The decrease in mean temperatures has shortened the growing season and has reduced the production of hay, Iceland's most important crop. Fishing may also have been affected by the colder weather and the increase of drift ice.

Many more studies and a great deal of weather data will be needed before the exact causes of temperature drops such as those in Iceland can be determined. It is not enough to know that the climate of the earth is constantly changing. We also must learn exactly how this happens. In our laboratory at the University of Wisconsin, we are working to trace the changing patterns of weather at many different locations throughout the world. By using computers to correlate available weather records and such related facts as the type of farming and local industrial development, we can define the weather pattern and project it into the future. There are some unpredictable factors that may change weather patterns, of course, so we also project what the development of new energy sources may do to local temperatures and how an increase in local pollution may affect local rainfall.

For the most part, nature produced these climatic changes in the past. Now it seems that man's activities are a significant cause. If so, then we are sure that airborne particles produced by man, particularly dust particles, are the most important reasons for this change.

The increase of field dust in the air, much of it the result of our mechanization, could change the climate enough to affect the world's capacity to produce enough food. The world relies heavily on the grain production of temperate regions. A lower mean annual temperature, including summer frosts, could result if dust levels continue to rise. Temperature and rainfall patterns might also shift. This would hurt the production of grain, particularly corn. And the moisture that is needed during the grain-growing season can be affected by these pollutants, as it was in India. The problem would be somewhat like the one Iceland now has with its declining hay crop, but it would take place on a worldwide scale.

It seems more evident than ever before that man has the power to damage or even destroy his environment. It would be a tragedy if we changed the temperature or rainfall patterns so much that we could not produce enough food, or if increased farming generated the dust that caused worldwide temperatures to decline. If we are to stem or reverse this process, we need to know more about what is involved in climatic change and how man can correct the effects of dust on the climate. Otherwise our dust may destroy us.

For further reading:

Bryson, Reid A., "The Climatology of Cities," *Saturday Review,*
April 1, 1967.
Landsberg, Helmut E., "Man-Made Climate Changes," *Science*, Dec. 18, 1970.
Man's Impact on the Global Environment, Report of the Study of Critical
Environmental Problems, M.I.T. Press, 1970.

Livestock From the Sea

By John E. Bardach

If experimental aquaculture techniques fulfill their promise, the fish farmers of the world may one day outnumber the fishermen

A race of "supertrout" inhabit special raising ponds at the University of Washington in Seattle. These spectacular fish grow at least 2 feet long and usually weigh well over 15 pounds—three times as long and 15 times as heavy as any of their wild rainbow trout ancestors when they are the same age.

The supertrout are the product of many generations of selective breeding by School of Fisheries Professor Lauren Donaldson. He uses the largest eggs (because they produce the largest fish) from the biggest, fastest-growing parents. Then he feeds the young a special, highly nutritional diet to guarantee that they will achieve their full growth potential. Although they are not being produced commercially, supertrout symbolize what can and is being achieved in the fast-growing field of aquaculture.

Aquaculture is the underwater counterpart of agriculture, and uses similar techniques. For example, the supertrout were developed by applying the same principles of genetics and feeding that man has used on domestic farm animals such as chickens, cows, and pigs. Another similarity is the practice of confining the livestock so that they can be easily tended and cannot wander off. The fields, barnyards,

Young stock roil the water in a frenzy of feeding on a catfish farm in Aliceville, Ala. Food can be seen falling toward the water, lower left.

Three 2-year-old trout reveal the potential of selective breeding in aquaculture. The big fish was bred at the University of Washington. The other two are typical wild trout.

and farm buildings of aquaculture are raising ponds or raceways large tanks, and cages placed in natural bodies of water.

In at least one way, fish are better suited to farming than are their land-based counterparts. Fish grow more flesh from the same amount of feed than do cattle, chickens, or pigs. This is primarily because the terrestrial animals use part of what they eat just to provide the energy to stand and move about. Fish, however, can hover "weightless" in water. This feed-conversion advantage is at least partly responsible for the comparatively rapid success and expansion of commercial aquaculture enterprises throughout the world.

In agriculture, total domestication—the control of all phases of the life cycle of the animals—is required for the best yields. In aquaculture, mating, egg hatching, and raising of the very young are fully controlled in only two fish, the trout and the carp. The eggs of both fishes are large, about 1/10 inch in diameter, and do not scatter and float. Trout deposit their eggs in gravel and carp lay their eggs on underwater plants. Thus, the eggs are easy to collect. In addition, the fry, or tiny young fish, that hatch out are raised with little difficulty.

On the basis of number and total weight, the carp is the most important cultured food fish in the world. It is usually grown in commercial ponds and in reservoirs, mainly in Asia and Europe. The specially bred fish produced have few scales, small bones, and an abundance of tasty meat. Such carp are not produced in the Western Hemisphere, mainly because of the variety and quantity of edible lake and river fish native to countries there.

Sewage is often used to fertilize carp ponds. This encourages the growth of small water plants and animals upon which the carp "graze." I have observed a particularly interesting variation of sewage fish culture in Java. I first saw it in 1959 in a stream near Bandung, and when I revisited the area in 1969, the technique had spread considerably. The stream drains a heavily populated valley in which the people's only toilet facilities are the outdoors. Bottomless wicker cages, each about 3 feet high, 3 feet wide, and 6 feet long, stand in the stream, their tops visible a few inches above the water. About 200 pond-raised carp fry, each 4 inches long, are put into a cage through a trap door in the top. They graze on bottom-living worms and insect larvae that

The author:
John E. Bardach is director of the Hawaii Institute of Marine Biology at the University of Hawaii.

thrive on the dissolved organic waste washed into the stream by the rain. For some reason, the grazing increases the number of worms and larvae that become established. The result is an incredibly efficient, self-regulating, automatic-feeding system that grows with the carp. Each square yard of stream bottom covered by the cages can produce 125 pounds of live carp per year.

The Javan method is particularly economical because neither the fertilizer nor the feed cost anything. Because of local cooking and eating customs, however, such fish culture is beset with public health problems. The method is possible only in warm water that runs fairly rapidly. Such water is found almost exclusively in the tropics, where people carry the greatest number of intestinal parasites. The fish pick up the parasites and harbor them temporarily, particularly in their internal organs. The Javanese eat these organs and, to add to the problem, do not cook the fish thoroughly. The result, of course, is an increased risk of infection for those who eat the fish.

Carp are not farmed in the United States. But one native fresh-water fish–the catfish–approaches domestication, although, so far, there has been very little selective breeding. The hardy channel catfish of the Mississippi River and its tributaries is spawned in captivity, fed cheap pelletized food, and raised in as little as a year. Each fish provides nearly a pound of meat.

Most commercially raised catfish in North America are grown in special ponds dug in the bottomlands of Arkansas, Louisiana, and Mississippi where the soil retains water and rainfall is dependable. In 1970, fish farmers harvested 15 million pounds of catfish and an additional 4 million pounds were predicted for the 1972 crop.

Attempts to domesticate salt-water fish are still highly experimental. For example, at the fish-rearing station at Port Erin, Isle of Man, in the Irish Sea, marine biologist James E. Shelbourne and co-workers raise flounder and other flatfish highly prized as food. There are many problems, however. Flatfish can be coaxed to spawn in confinement by manipulating their water temperature. But many of the fry die because they are living in unnatural conditions. Furthermore, the scientists do not know exactly what the fish eat. Yet Shelbourne and his crew have been able to coax a third or more of their hatchlings through the critical first two or three months.

The fry, which up to this time had been swimming at all levels, then take on their typical flat form and begin to live at the bottom. They require very little space–1 square foot of bottom per fish–for the rest of their lives. They are quite content to lie quietly most of the time if they are fed enough to keep them from having to forage. Considering these space requirements, Shelbourne has speculated that shallow aquaculture ponds covering less than 2 square miles could produce all the flatfish now caught in the North Sea by British commercial fishermen. Similar flatfish farms scattered throughout the world could easily match the world flatfish catch.

Aquafarmers examine two breeding-size fish, *above,* on catfish farm in Alabama. Young stock churn up water, *above right,* as they gulp down food thrown from the platform. Workers harvest catfish, *right,* by slowly pulling in net.

Unfortunately, in 1971 such an operation would be prohibitively expensive. However, experiments that promise commercial flatfish farming at lower cost have been underway at the Scottish nuclear power plant in Hunterston since 1967. Cold water from the Irish Sea is pumped through the Hunterston power plant to carry off excess heat generated during power production. The water, warmed at least 15 to 20°F., is channeled into tanks. The flatfish not only adjust to the warm water, but they grow faster in it. Thus, marketable fish are raised in two years rather than the four or five years required in the sea.

Comparable experiments with water heated in conventional and atomic power plants are underway in Florida and New York as well as in Russia and several other countries. In 1971, some of these projects began operating as commercial aquafarms.

Desirable as it may be to totally domesticate all the aquatic animals we farm, it is sometimes surprisingly efficient to raise aquatic animals from young collected in the wild. The yellowtail is an example. Relished raw by the Japanese, this fish was netted from the seas around Japan for generations. In the mid-1950s, researchers worked out a way to raise young yellowtails netted under floating patches of seaweed, where they normally live from their late fry stage on. Fish farmers put 100 to 200 of the 3- to 4-inch young in large bags of nylon netting suspended in the sea from floating platforms. They then feed the yellowtails on trash fish.

In nature, yellowtails are wide-roaming predators capable of bursts of speed of from 9 to 15 feet per second, a respectable velocity for any fish. Yet, they seem content to swim slowly and continuously in the relatively crowded, confining nets, which are typically about 50 by 30 by 20 feet. This method has become so widespread that more net-raised than wild fish are now sold in Japan.

Unfortunately, not enough young yellowtails inhabit Japan's coastal waters for economical expansion of their culture. The obvious solution is to raise the young from eggs under controlled conditions. However, this is very difficult with yellowtails and other seasonal spawners. Their eggs can be fertilized only when ripe and about to be shed. A number of environmental factors influence the eggs' ripeness, and removing them at any but precisely the right time is useless.

Scientists in South America discovered, in the 1940s, that an egg-laden fish injected or implanted with material from the pituitary gland of another fish of the same species will hasten egg maturation. This method is called hypophysation, because the pituitary gland is also called the hypophysis. When the eggs are mature, a bulge is visible low in the female's belly. In some species, the eggs then begin to come out spontaneously. In others, slight pressure on the belly is necessary to eject the eggs. Hypophysation has been successfully used on the mullet, a fish with great promise for commercial growers. Japanese experimenters are confident that in some form it will also work with yellowtails.

Brackish-Water Aquaculture

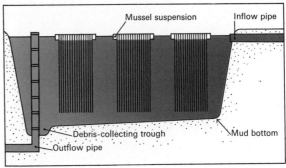

Several creatures, such as crayfish, mussels, mullet (at top), and *Tilapia* (bottom), could be raised together in brackish, or slightly salty, water. Natural foods, such as algae, would be added. Color in cutaway indicates lowered food concentration as water flows through pond.

Even if the eggs could be removed and fertilized easily, they would still have to be coaxed through hatching, and the delicate fry raised. Both eggs and fry, tiny as pinheads, normally float among the *plankton* (microscopic living organisms) on the high seas. There they live a very complex life, choosing certain types and amounts of food organisms and growth substances from the plankton. So far, nobody has been able to analyze, no less duplicate, the conditions. In spite of the work of skilled scientists, by mid-1971 only a tiny fraction of 1 per cent of the fry obtained from the eggs of hypophysized mullets had ever survived more than three weeks after birth.

Clearly, extensive marine aquaculture is not yet possible when man cannot control reproduction and early growth stages, or where nature does not provide great numbers of spawn and easily raised young. Oysters and mussels, which do produce a superabundance of young, are the most successfully farmed of all water creatures. In the wild, each female yearly sheds millions of eggs that hatch out pinhead-sized, free-swimming larvae. In about two weeks, the larvae sink to the bottom, where only those that find and attach themselves to a rough surface survive. Because such surfaces are very scarce, only a fraction of 1 per cent of the larvae survive. This is why evolution has endowed these mollusks with such large reproductive potential.

The mussel farms in northern Spain have become very successful by providing the larvae with an

Mussel-growing rafts
dot Vigo Bay in Spain,
right. The mussels
are attached to the
ropes that are hanging
from each raft, *below.*

abundance of suitable places to attach themselves. The world's most prolific natural mussel grounds are in the bays of Galicia. The upper waters of the bays are fertilized by wastes carried in by rivers and run-off from the land, and by nutrient-rich deep waters plowed up by the mixing action of tidal currents. This makes the bays' waters exceptionally rich in plankton and other suspended organic matter that the mussels eat. In addition, the water is about 65°F. during most of the year, an ideal temperature for mussel growth.

The mussels spawn from early spring to late fall, and billions of larvae swim about searching for a place to anchor themselves. Spanish mussel farmers use a technique similar to one devised by the Romans at about the time of the birth of Christ. They build large, latticed, wooden rafts from which they suspend as many as 500 long, thick ropes almost to the bottom. The larvae attach themselves to the ropes by secreting a sticky thread called a byssus. The mussels often crowd onto the ropes and each other in such great numbers that wooden pegs must be inserted in the ropes at intervals of 12 to 18 inches so that clumps of mussels do not slide down and drop off as they gain weight. The mussels grow to harvestable size—3 to 4 inches long—in 12 to 18 months. The 8 to 10 rafts per acre of surface in the best locations yield about 300,000 pounds of shelled mussels. Spain's mussel production is enough to supply the needs of the entire European market.

In Japan, the oyster growers of the Inland Sea use a similar technique. Each spring, they string thousands of scallop shells, barely 1/2 inch apart, on wires suspended from bamboo frames that are built where plankton abound. Here, too, the larvae, or seed oysters, float in great numbers. After only a few weeks, an average of about 200 larvae have attached themselves to each shell. About 500 of the wires are suspended from each of many 50- by 75-foot oyster rafts. As the oysters grow, some of the scallop shells are strung onto new wires, and all the shells are spaced more widely by inserting longer hollow bamboo pieces between them. In the fall, the wires are hauled up on a hoist by a boat-mounted winch, and the oysters are harvested.

The Inland Sea is so fertile that in some places an oyster raft kept afloat by 6 buoys at the beginning of the growing season requires 18 to 20 just before harvest. One such raft, in one year, may produce 24 tons of whole oysters, which yield 4 tons of oyster flesh.

Raft culture offers important advantages beyond such obvious ones as easy care and harvest. For example, the entire volume of food-rich water is used when the animals are stacked in three dimensions. Also, by preventing the ropes from touching the bottom, the aquaculturist controls predators. Mussel and oyster predators such as the starfish, crawl on the sea bottom to locate their prey. The raft technique prevents them from reaching the well-stocked larders suspended so temptingly above them.

Trout, carp, catfish, mussels, oysters, and a growing list of other successfully farmed water creatures are proving that aquaculture can

Scallop shells implanted with young oysters, *above,* are kept the proper distance apart with bamboo spacers on ropes suspended from rafts in Japan's Inland Sea. Ropes are raised and lowered, *right,* by a ship's winch.

Salt-Water Aquaculture

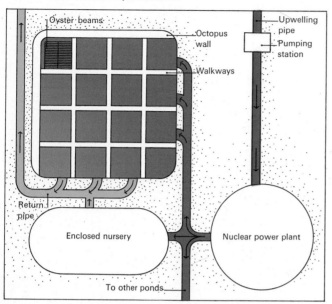

Species such as oysters and octopuses might someday be raised together. A diver could feed octopuses trash fish. Oysters would eat plankton. Wall of bubbles would frighten octopuses, keeping them from eating oysters. Pumping nutrient-rich water up will be more practical if it is first used to cool power plants, *above.* Color in diagram indicates decreasing nutrient concentration as seawater flows through the ponds.

be as effective as land farming. But what is aquaculture's potential? The question is critical, for the world's population, much of it now undernourished, is likely to double and double again well within the next century. Natural sources of fish and mollusks, places where they are free for the taking, are inevitably overfished. As a result, their numbers decline. For example, the average yield of U.S. public oyster grounds, such as those in Maryland, has dropped from about 16 pounds per acre in 1920 to 8 pounds per acre in 1970. Several important fisheries are also declining because of overfishing. But, even if we did not overexploit them, natural marine sources could at best provide only double or triple what they do now.

Fresh-water bodies cover less than 1 per cent of the earth's surface, but more than 70 per cent is covered with salt water. Our hopes, then, must rest on extensively farming the seas. Years of experience

Nutrient-rich water can
be pumped from
the depths of the sea
into ponds where it
can be used to grow
plankton for use as
food in aquaculture.
This experimental
system has been proved
successful on St. Croix,
U.S. Virgin Islands.

Plankton
pond   Upwelling
Pumping   pipe
station Sea
wall
                        Depth
                        0 ft.

650 ft.

1,312 ft.

1,969 ft.

2,625 ft.

3,281 ft.

have proved, however, that fertility varies as much in the seas as it does on land. Vast stretches of ocean show by their very color—the blue of clear water—that they are all but devoid of plant life and, consequently, can produce little seafood. Naturally fertile areas, those most suitable for aquaculture, cover only about 10 per cent of the world's oceans. These spots include the shelves and banks near coasts, where run-off from the land fertilizes the ocean. There are also the so-called zones of upwelling, where nutrient-laden deep water is constantly plowed to the surface by peculiar hydrographic conditions. Not surprisingly, these areas already support large populations of wild marine life.

.The most prominent zones of upwelling are in the Antarctic Ocean and on the west coasts of Africa and South America. Such areas furnish nearly half of man's ocean harvest. Their tremendous productivity has led to attempts to create upwelling in miniature for use in marine aquaculture.

One way is to sink a pipe and pump the deep water up. In tropical seas, the water 1/4 to 1/2 mile down contains much more of the fertilizing elements—10 to 15 times as much nitrogen and 6 to 25 times as much phosphorus—than does water in the upper layers, where most aquatic creatures live. Marine biologist Oswald Roels and co-workers of the Lamont-Doherty Geological Laboratory of Columbia University in New York City have demonstrated that this can be done. At their pilot operation for aquaculture on the island of St. Croix in the Virgin Islands, a pump brings water up from about 1/2 mile down through a 3-inch-diameter pipe. The water is spilled into small (16,000-gallon) ponds. Some algae and other plankton grow about 25 times as fast in this water as they would in normal surface water. Oysters have been planted in the ponds to feed on the plankton. Data on the oysters' growth obtained in 1970 encouraged Roels to plan experiments with other sea creatures.

One roadblock to this method is the prohibitive cost of pumping. Commercial operations would require large pumps that are expensive to buy, operate, and maintain. A solution is to use the deep water for some other profitable purpose in addition to aquaculture. One of the very promising uses is as a coolant for power plants. The fertile deep water is particularly useful for cooling because its temperature is 40°F. colder than surface water. The flow through the plant could easily be regulated so that when water flows into the aquaculture tanks or ponds, its temperature could be the same as that of the tepid surface waters in which the creatures being raised grow best. Coincidentally, this would assure that the water draining back into the sea from the power plant and the fish farm would not be warm enough to cause thermal pollution of the seawater.

However, there is also a pollution potential similar to that caused by sewage. The nutrients from the deep water must be completely used as the water flows through the aquafarm, or they will fertilize the waters into which they are returned. If these waters are shallow and are not being mixed by winds and tides, this can result in excessive algae growth that ultimately chokes out all life.

Such problems can be solved, clearing the way for the aquaculture of the future. And, although aquaculture is no panacea, as an industry, it can grow more rapidly than can commercial fishing. The Food and Agriculture Organization (FAO) estimated the 1969 world aquatic catch at over 60 million metric tons, about 4 million of which came from aquaculture. Sidney Holt, then chief of the biology branch of FAO Fisheries Division, further estimated that man can increase the aquaculture harvest 10 times within 30 years.

Holt's forecast is conservative. Unfortunately, however, I fear even this figure will not be met. Aquaculture cannot be expanded without financial support for research. Such support has been difficult to get because, despite the depletion of our natural fisheries, it is still usually cheaper to catch fish than to grow them. Furthermore, the bulk of man's food has always come from the land, so agriculture will continue to receive research funds that might otherwise go to aquaculture. No field of science or technology can, of course, achieve on a tight budget what it might with adequate funding.

But the greatest threat to the ultimate expansion of aquaculture is pollution. Man has used the earth's waters, especially its seas, as a universal septic tank. The practice is based on age-old patterns first developed when our relatively few wastes were natural ones that the animals and plants of the sea could cope with. However, now we not only dump greater amounts of natural wastes into the sea, but we also add industrial poisons with which living organisms cannot cope. The spread of modern technology will probably increase the problem of pollution greatly.

The sea life we attempt to exploit live fully in the slow time frame of natural organic evolution. If ocean pollution continues at its present rate, these creatures may die in greater and greater numbers. If so, advances in caring for those we have domesticated will be like trying to fill a leaky barrel.

The way out of the dilemma is better management of world ecology. Only then will there be hope that our waters will support a full complement of living creatures, both in the wild and on our aquafarms.

For further reading:

Bardach, John, *Harvest of the Sea,* George Allen and Unwin Ltd., London, 1968.

Bardach, John, "Aquaculture," *Science,* Sept. 13, 1968.

Bardach, John, and Ryther, John, *The Status and Potential of Aquaculture, Particularly Fish Culture,* American Institute of Biological Sciences, 1968.

Nash, Colin E., "Power Stations as Sea Farms," *New Scientist,* Nov. 14, 1968.

Pinchot, Gifford B., "Marine Farming," *Scientific American,* December, 1970.

Drilling into The Ocean Floor

By N. Terence Edgar

Scientists go to sea in a special ship to sample the earth's history from the sediments below

Early in the morning of Dec. 2, 1970, I stood with half a dozen other scientists on the dock at San Juan, Puerto Rico, and watched a unique vessel move slowly but steadily across the city's skyline to its assigned berth. The ship hesitated alongside the narrow space like an automobile edging up to a small parking place. Then, without help from tugboats, it began to move broadside toward the dock. Within minutes, lines were thrown ashore and secured. The *Glomar Challenger* had completed a sea floor drilling expedition that had begun in Lisbon, Portugal, 56 days earlier. It was the 14th cruise for the ship, which had been drilling in both the Atlantic and Pacific oceans for two and one-half years. I was about to join it for the 15th cruise.

Soon, the dock was teeming with activity. Scientists leaving the ship chatted with those of us waiting on the dock, telling of their discoveries and their obvious joy at seeing land. Yet, their enthusiasm was mixed with regret that the cruise, a highlight in their careers, was over.

In only three days, the ship's 45-man crew would be replaced, and fresh provisions and supplies hoisted aboard. Drilling tools, lubricants, miles of wire cable, and tons of mud and cement to plug the drilled holes would be loaded for the next two-month cruise. At the same time, the thousands of feet of cores cut and lifted from the deep sea bottom during the cruise just ended would be unloaded and shipped to laboratories in the United States for further study. Within hours after docking, the loading and unloading were well underway.

The *Glomar Challenger*, its deck neatly stacked with drilling pipes, slices through the Caribbean waters on its way to its next assignment.

The *Glomar Challenger* is the primary instrument and laboratory of the Deep Sea Drilling Project. The project's mission is to determine the age and history of the crust of the earth that lies beneath the ocean floor. Conceived in 1966, the project is managed by the Scripps Institution of Oceanography, La Jolla, Calif., under the direction of the Joint Oceanographic Institutions for Deep Earth Sampling (JOIDES). The latter was formed by four of the leading oceanographic laboratories in the United States, the Lamont-Doherty Geological Observatory of Columbia University, the Scripps Institution, the Rosenstiel Institute of Marine and Atmospheric Sciences of the University of Miami, and the Woods Hole (Mass.) Oceanographic Institution. Subsequently, the University of Washington at Seattle joined JOIDES.

What lies at the ocean bottom? Scientists began to seek the answer to that question almost 100 years ago. On Dec. 21, 1872, the corvette H.M.S. *Challenger* sailed from Portsmouth, England, and began a systematic study and mapping of the oceans. For the next century, scientists continued to map, sample, and photograph the ocean floor. But it was always the floor as it exists today. Dredges and core pipes could sample only near-surface sediments, the debris that had settled to the bottom during the very recent geologic past.

To learn the nature and age of the older and more deeply buried sediments, scientists would have to bore deep into mud, limestone, various types of shale, and hard rock. They would need technology developed by the petroleum industry for offshore exploration and a very special ship.

That ship was launched in 1968 at Orange, Tex., and christened the *Glomar Challenger* after the H.M.S. *Challenger*. Glomar is a contraction of Global Marine, Inc., of Los Angeles, the company that owns and operates it. The 10,500-ton ship is 400 feet long and 60 feet wide. True, a 142-foot derrick straddling it amidships looks a little unusual. But in this respect, it resembles other drilling ships used to find offshore oil. What makes the *Glomar Challenger* unique is that it can hold to a 50-foot circle for days at a time, even in rough seas. This is a prerequisite for drilling in ocean depths—sometimes greater than 3 miles —where anchors are useless.

The ship maintains its position by a process called "dynamic positioning." Three hydrophones mounted in the hull pick up signals from a beacon placed on the sea floor. A computer records the arrival time of each signal and calculates the position of the ship in relation to the beacon. If the ship begins to drift, the computer automatically activates the appropriate screws (propellers) to hold the ship in position. To achieve precise maneuverability, special screws are mounted in tunnels that lead from one side of the hull to the other. These "tunnel thrusters," two in the bow and two in the stern, enable the ship to move sideways and to rotate within its own length.

The *Glomar Challenger* is also equipped with the most modern navigational aids, including a receiver tuned to a navigation satellite,

The author:
N. Terence Edgar is Chief Scientist of the Deep Sea Drilling Project and a marine geologist with the Scripps Institution of Oceanography. He was co-chief scientist on the 15th cruise of the *Glomar Challenger.*

which locate the ship's position to within one-tenth of a mile. A radio facsimile system receives weather maps and photographs of large areas of the earth's surface taken by weather satellites.

The ship has a number of scientific laboratories on board. For example, one is equipped to handle chemical and physical studies of ocean sediment samples. Another laboratory is for studying the microscopic fossils found in the sediments. There is a photographic laboratory, as well, that is used for developing pictures taken of the core samples when they are fresh from the bottom.

Spacious living quarters on the *Glomar Challenger* include a large lounge, with a library and stereo tape player, that doubles as a movie theater in the evening. The food is good and plentiful, and the ship also has a gymnasium for those who cannot resist the pastry.

Drilling into the ocean floor began in the summer of 1968. During its first two and a half years, the *Glomar Challenger* drilled 230 holes at 144 sites in the Atlantic and Pacific oceans. Scientists have obtained cores from sedimentary layers under water more than 20,000 feet deep. In fact, the ship has assembled the longest length of drilling pipe, or drill string, ever suspended from a floating platform—20,760 feet, almost 4 miles. Several holes have been drilled more than 3,200 feet into the ocean bottom. The drillings have produced enough significant information to mark the Deep Sea Drilling Project as one of the most successful scientific expeditions of all time. For instance, core samples taken on the second and third cruises of the *Glomar Challenger*, in 1968, demonstrated that the continents of North and South America are drifting away from Europe and Africa.

The scientists who moved their personal belongings on board for this cruise represented several countries. John B. Saunders, an English micropaleontologist from Texaco of Trinidad, was co-chief scientist with me. The geologists were American Thomas W. Donnelly, Israeli Nahum Schneidermann, and Haitian Florence Maurasse. The paleontologists were Saunders, Hans Bolli from Switzerland, Isabella Premoli-Silva from Italy, William Riedel from Australia, and William Hay, an American.

We immediately familiarized ourselves with the laboratories that would become our homes for the next couple of months, checking the microscopes, chemicals, sieves, slides, ovens, hot plates, and other equipment. When we were satisfied that we had what we needed, we made a field trip ashore to examine the rocks of Puerto Rico. These could have been deposited or formed at the same time as those in the deep ocean where we planned to drill.

At five minutes past midnight, on December 5, the docking lines were pulled aboard the *Glomar Challenger*, and our cruise began. Saunders and I spent the night outlining our plans for the approach to the first drilling site, a small rise in the Caribbean Sea floor about 300 miles southwest of Puerto Rico. Meanwhile, technicians were preparing a seismic profiler to be lowered over the side and towed behind

Glomar Challenger Deep Sea Drilling Sites

| • | Completed | → | Planned track for future drilling |

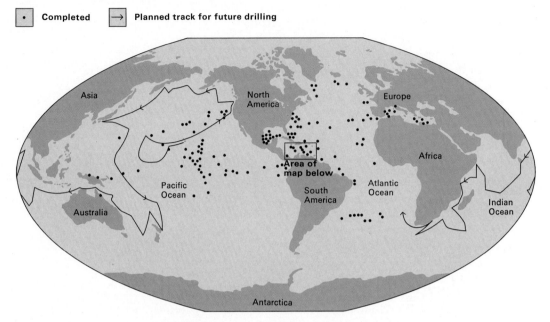

Cruise 15 Drill Sites in the Caribbean Sea

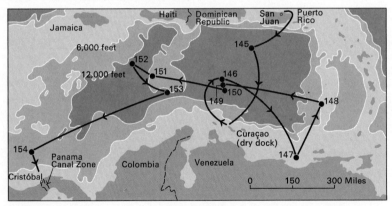

On Cruise 15, the _Glomar Challenger_ drilled holes 145 to 154. The trip began at San Juan, Puerto Rico, and ended at Cristóbal, Panama Canal Zone, with an unscheduled visit to a Curaçao dry dock in-between. The ship then set sail for Hawaii to begin a series of drillings in the Pacific and Indian oceans.

the ship. This instrument and an echo sounder mounted on the hull would locate the site and give background data for interpretation of the bottom cores.

The echo sounder emits a high-frequency signal, then listens to the echo from the ocean floor. Sound travels through the water at about 4,800 feet per second. If the echo arrives in one second, the depth is 2,400 feet (the signal must travel twice the distance). In oceanic depths of 15,000 feet the echo sounder is accurate to within 6 feet.

The seismic profiler operates in much the same way as the echo sounder, except that it emits a lower frequency signal. The low-frequency sound penetrates the ocean floor and reflects back from denser layers of sediments and rocks beneath. The reflected signals are received at slightly different times by the hydrophones. The seismic profiler thus calculates the exact depth of reflection of each returning signal and produces a cross-section of the sedimentary layers along the path of the vessel.

Our first site (Site 145 by total count), was a rise in the ocean floor. It had been discovered accidentally several years before by another ship that passed over it with a seismic profiler. The record indicated that hard rock, perhaps the ancient ocean crust, protruded to within a few hundred feet of the top of the rise. This offered us an opportunity to sample the crustal bedrock of the Caribbean with a minimum of drilling. We first wanted to determine the shape of the rise—it could be a circular formation, perhaps an underwater volcano, or elongate, like a mountain ridge.

When we arrived at the specified location, we couldn't find the rise. We headed west; no luck. Next we went north, and then east, without finding it. Finally, we decided to make a wider search, steamed across the area the ship had already circled, and, quite by chance, ran right over the rise. We had apparently steamed all around it. This suggested it was a circular pinnacle. We surveyed the pinnacle with the seismic profiler, found a spot where the sediments were thin, and stopped the ship directly above it.

We dropped the positioning beacon by cable. After the beacon had settled on the crest, the computer began to hold the ship in position over the beacon. During the 45 minutes it took for the beacon to reach bottom, 14,000 feet below, the drilling crew assembled the drill bit and screwed it onto the first length of pipe. They then began lowering the entire assembly through a shaft, called the "moon pool," in the center of the vessel under the derrick. The 8-inch drill pipe is in 30-foot sections, which are linked in sets of three to make 90-foot lengths. Several hundred 90-foot lengths are stored on a pipe rack forward of the derrick. A huge traveling block clamps onto the near end of a length

At Site 145, the *Glomar Challenger* hovers over the positioning beacon 14,000 feet below on the floor of the Caribbean.

Standing on drill collar in drilling rig, *right*, the workmen add a length of pipe to drill string that passes through a hole in the center of the ship. A movable crane brings each 30-foot section of pipe, *below*, to the drilling rig.

of pipe and pulls it up to hang vertically in the derrick. The pipe is screwed into the top of the drill string, which is then lowered far enough that the next length can be screwed into it. It takes two to three minutes to add each 90-foot length.

The drill pipe has rubber rings, or collars, around it, spaced at 5-foot intervals. These reduce snakelike motions of the long drill string caused by motion of the ship. The collars are colored so that the pipes can be counted. The color-coding system was devised in 1968 by the father-and-son team of Jim Dean, Sr., the project's operations manager, and Jim Dean, Jr., then a sixth-grade student. Jim, Sr., wanted to number each 30-foot section of pipe permanently without scratching its surface. He and Jim, Jr., resolved the problem by applying "new math," using base 4. The collars are in four colors—black, brown, green, and red. Since there are five collars per pipe, these four colors provided enough combinations to identify 4^5, or 1,024, lengths of pipe. As a result, the drill team knows how many feet of pipe are beneath the water line at any time merely by checking the color code on the above-board pipe.

The drill had not even reached the ocean floor at Site 145 when Captain Joe Clarke informed us that the tunnel thruster screws were not working properly. "One thruster is losing oil; we can hardly turn another," he said. "We'll try to drill this hole, but after that we'll have to return for dry-docking." As we talked, the chief engineer telephoned to report that we were now using twice as much oil as we had half an hour before. That was it. We had to abandon the site and leave immediately for a dry dock.

Telegrams were sent to ports throughout the Caribbean, Gulf of Mexico, and as far north as Norfolk, Va., inquiring about the availability of a dry dock. Fortunately, a dry dock was available at nearby Curaçao, an island off the north coast of Venezuela.

The 5-ton re-entry cone splashes into the water and is lowered with the drill string. With this equipment, worn drill bits may be replaced and the drill reinserted in the underwater hole.

We headed for that port and soon had the ship high and dry. Repairmen discovered that the thruster blades were chipped and bent, indicating that they had struck some large objects, presumably in San Juan harbor. The *Glomar Challenger* would use tugs for docking and departures from that time on, saving the thrusters for their more important service of keeping the vessel on station. The repair work on the thruster took five days, during which time the scientists examined sedimentary rock structures on Curaçao that might be of value when compared with drilled samples taken far out at sea.

We went to sea again on December 14, but because of the delay, we had to drop the first site from the program. We steamed, instead, toward a new location, Site 146, the key site of the entire cruise for both the scientists and the engineers.

For the scientists, this site was important because the sea floor there had accumulated a thick layer of sediments. Our prime objective was to accurately date the formation of the crust in this area by identifying fossils entombed in the layer of sediment directly on top of the crustal

Paleontologist John Saunders collects particles that stick to the drill bit center, *above*, before workmen replace it with a sonar scanner, *right*.

rocks. Once we established that date, we could begin to resolve the mystery of how the Caribbean Sea fits into the giant jigsaw puzzle of the earth's drifting continents.

Geological and geophysical studies indicate that the earth's crust beneath the oceans is being created and destroyed continuously and that the continents are moving relative to one another. Evidence gathered from ocean and surface rocks found on the continents indicates that about 200 million years ago all the continents were part of one great supercontinent. Then, North America began to separate from Europe and Africa, creating the North Atlantic Ocean, which has grown steadily wider ever since. Perhaps 50 million years later, South America broke away from Africa, and the South Atlantic Ocean was formed. If we try to fit the continents back into their original position before they broke up, we find that the present Caribbean Sea cannot be accounted for. The date of its origin—when North and South America moved apart—is therefore of great interest to geologists.

Re-Entering the Drill Hole

Unique arrangement of special propellers (A) are driven by hydrophones (B) to guide ship and drill string (C) over the beacon. The pipe is designed so that water jets can shift the drill bit into place over the submerged re-entry cone.

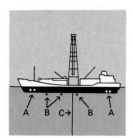

For the drilling engineers, Site 146 was to climax more than a year of planning, designing, construction, and testing to achieve a seemingly impossible task—to pull a worn bit out of a hole in the ocean floor and put a new bit back in the same hole. The water depth was 13,000 feet—nearly 2½ miles. The hole probably was less than a foot in diameter.

Most of the drill bits used on the *Glomar Challenger* are complex devices composed of several wheels of jagged teeth that form a ring. A solid plug fits in the hole formed by the rest of the bit. This center section can be removed whenever the drilling crew wishes to retrieve a sample of the sediment or rock. In the past, whenever the drill bit encountered hard rock, such as flint or flintlike chert, the teeth were quickly dulled and broken. Even diamond bits were demolished before they penetrated far into these hard formations. Generally, the older the sediment, the greater the chance of encountering chert. On its fourth cruise, late in 1968, the *Glomar Challenger* had drilled four holes about 20 miles south of our Site 146. But after these attempts to penetrate the chert had failed, the scientists gave up hope of retrieving any of the oldest sediments from the Caribbean Sea.

The problem of how to drill through flint to the ancient sediments was approached by the engineers of the Deep Sea Drilling Project in two ways. First, they improved the toughness and cutting power of the bits in chert. Then, they developed a means of replacing a worn-out bit.

The project engineers designed a steel cone, or funnel, that was to be placed on the ocean floor as a guide for the drill bit. To find the cone with a new bit, a sonar scanner is lowered through the drill string and extended through the hole at the center of the bit created when the central plug is removed. On command, the scanner rotates 360 degrees, sending and receiving signals much as does a

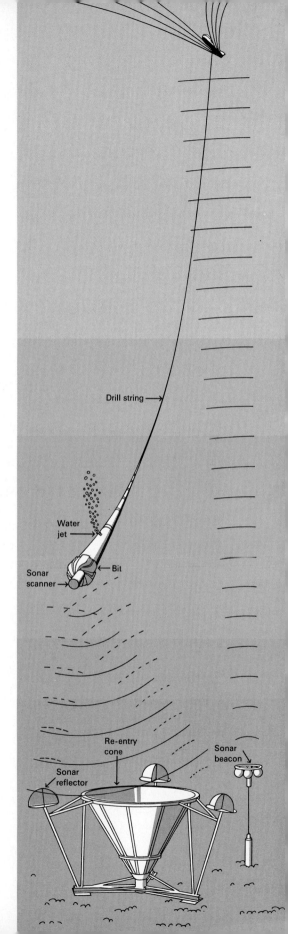

Drill string

Water jet

Bit

Sonar scanner

Re-entry cone

Sonar beacon

Sonar reflector

Taking a
Core Sample

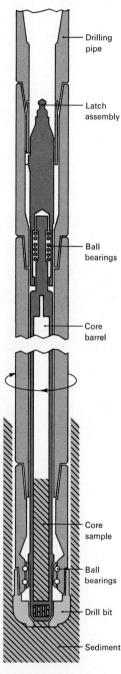

— Drilling pipe

— Latch assembly

— Ball bearings

— Core barrel

— Core sample

— Ball bearings

— Drill bit

— Sediment

Removing the center of the drill bit permits the remainder of the bit to core sediments that are forced up drill barrel.

radar antenna. On the ship, the reflected signals show as a blip on a sonar screen. Once sonar reflectors on the cone are spotted, the ship moves toward it. The bit is then positioned directly over the cone by "jetting," or pumping water down the center of the drill pipe and out a hole in its side. By simply turning the pipe, it can be aimed in the desired direction. Once the bit re-enters the hole, the sonar scanner is removed, and a latching tool lowers the central plug and locks it in position. Then, drilling resumes.

The system was tested in June, 1970, in relatively shallow water off Cape Hatteras. Many problems developed, and it took two weeks to accomplish the re-entry. One problem was that the jetting system did not work, and the bit finally had to be positioned over the cone by maneuvering the ship.

When we arrived at Site 146, on December 15, the beacon was promptly lowered and the vessel locked onto its position. Before we could lower the drill string, we had to prepare the re-entry system. The 14-foot diameter, 10,000-pound cone was lowered over the side of the ship and moved under the keel until it was directly below the moon pool. In calm seas this operation is fairly simple, but that day the drilling crew had to contend with 10- to 12-foot waves and a 20-mile-per-hour wind. Despite this handicap, the operation went smoothly until a line fouled on one of the sonar reflectors and the cone had to be raised out of the water. The combined motions of the ship, the sea, and a funnel full of water put excessive stress on the crane cable, and suddenly it broke. Members of the drilling crew, scientists, and technicians stampeded across the deck to avoid the wildly lashing cable. The cone dropped into the sea and was lost— along with our morale. First the thrusters, and now the cone.

The engineers then regrouped and devised a plan for lowering our second cone (we had three). It would be launched on its side so that it would not fill with water as it was lowered. This was successful, and the cone was soon in place under the keel. Workmen thrust the drill bit and pipe through the cone, and latched them to its bottom. They then attached the first of more than 150 lengths of drill pipe to lower this entire assembly to the sea floor, a process that took more than eight hours. When the cone had been pressed into place in the soft bottom mud, the bit was detached from the cone. We were finally ready to drill, and could continue until the bit was worn out. Then, hopefully, we would be able to replace the bit and reinsert it in the same underwater hole.

In order to take a core sample, the latching tool is lowered through the drill string and the center of the bit is pulled out. If the drill string is 3 or 4 miles long, this task may require one and one-half to two hours. A core barrel, 30 feet long and lined with a plastic tube, is then dropped through the hollow drill pipe into place over the hole left in the bit. The drill is started again and the bit cuts a ringlike hole, trapping a central cylinder of sediment in the core barrel. When

30 feet of sediment is cut, the drillers pull the core barrel up to the surface.

Our prime objective at this location was to recover samples of the oldest sediments, so we planned to drill the upper few hundred yards down to the first hard layer of rock taking only a few cores. Then, we would continuously core the remainder of the hole down to the very bottom of the sediments. Drilling went smoothly and rapidly through the first 200 feet of very soft sediment. When the first core arrived on deck, the impatient paleontologists scooped small samples from the end of the barrel and scurried off to their laboratories to determine the age of microfossils in the drab gray clay. It proved to be late Miocene (approximately 10 million years old). Each 30-foot core encased in the plastic liner is taken from the bottom, pushed from the barrel, labeled, and cut into 5-foot sections. Each section is bombarded with gamma radiation to determine the porosity and density of the sediment. Instruments measure the rate at which sound passes through it. Technicians then cut the entire core in half lengthwise. One half is photographed and stored for future reference. The other is used for more sophisticated shipboard investigations and also for later studies on shore. Some of the material is examined under a microscope by geologists to determine the mineral content. A geochemist takes a sample to be squeezed dry in order to examine the pore water. Paleontologists examine the sediment in detail, noting changes in the fossil assemblages and their preservation. Almost two hours of shipboard study may be spent on each 5-foot section.

We recovered an early Miocene core (approximately 25 million years old) 840 feet into the sediment. At 1,332 feet, the bit struck hard rock. After a few minutes, it broke through this layer of rock and then hit another, and another. Clearly we were drilling through the chert layers that had stopped the scientists on the fourth cruise.

We began coring the hole continuously, dating and describing limestones containing fossilized worm burrows and other evidence of bottom dwellers in this ancient sea. Particles of pumice and ash reminded us that the Caribbean was once an active

A 30-foot length of ocean-bottom core, still in its plastic case, is photographed on deck before being cut up into 5-foot sections for laboratory study.

volcanic area. The drill probed back in time—40-million, 50 million, and 60 million years. Then, as 80-million-year-old sediments were being removed from the core barrel, the engineers reported that the bit should be replaced. At that point we had drilled 2,296 feet into the ocean sediments.

The crew brought the entire drill string back up to the surface, installed a new bit, and then lowered the drill string again to about 50 feet above the sea floor. This process took 16 hours. The re-entry team under Roy Anderson, the operations manager, then lowered the sonar scanner through the drill pipe into its place in the center hole of the bit. Immediately, the reflectors on the cone showed up clearly on the sonar screen. The jetting system again failed to move the pipe, however, so the 10,500-ton vessel had to be delicately maneuvered to center the bit over the 14-foot diameter cone, 2½ miles below.

This posed a problem because the computer was holding the ship steady relative to the beacon. The small shifts of position required to center on the drill hole had to be made by feeding an incorrect water depth to the beacon. The computer would then automatically shift the ship's position. We also achieved small shifts through electrically switching among the three hull-mounted hydrophones. In switching from any one pair of hydrophones to another, the vessel responds by centering the new pair over the beacon.

The crew maneuvered the ship for six hours, using offsets, changes in depth settings, and switching hydrophones. Whenever the bit approached the cone, oscillations in the drill string caused it to veer away at the last moment. Frustration mounted. Finally, as the first glimmer of dawn appeared on December 23, the bit apparently centered over the cone. The order was given to lower the pipe. Down it went, and we happily went to breakfast.

Before we could start drilling again, the drill crew had to close the side hole used in the jetting system so that normal circulation of drilling mud could be maintained around the bit during drilling. Then, the drillers plugged the center hole in the bit with a core barrel, and started the pumps to test whether the drill shaft would now hold pressure.

Hans Bolli, left, William Riedel, and Isabella Premoli-Silva examine an 85-million-year-old core segment for microfossils in ship's paleontological laboratory.

If so, the jet must be closed. But apparently the tool lowered to close the hole failed to do so. The drill shaft did not hold pressure, and the core barrel came up full of clay. This indicated that the bit had not re-entered the hole. A second core, also full of clay, confirmed it.

The team then lowered the acoustical scanner for a second re-entry attempt, but mud had clogged the lower few feet of drill pipe. A discouraged drill crew then began to pull up the entire drill string so they could clean out the pipe and bit. We all waited impatiently for the next 16 hours until the assembly had been lowered once again.

When the sonar scanner was in place, the cone was sighted almost immediately. Only three precise and well-planned maneuvers, based on our growing experience, were needed to center the bit over the cone. Anderson yelled, "Stab it!" and the massive brakes screeched in protest as the pipe slid through the moon pool. The drilling crew quickly added additional lengths of pipe, but there was no resistance, no indication of pushing through sediment. The drill bit had successfully re-entered the hole.

The ship's scientists and most of the crew celebrated Christmas Eve, the night of the drill re-entry. The dinner feast included cornish hen, turkey, ham, and even frogs' legs. The food was so good that everyone who attended was incapacitated for a few hours from overeating.

Coring resumed through limestones, volcanic sands, and black muds to a layer of diabase, a rock similar to basalt, less than 3 feet thick. A thin limestone layer lay beneath that and then another section of diabase. The paleontologists noted with delight that the limestone between the two diabase layers contained fossils that were 85 million years old. We cored almost 45 feet into the lower diabase layer before we stopped drilling. This diabase was probably the crustal rock of the Caribbean. By that time, the second bit was demolished. Altogether, we had pierced a hole 2,407 feet into the ocean bottom.

A few weeks later, a second hole was drilled at Site 146 to recover the soft upper sediments so sparsely cored in our desire to get the record of the earliest years. The sediments collected from these two holes have yielded an unprecedented continuous history of sedimentation and evolutionary changes of flora and fauna over the past 85-million years in the Caribbean area.

To many geologists and geophysicists, the most important discovery was that the sediments overlying the lower diabase layer were only 85-million years old. The age hardly fits into our concepts of sea-floor spreading. Scientists aboard the *Glomar Challenger* on its third cruise had clearly demonstrated that the spreading separated North America from Europe and Africa about 180 to 200 million years ago, and that South America probably parted from Africa sometime later, about 150 million years ago. If the Caribbean basin developed as North and South America pulled apart during their dominantly westerly drift, then shouldn't the sea floor that was created be considerably older than 85 million years? The simple answer is "yes," but the earth doesn't always respond in simple terms and so the process of sea-floor spreading may have been much more complex. The Caribbean basin could conceivably be older. For instance, the crust may have started forming at the same time that South America separated from the supercontinent, 150 million years ago. In this case, the diabase where we stopped drilling could represent only the last phase of 65 million years of igneous and volcanic activity. Another possibility is that the Caribbean crust may have formed quickly 150 million years ago, followed by little activity until new lava flows covered the older sediment 85 million years ago, deeply burying it under the diabase. Further study of other samples of the diabase may reveal if either concept is right.

Some geologists offer yet another theory—that the Caribbean was once a land mass that sank to oceanic depths. There is no indication of this in the sediments recovered from the deep basins. However, later in our cruise we found evidence in cores suggesting that at least part of two submarine ridges, the Aves Swell and the Beata Ridge, were once above sea level. These may have been isolated volcanic peaks that slowly sank beneath the sea. In January, 1971, at four other drilling sites 200 to 300 miles from Site 146, we cored diabase of approximately the same 85-million-year-old age. This established the great extent of this type of rock in the Caribbean. We also cored at two other sites in the Caribbean during the 15th cruise. Scientists are planning a deeper probe into the crust in 1972 to see if the thick diabase zone has sediments beneath it.

On February 2, the *Glomar Challenger* docked at Cristóbal, a port city of the Panama Canal Zone, having drilled a total of nine holes. As we carried our gear from the ship that afternoon, we found the scientists who would conduct the 16th cruise in the Pacific Ocean waiting on the pier. Our team went to a Panama City hotel where we reviewed our findings for the next two days, and prepared a report for the scientific community. Then, we returned to our separate laboratories to begin a detailed analysis of the many rocks we had collected from beneath the bottom of the sea.

In the months and years to come, our understanding of the ocean and, indeed, of the entire earth should be greatly enhanced by the explorations of the *Glomar Challenger*, as it continues to bring up thousands of feet of sea-bottom core for study. With the success of re-entry, we took a giant step toward a deep crustal probe. The more we learn, however, the greater the demand becomes for improved technology, better bits, faster drilling, more samples, deeper penetration, and new laboratory equipment to improve core analysis. The voyages of the *Glomar Challenger* have opened wide the future for oceanographers.

For further reading:

Taylor, D. M., "The Challenger's Adventure Begins," *Ocean Industry,* October, 1968.

Peterson, M. N. A., and Edgar, N. T., "Deep Ocean Drilling with Glomar Challenger," *Oceans,* June, 1969.

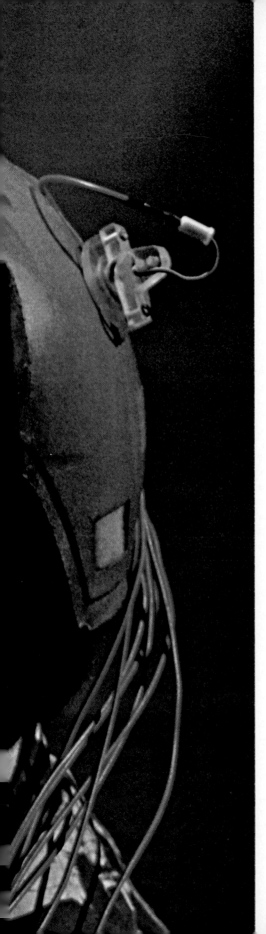

Attempting to Understand The Brain

By Lee Edson

The first steps are being taken to find out how the brain sorts and codes information from our senses

For thousands of years, man's brain has intrigued his mind. About 3 pounds of jellylike tissue that you could hold in the palm of your hand, the brain is the director of almost all human behavior, and indeed, of human society, culture, and civilization.

How did this pinkish-gray, convoluted organ reach such an unmatched pinnacle of achievement? How does it help us remember and learn? And how does it think and guide our behavior?

Until recently, there seemed to be no way to answer such difficult questions. But, sophisticated new instruments used in a host of new and exotic disciplines, such as molecular psychobiology, communication biophysics, and neural biochemistry, are bringing the understanding of the brain into a sharp new focus.

Neurophysiologist David Hubel, a leading brain researcher at Harvard University, put it this way, "For the first time, we may be on the verge of an understanding of the brain—and, ultimately, it is hoped, of solving the mystery of behavior—in terms of the brain's physical mechanisms." Not all scientists, however, agree with Hubel's view. Many feel that the brain cannot be understood by reducing it to its component parts through observation, surgery, and chemical and physical analysis. Some psychologists, for instance, tend to view the mind as more than the sum of the brain's parts. To try to understand the mind, these researchers analyze its most complex products, such as language. Some of them are convinced that to fully understand the brain would take a thinking device of a higher order than the brain itself.

If, however, the brain is strictly an electrochemical machine, it is as subject to analysis as a rocket or an automobile. True, its 10 billion neurons, or nerve cells, each with 60,000 connections, make the brain the most complicated mechanism known. But, despite its complexity, researchers throughout the world are bravely taking the brain apart, tracing its electrical and chemical pathways, and searching deep inside its molecules.

Most modern experimenters approach the brain via three main routes. These are: (1) electrophysiology, in which electrical signals in the brain are studied, (2) biochemistry, in which the chemical activity of brain cells is monitored, and (3) neuroanatomy, a surgical method related to the ancient lesion approach in which damage to the brain was correlated with changes in perception and behavior. From this neuroscience trinity and a little psychology, researchers hope to show at least the first stage of the brain's function: exactly how information from our senses is sorted and coded. And they have made remarkable progress in some areas.

In electrophysiology, one of the first big steps was taken in 1959 when biophysicist Jerome Lettvin and co-workers at the Massachusetts Institute of Technology published a classic paper, "What the Frog's Eye Tells the Frog's Brain." This article describes a "bug detector" in the frog's eye. When an insect flies in a certain path in front of the frog's eye, the detector passes an impulse to the brain. The brain responds with a signal to the tongue, which flicks out and snags the insect. The researchers discovered that the bug detector reacts only to a small dark object in flight. This highly specialized information-processing feature explains why a frog can starve on a carpet that is covered with dead flies.

Also in 1959, Hubel and his assistant Thorsten Wiesel began experiments that showed that mammals also had feature detectors. But the mammals' detectors are even more specialized than those of the frog, and they are in the brain, rather than the eye. The researchers used a microelectrode probe, a wire less than 1/10,000 inch in diameter. They connected it to both an oscilloscope, which displays voltage variations on a screen, and a device that clicked when the tiny voltages detected by the microelectrode exceeded a preset amount. The probe was inserted into the visual cortex of the brain of an anesthetized cat whose eyes were open and staring fixedly at a blank screen. Hubel and Wiesel then projected and moved geometrical patterns of light on the screen. At the same time, they shifted the microelectrode about in the cat's brain. When the probe encountered a neuron that was activated by what the cat saw, loud clicking was

The author:

Lee Edson is a science writer whose articles have appeared in numerous magazines and books. He is also a former contributor to *Science Year.*

heard. Simultaneously, a spike, representing a voltage peak, appeared on the screen of the oscilloscope.

Many cells of the visual cortex were monitored in this way to define what pattern the cells could "see." The nerve cells of the cat's visual center responded to straight lines—each cell to a line having a unique length, orientation, and position. One neuron, for instance, responded only to a horizontal line in a certain position. Another fired only when an appropriately positioned vertical line was projected. In addition, some of the cells responded more strongly when the line was moving. Thus, Hubel and Wiesel reasoned, when a cat is confronted with a complex object, different brain cells probably register each of the object's linear components. "The cells in the brain of a cat that judge shape and movement," Hubel explains, "continually converge upon other cells that are probably able to combine the simple information into a more complex picture."

Scientists question whether the cells of the visual cortex are preprogrammed at birth, or whether they "learn" what to respond to. Hubel and Wiesel discovered in 1963 that there is a critical period of life when environment is all-important for the development of the visual cortex. In one series of experiments, they kept a patch over one eye of a cat from birth to maturity. They found that certain cells in the brain's visual cortex had died, apparently from lack of use. "Sensory processing is normally present at birth," says Hubel, "but for the cells to remain effective throughout life, they have to be used. If they're interfered with for too long a period—say somewhere between four weeks and four months—the cortex may be changed irreversibly." Hubel found that this was also true in infant monkeys. However, adult animals showed no deterioration in the visual center, even when one eye was kept covered for as long as two years.

In related work, psychologist Helmut Hirsch and physiologist D. Nico Spinelli at Stanford University, as well as physiologists Colin Blakemore and G. F. Cooper of Cambridge University in England, found similar results when they raised cats in very limited visual surroundings and then tested the animals.

For example, in 1969, Hirsch and Spinelli raised cats from birth wearing goggles in which one eye saw only three horizontal lines, and the other eye saw only three vertical lines. Then, with a micro-electrode-probe technique they checked cell response in the visual cortex. Cells activated by the eye restricted to the vertical lines responded only to lines that were vertically oriented, and cells activated by the eye restricted to the horizontal lines responded only to horizontally oriented lines.

To test further the effects of the pattern of lines, Hirsch and Spinelli monitored the cells while the cats stared at a blank screen across which a spot moved. When a cell fired, the position of the spot was recorded. Each of the cells responded only when the spot passed across a part of the screen that represented portions of the three lines to which

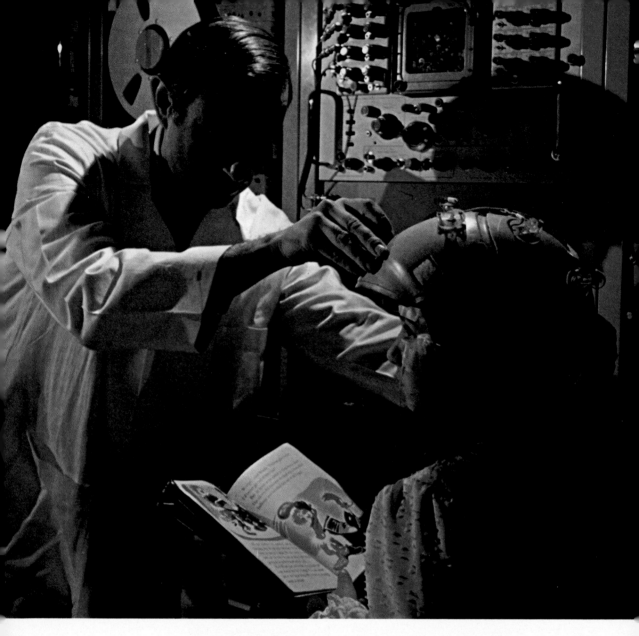

Electrical engineer Bernard Sklar of UCLA adjusts cap studded with electrodes that monitor brain waves. The brain waves recorded while a child reads can be analyzed by computer to determine if the child has dyslexia, a neurological disorder that seriously impairs learning to read. In 1971, Sklar designed a special alphabet for these children that has marked differences in the shapes of each of its letters.

 Ducks eat fish.
Cows eat grass.

the eye had been restricted by the goggles. The points at which some cells fired produced a pattern that re-created all three lines. Other cells seemed to "remember" only parts of the stimulus, perhaps only a single line or a part of a line.

The scientists wondered what would happen with a more complex stimulus, so they raised cats wearing masks that allowed them to see only a dot surrounded by concentric circles—a bull's-eye shape. Visual cortex cells of those animals re-created only small parts of the bull's-eye shape when they were tested.

The work of Hirsch and Spinelli may indicate that cells in the visual cortex are not programmed, but rather learn what patterns to see. "Some cells may not be specified at birth," Spinelli says, "but may develop their 'shape' response through stimulation by the environment." To prove it, of course, is difficult. Scientists will have to follow single cells and trace the development of their patterns of sensitivity throughout the animal's growth.

In 1970, Blakemore and Cooper put kittens inside a cylinder marked only with vertical lines. The animals wore wide collars so that they could not see their own bodies. After five months, the scientists tested the animals for their response to vertical and horizontal lines. The cats responded to vertical lines, but ignored horizontal lines. In fact, they seemed not to see horizontal lines at all. For example, they dodged a black rod held vertically and swung at them, but did not seem to see the rod when it was swung at them in a horizontal orientation. Surprisingly, after about 10 hours of disorientation, in which they seemed to guide themselves mainly by touch, the animals learned to behave normally. Apparently, their brains began to extract enough information about horizontal lines from the vertical lines they saw to get along in the complicated geometry of the real world.

Two months later, the researchers checked the cats' brains electrophysiologically, using a microelectrode probe. They monitored the discharges of 125 randomly selected cells in the visual cortices of two cats. Not one cell responded to any line that was slanted within 20° of the horizontal. This indicates that early visual experience permanently modifies the brain.

Processing the input of the senses is only the initial stage of the activity of the mind. At New York Medical College, E. Roy John, director of the Brain Research Laboratory, has tackled the difficult problem of analyzing the effects of sensory experience. "Learning and memory involve not specific nerve cells," he says, "but the statistics of populations of cells. The statistics are altered during the process of learning and that alteration can be seen in the electrical wave patterns that come from the brain."

Using sensitive recording devices, tiny electrodes, and sophisticated computers, John has followed the subtle electrical waves that accompany thought. In fact, the waves can actually tell you accurately what is on an animal's mind.

Experimenters at Stanford University raised cats wearing
goggles, *above,* that allowed them to see only simple patterns.
(The cardboard collar keeps the animal from knocking goggles
off.) After several weeks, the cats were ungoggled, and
drugged to keep them inactive and staring straight ahead.
Then, they were placed in front of a blank screen, *right.* A
microelectrode probe monitors a single visual cortex cell as
a light circle moves slowly back and forth across the screen.
The cell fires each time it "sees" the circle. A record of
where the circle was at each firing, *below,* is a partial
reproduction of the horizontal lines seen in the goggles.
This indicates that the cell is associated with the cat's
right eye and, most important, that the cell sees only in
the area of the original pattern. Similar records of visual
cortex cells associated with the left eye are partial
reconstructions of the bull's-eye pattern. Such results
suggest that cells of the visual cortex are not preprogrammed.
Rather, they "learn" what to see by early visual experience.

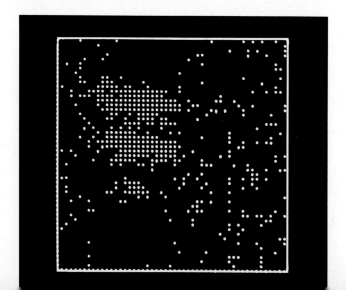

On a recent visit to John's laboratory, I saw a motion picture of one of his experiments. The star of the film was a cat wearing a headdress of 34 electrodes so thin as to be almost invisible. Such electrodes, I learned, have been kept in a cat's head for as long as eight years without ill effect. The cat in the film had been trained to move toward a certain door when a light flashed once each second, and to choose another door when the light flashed three times per second. The different electrical waves reproduced by the brain in response to these signals were displayed on a recorder.

John then flashed the light two times per second. Not surprisingly, the animal looked puzzled at first. Moments later, however, the recorder showed a brain wave characteristically produced by either one of the two training signals. From the shape of the wave, I could reliably predict which door the cat would choose. John feels that the wave shapes that permit such predictions not only represent memory, but also decision making on the part of the cat.

In 1968, electrophysiologist Dennis Willows of the University of Washington accomplished an even more remarkable feat. He recorded a wave associated with a specific activity, and then triggered the activity at will by playing the wave back into the brain. He did this with a sea mollusk, a primitive animal with a simple nervous system.

Willows sank microelectrodes into the brain of the sea mollusk. Then he placed a predatory starfish in the creature's tank, and monitored the mollusk's neurons while it tried to escape. He was able to trace the pathway of the electrical signal back to a specific neuron. Willows then took the starfish out of the tank, and played back into that neuron the same series of electrical bursts. The mollusk again tried to escape—from a nonexistent starfish. This demonstrated that, for at least one form of behavior, the animal's brain is structured to operate through a very rigidly specified wiring system.

Willows believes that results from experiments working with this simple creature can be applied to higher animals. He bases this belief on the theory that the basic operation of the nervous system has not changed significantly through evolution.

Although such electrical studies have shed new light on the processing of information in the brain, most scientists agree that chemical change is the ultimate basis of learning and memory. In the early 1960s, molecular biologist Holgar Hydén of the University of Göteborg in Sweden tested whether or not learning and memory result from the brain cells' production of new forms of ribonucleic acid (RNA). This is an important substance used to make proteins. Hydén performed a series of spectacular experiments in which rats were trained to balance themselves on steel wires or to transfer handedness—that is, to use their left paws instead of their right paws to reach for food. After the training, Hydén extracted a broth of large nerve cells from the medulla, a part of the rat's brain that was very active electrically during training. Working with only a millionth of a gram of extract and using an incredibly sensitive analytical method, he measured the chemical changes in these cells. He did this by comparing his results with data from identical analyses of the brain cells of untrained rats. Hydén found a change in the chemical composition of the RNA and an increase in the number of RNA molecules. He concluded that RNA and the proteins it helps to produce could well be the chemical basis of memory and learning.

Most scientists applaud the elegance of Hydén's microtechniques. But they are not yet ready to believe with Ralph Gerard, the former dean of the University of California, Irvine, that "behind every twisted thought lies a twisted molecule." Despite the doubts of his peers, Hydén is unfazed. In 1970, he and his associate Paul Lange reported that they had identified a specific protein made in large quantities in the rats' brains while the animals were learning to change paws. The protein is called S 100, and it is found exclusively in the brain of several species of animals. What it does there and how it may be related to learning and memory is not known, but Hydén found a large amount of it in the hippocampus, a part of the brain where most scientists suspect certain kinds of learning take place.

Other scientists think that protein synthesis may also explain why some memories are short while others last a lifetime. Several biochemists have devised ways to interfere with protein synthesis, thus presumably interrupting the consolidation of learned information. During the early 1960s, anatomist Louis B. Flexner of the University of Pennsylvania injected mice with puromycin, an antibiotic known to inhibit protein synthesis, and showed that the chemical also inhibited learning. Biochemists Roger Davis and Bernard Agranoff of the University of Michigan trained 500 goldfish to swim over an underwater hurdle by giving them electric shocks every time they went in the wrong direction. When the fish had learned the task, the scientists injected them with puromycin. If injected immediately after the training, the fish seemed to forget what they had learned. But, if injected after an hour, which would be long enough for a great deal of protein synthesis, the fish remembered the task.

These scientists readily admit that their work does not absolutely link the interference of learning with the blockage of protein synthesis. However, they have drawn a tentative conclusion from biochemical and electrophysiological work. Long-term memory is essentially chemical, and, unlike the presumably electrical nature of short-term memory, depends on the synthesis of new proteins.

The most dramatic chemical route to the secret of memory would be to find a "memory molecule." A scientist might then isolate it from the brain of one animal, inject it into another, and find evidence that the second animal has the memory. In 1961, biochemist James V. McConnell of the University of Michigan trained flatworms known as planaria to contract in response to a flash of light. Then, he cut up the trained worms and fed them to other planaria. After the canni-

Goldfish can be taught to avoid the blue side of a compartmented fish tank, *above,* by giving them an electric shock each time they enter that side. When biochemist Georges Ungar of Baylor College of Medicine, Houston, injects extracted brain material from such a trained fish into an untrained goldfish, *right,* the untrained fish avoids the blue compartment without being shocked, *far right.* Ungar claims that this is memory transfer, and he is now trying to isolate the chemical that carries the memory. He has already started training some of the thousands of goldfish he feels he needs to do the job.

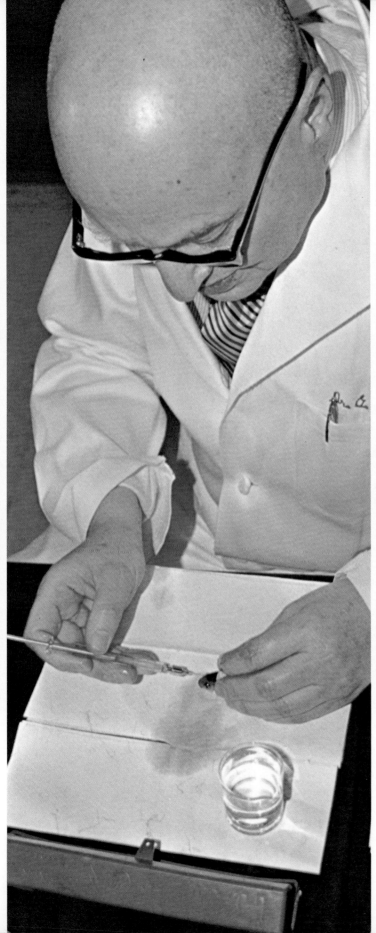

balistic worms had gorged themselves on their learned brothers,
McConnell found that they contracted in the light as if they had been
trained. This remarkable experiment created a sensation among
scientists and laymen. A standing campus joke was that now the
university could cut up the professors and feed them to the students.
McConnell himself began to talk of education by injection, and a
popular book later made into a motion picture, *Hauser's Memory*, was
based on this concept.

McConnell's work inspired a flurry of research on memory transfer.
Several scientists reported success. But it was not until 1970 that
anyone claimed to have taken the next logical step. Biochemist
Georges Ungar of Baylor College of Medicine, Houston, claimed not
only to have transferred a memory from one group of rats to another,
but also to have isolated the chemical responsible. The memory was
fear of the dark. Rats normally prefer darkness to light, but Ungar
soon taught his rats to reverse this preference by giving them an
electric shock when they went into a dark area. When they had
learned to avoid the dark, the rats were sacrificed and their brains
removed. Other rats injected with material from these brains avoided
the dark without any training. Ungar then isolated from the brain
material a single substance that could cause untrained rats to fear the
dark. The substance was a peptide, a small proteinlike molecule, that
Ungar named scotophobin, from the Greek words for dark and fear.

By mid-1971, the scientific community was still very cautious about
Ungar's findings. Some critics argue that fear of the dark is not a
memory at all. Moreover, most brain scientists want more definitive
experiments before they accept any memory transfer as valid. In general, biochemists seemed to agree that learning capacity might be
transmitted by injection, but specific content cannot. Ungar, however,
continued his research, and, in 1971, he reported the results of preliminary experiments in which he claimed to have transferred learned
color avoidance in goldfish. To isolate the peptides for blue and green
avoidance, he has begun training the first of the 20,000 goldfish.

In addition to electrophysiology and biochemistry, there is the third
and oldest highway to understanding the mechanisms of the brain—
the neuroanatomical approach. Here scientists examine the fundamental functioning of the brain by altering it surgically and observing
behavioral changes. A leading figure in this research is psychobiologist
Roger Sperry of the California Institute of Technology.

Sperry began brain research in the 1930s when scientists were largely influenced by the behaviorist theories of psychologist John B. Watson. Watson believed that environment controlled all behavior. Thus, nerve cells could be taught how to behave. This notion was so pervasive that neurosurgeons often connected nerves severed in accidents without concern as to whether the two ends came from the same nerve. The surgeons claimed that the patient could restore the original function of the nerves with practice. One patient who suffered from such an operation was a woman whose face was injured in an accident. Her surgeon had connected severed salivary gland nerves to tear duct nerves. After that, the woman salivated when she wanted to cry, and, on at least one occasion, she cried when she looked at a T-bone steak.

Suspecting Watson's theories to be wrong, Sperry began a series of experiments with newts that proved that the nerves connecting the eyes to the brain transmit only specific information and cannot be taught to transmit new information. The scientific world did not accept Sperry's conclusions at first. So he went on to perform a number of additional experiments on a variety of animals to see whether severed nerves grow back to their old connections, or whether they join at random. He found that severed nerve fibers grow back only to their original place "with utmost precision and selectivity in accordance with the animal's genetic wiring diagram."

Sperry next turned to the problem of why the brain has two hemispheres—why we have, in effect, two brains. Each hemisphere controls movements in the opposite half of the body, but how does one hemisphere learn what the other hemisphere knows? How is information transferred? This could not really be understood until after Sperry and his associates cut all contacts between the two hemispheres in experimental animals in 1951. These split brains opened up a new and very informative area of brain study.

Sperry's continued experiments with split-brain animals have shown that the two hemispheres are independent when not joined by the corpus callosum, the major connection between the hemispheres and the largest bundle of nerve fibers in the body. Sperry's first chance to investigate the effect of brain splitting in a human being came when he was called upon to treat a World War II paratrooper who suffered epileptic seizures because of a brain injury suffered in a fall. The injury was confined to one hemisphere, and Sperry knew that the electrical discharge typical of epileptic seizure is transferred across the corpus callosum. He suggested split-brain surgery to try to stop the seizures.

The operation halted the seizures. However, Sperry and his associate Michael Gazzaniga found that while the patient was normal in nearly all respects after the operation, he could not use his hands simultaneously to work a jigsaw puzzle. Stranger still, one hand often did the opposite of the other hand. For example, the man would use his right hand to button a shirt only to find that his left hand was unbuttoning it at the same time.

Experiments showed that the halves of the man's visual field were perceived and remembered separately. In one such experiment, the man stared straight ahead as pictures were flashed to the left and right sides of his field of vision. A picture flashed to the right side of the field was perceived and remembered only by the left hemisphere, and one flashed to the left was seen and remembered only by the right hemisphere. Even smell was divided. An odor identified through one nostril could not be recognized or remembered a moment later when smelled through the other nostril.

Sperry says that each cerebral hemisphere in the normal human brain has its own domain of consciousness and its own chain of memories. And each also tends to have its own specialty and to handle problems with its own built-in strategy. The left hemisphere of most persons tends to specialize in language and operates in logical computerlike fashion. The right hemisphere can best put together bits and pieces into whole images, but it must pass the images to the left hemisphere if they are to be described verbally.

By 1971, the split-brain technique was being used as a research tool in laboratories around the world to shed some light on the higher processes of learning and memory. So far, the major finding that holds promise is that the right hemisphere seems to have potential in areas with which it is not normally associated. In 1971, Gazzaniga, working with psychologist David Premack, a visiting professor at Harvard, showed how this potential might be put to use. Premack startled his colleagues in 1970 by showing that he can conduct a conversation with a chimpanzee. Using symbols that refer to items in the environment, Premack and the chimpanzee converse, with the animal showing such versatility in manipulating the symbols that humanlike thought seems to be involved. Gazzaniga and Premack have recently applied the method to helping victims of aphasia, a lesion of the left hemisphere that results in loss of the ability to talk or use words in any other communicative way. They reasoned that the undamaged right hemisphere, which is perfectly capable of using symbols, might help these people communicate. Initial reports indicated good results in communicating with these persons.

The portrait of the brain emerging from this honeycomb of findings in the last few years shows a structure of remarkable specialization. Scientists are just beginning to understand the brain, however. Yet, their continued research promises an era in which science will at last be able to learn what fundamental physical processes are behind the workings of the mind.

For further reading:

Calder, Nigel, *The Mind of Man: An Investigation into Current Research on the Brain and Human Nature,* Viking, 1971.
Rose, Steven, "The Future of the Brain Sciences," *New Scientist,* June 25, 1970.
Tiplady, Brian, "The Chemistry of Memory," *New Scientist,* June 25, 1970.

The Unwelcome Badge of Adolescence

By Marion B. Sulzberger

A disease that can scar the psyche as well as the skin, acne is beginning to yield to medical research

Man's skin protects him from a hostile world. It resists excessive heat, cold, dirt, chemicals, and living adversaries—bacteria, fungi, and insects. It endures infections, burns, scrapes, scratches, cuts, and bruises. One of the body's largest organs, the skin encloses all the others, providing them with a germfree, shockproof environment of constant temperature and humidity. But sometimes these internal organs harass their protector with rashes, blotches, and swellings. Moreover, an imbalance of such substances as enzymes or hormones, or the body's reactions to certain foods or drugs, can produce diseases of the skin itself.

Because the skin is assaulted both from within and without, its diseases are caused or aggravated by a complex mixture of influences. This sometimes makes them difficult to diagnose and treat. Skin diseases are usually visible, so they have profound psychological effects, even though most cause little discomfort and do not threaten life.

Perhaps nothing makes more people feel more insecure than an unsightly skin problem called *acne vulgaris* (common acne). The disease comes at an uncertain time of life—adolescence—when a person is making the mental and physical transition from childhood to adulthood. Furthermore, acne usually appears on the most visible parts of the body—the face, shoulders, and chest. From 80 to 90 per cent of all adolescents suffer from the disease to some degree. Probably no one escapes a few pimples or blackheads.

The problems of acne and attempts to cope with them go back many centuries. Over 1,600 years ago, Marcellus of Bordeaux, court

physician to Emperor Theodosius I of the Roman Empire, advised: "Watch for a falling star, and then instantly, while the star is still shooting from the sky, wipe the pimples with a cloth or anything that comes to hand. Just as the star falls from the sky, so the pimples will fall from your body; only you must be very careful not to wipe them with your bare hand, or the pimples will be transferred to it."

Acne seems to develop from a number of external and internal factors affecting the skin. Thus, to understand acne and how it begins, we need to examine the structure and function of normal human skin.

The skin weighs from 6 to 9 pounds in an average adult. Spread out, it would cover from 18 to 20 square feet. The structure of the skin varies greatly from place to place because it must perform different duties. For example, the heavy hairs of the scalp protect the brain from blows and too much sunlight. The thick-skinned palms can grasp sharp objects and resist pressures without damage. The fingernails—also a type of skin—help us pick up tiny objects.

In most animals, the skin is covered with hairs that insulate from excessive cold and heat. Man gradually lost much of his body hair during evolution, and he learned to keep warm by clothing himself in animal skins. His main protection against heat has resulted from another evolutionary development—sweat glands. Their purpose is to cool the body through evaporation of fluid from the skin's surface. Each person has more than 2 million sweat glands, an average of 860 to 1,200 per square inch of skin.

On the skin's surface, sweat combines with *sebum* (an oily substance), which is secreted by the *sebaceous* (oil) glands. This mixture discourages the growth of harmful bacteria and lubricates the skin so that it will not become brittle and scaly. Sebaceous glands are distributed very unevenly, however. For every gland on your leg, there are about 20 on your trunk and 100 on your face.

The skin has several layers of cells. The number of layers depends on the part of the body it covers, the person's age, and other factors. It consists of two distinct structures—the outer epidermis, and the inner supportive structure called the dermis. The dermis contains the skin's blood vessels, nerves, and connective tissue. In the epidermis, the cells of the outermost layer—called the horny layer—are dead. They interlock to form a tough protective sheet. Over most of the body, the horny layer is only a thousandth of an inch thick, no thicker than a sheet of very fine tissue paper. Yet this layer helps to keep in vital body fluids that otherwise would evaporate. Below the horny layer is the living epidermis. Its cells rise toward the surface as they mature and become the horny layer when they die.

In millions of places on the body, the epidermis folds in, forming finger-shaped invaginations that protrude deeply into the dermis. (You can picture them as the fingers of a glove pushed back inside the hand.) These invaginations form the hairs, nails, sebaceous glands, and sweat glands. The invaginations involved in acne are the hair

The author:
Marion B. Sulzberger
is professor emeritus of
dermatology at New York
University and former
technical director of
research at Letterman
General Hospital
in San Francisco.

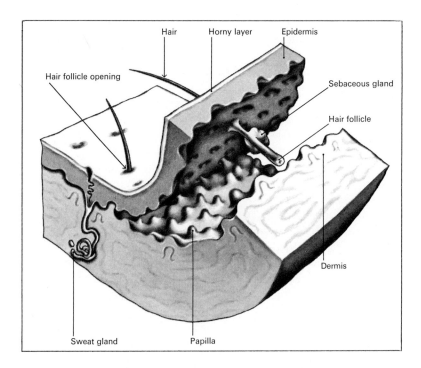

Hair Horny layer Epidermis

Hair follicle opening

Sebaceous gland

Hair follicle

Dermis

Sweat gland Papilla

The Skin's Architecture

The dermis and epidermis form the main layers of the skin. The papilla contain nerves and blood vessels. They fit into pits in the bottom surface of epidermis, helping to connect the two skin layers.

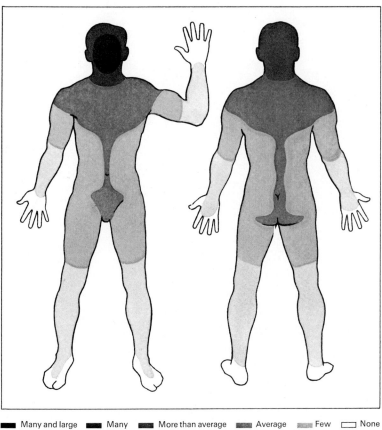

The Variable Sebaceous Glands

The concentration and size of sebaceous glands vary greatly over the surface of the body. There are no glands on the palms of the hands or the soles and sides of the feet.

■ Many and large ■ Many ▥ More than average ▥ Average ▥ Few ☐ None

follicles and the sebaceous glands. At the lining of each gland, the sebaceous cells constantly divide and grow larger as they are pushed toward the center by new cells. When these sebaceous cells reach the gland's duct, their walls disintegrate. The contents—sebum—reach the skin's surface through the hair follicle.

When a person reaches puberty, his anterior pituitary gland begins to secrete new kinds of hormones into the blood stream. Most of these hormones stimulate other glands, such as the sex glands—the ovaries and testicles. These glands, in turn, send out their own hormones, which cause, for example, voice change in boys and breast development in girls.

One type of male hormone, the androgens, is produced by both sexes. At puberty, the androgens cause the epidermis and its horny layer to thicken. In certain parts of the skin, the sebaceous glands become very large and produce great amounts of sebum. Although the androgens cause many body hairs to grow sturdier and thicker, for some reason the hairs in certain follicles into which the large sebaceous glands empty fail to grow; they remain just as small as they were before puberty.

One of the reasons I became a dermatologist was my fascination with the complexities of the skin. I recognized the importance of its

The scanning electron microscope shows the sheathlike structure of a hair follicle. This 200-times magnification also reveals its cut hair and sebaceous gland, which resembles a tiny cluster of grapes.

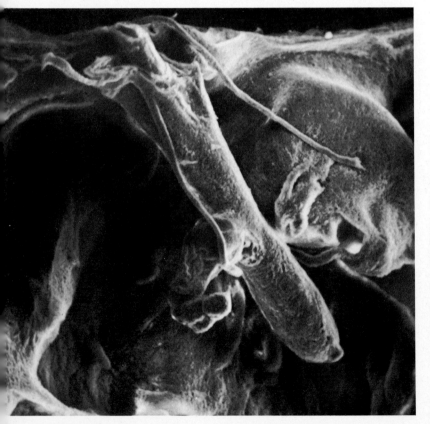

A very fine hair, magnified 320 times, extends from its follicle opening, *below.* The same area, *right,* magnified 3,200 times, reveals debris in the follicle opening.

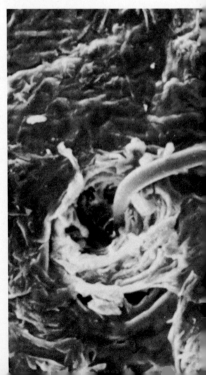

protective functions and the effect of its changes on a person's emotions. And I was also attracted by the accessibility of the skin for study and research. But perhaps my greatest motive was that my younger sister was severely affected by acne. Coming when she was in her early teens, a sensitive and impressionable period, it almost ruined her life. In fact, one day in 1920, in a doctor's waiting room, she pressed a pistol to her forehead and pulled the trigger. Fortunately, the gun did not go off.

I believe that acne has evolved in man as a consequence of the gradual loss of his skin's heavy hairs, without a parallel reduction in the size of his sebaceous glands. Some of my colleagues do not agree with this hypothesis. But it does fit many facts about acne. Man is the only animal that gets acne. Also, acne is generally confined to the follicles that, during puberty, develop large sebaceous glands but weak hairs. Weak, curled-up hairs are often found in plugged follicles.

Large, strong hairs, such as those on the scalp, move constantly as the skin moves or is rubbed by clothing and other objects. Thus, these hairs would act not only as wicks to draw the sebum to the surface but also as stirring rods to keep the follicles open. Small, feeble hairs, such as those on the face, would not do this. As a result, the horny cells, which are constantly being produced at the skin's surface and in

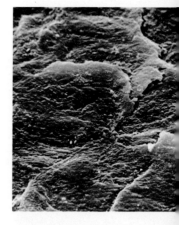

Nubby texture of normal fingernail is revealed by magnifying its surface about 2,000 times.

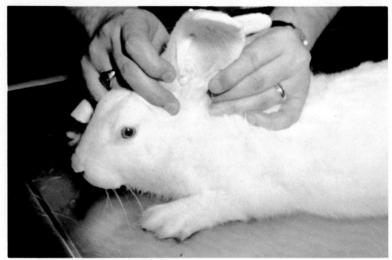

Rabbit ears are used in acne research at the University of Pennsylvania School of Medicine, *right*. Coal tar applied to inside surface of both ears plugs the hair follicles. One ear is then treated with vitamin A acid.

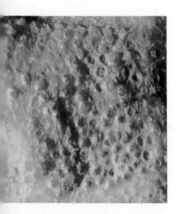

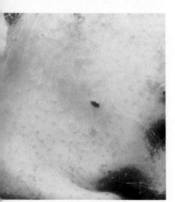

The follicles in the untreated ear, *top*, remain plugged, but the treated ear, *above*, is cleared of plugs.

the lining of the follicle neck, would accumulate with sebum and block the follicle openings.

Some researchers believe that chemicals in the sebum can stimulate an overproduction of horny cells, thus increasing the plugging. Moreover, Dr. Albert M. Kligman and his co-workers at the University of Pennsylvania found in 1970 that some of the masses of horny material and sebum may be denser, more compact, and stickier than others. The thicker masses would have a greater tendency to plug the follicle and sebaceous gland.

When sebum, horny cells, feeble hairs, and bacteria (particularly the species called *Corynebacteria acnes*) become impacted, a plug forms in the follicle. If the plug is large and its top is dark, it is called a blackhead. The blackness results from oxidation, when the horny material and the accumulation of skin pigment come into contact with air. If the plug is microscopic and pushes the skin surface up into a little translucent, whitish mound, it is called a whitehead.

Blackheads and whiteheads are the basic symptoms of acne. Why the acne process often stops at this stage and why it sometimes goes on to form the inflammatory lesions called pimples, no one knows. Some dermatologists contend that only whiteheads turn into pimples. But others, including me, believe that both blackheads and whiteheads can become pimples.

Whichever the case, almost all researchers agree that various enzymes act upon the stagnant mass of sebum, fats, and oils trapped beneath the plug to form substances called fatty acids. During the 1960s Kligman, Dr. Robert E. Kellum of the University of Oregon, and Drs. John S. Strauss and Peter E. Pochi of Boston University showed that fatty acids are extremely irritating to the dermis, and can produce severe inflammation.

One important aspect of this story that must still be clarified is where the enzymes come from. Some investigators have shown that the bacteria trapped in the follicles produce them. Other investigators have proved that as cells produced in the sebaceous glands mature and die, they, too, liberate enzymes. I suspect that bacteria play the major role in producing the enzymes, but the sebaceous cells must also contribute.

In some patients, the inflammation of pimples spreads and burrows into and under the skin. Big, abscesslike structures form, sometimes with a maze of interconnecting tunnels, like the passageways of a rabbit warren. This is called acne conglobata or cystic acne. It is the most severe kind of acne, causing much tissue destruction and scarring. It is particularly difficult to treat.

Many stimuli can aggravate or trigger acne attacks but they vary from person to person. For example, some skin specialists believe that foods such as chocolate, nuts, shellfish, pork, and milk can make some cases of acne worse. Others believe that foods have no effect. Most specialists agree that certain industrial and household chemicals such as synthetic waxes, coal tars, and oils bring on acne attacks when they are inadvertently transferred from the hands or clothing to acne-susceptible areas of the skin. Medicines such as iodides and bromides can also produce acne or aggravate it. And some of my colleagues think that emotional and other stresses are important stimuli. I am not sure, however, that this is always true.

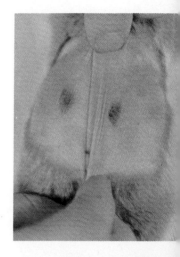

Although we know much about how acne starts, we know practically nothing about how and why most people get over it in their early 20s. Sex hormones continue to circulate in the blood stream in full strength and quantity throughout adult life. During World War II, I developed a hypothesis to explain these spontaneous cures. While on duty with a naval medical research unit on Guam, I noticed that many men in their 30s got acne conglobata. This was, of course, many years after their adolescent acne had cleared. The so-called "tropical" acne was not primarily on the usual acne sites, but on the neck, upper arms and thighs, and buttocks. The same thing has happened to serv-icemen in Vietnam. Severe acne exceeds all other skin diseases as a cause for evacuation from the war zone in Southeast Asia.

Two dark spots on back of a hamster, *top,* are areas where sebaceous glands are concentrated. The use of estrogen and androgen changes the size of glands, *above.* Such tests can help find ways to treat acne with hormones without adverse side effects.

I reasoned that perhaps the acne-susceptible follicles that are the sites of adolescent acne had either become resistant to the acne-pro-ducing stimuli, or had actually been destroyed by scarring from the earlier lesions. It also seems likely that the stresses of military service in the tropics constituted new stimuli so strong that even the usually nonsusceptible areas became the sites of tropical acne.

Because more girls than boys come to the physician to treat the blemishes of acne, doctors long thought that the disease was more common in females. But in 1935, Professor Bruno Bloch of the De-partment of Dermatology at the University of Zurich in Switzerland showed this to be an error. Bloch examined nearly 4,200 16- to 19-year-old students and found that 69 per cent of the males, but only 60 per cent of the females, were affected with some degree of acne. It is note-

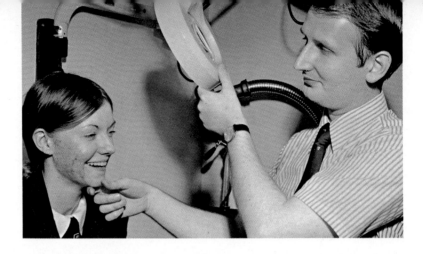

A patient's acne is examined through a magnifying glass by Otto Mills, a skin researcher in the acne clinic at the University of Pennsylvania School of Medicine.

Researcher scrubs scalp to remove bacteria, *left,* and rinses hair with ether to remove sebum, *right,* so these can be analyzed in the laboratory.

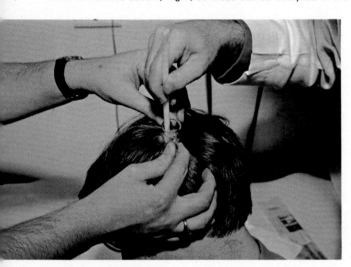

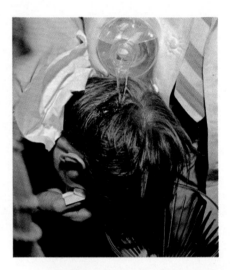

Mills treats severe cases of acne with the extreme cold of a cryosurgery probe, *below left,* or steroid injections, *below right.* Research indicates that these treatments reduce the inflammation.

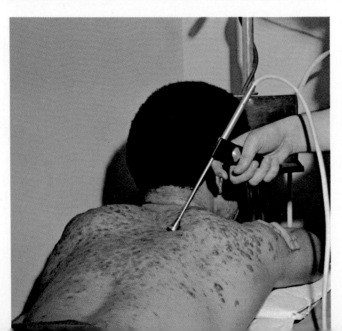

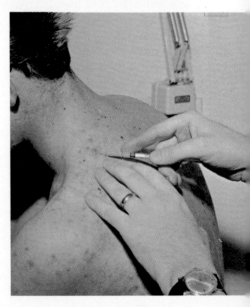

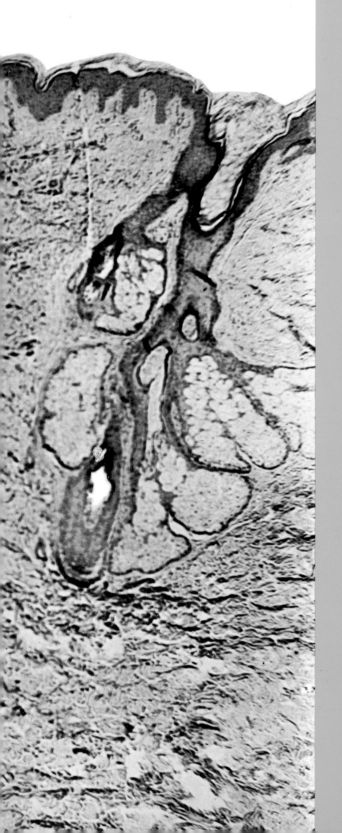

Acne Attacks

How to Use This Unit

To follow changes in the skin that lead to
acne, turn back this page. Then, place the
three acetate overlays on the right-hand page
so that the page entitled "A Vulnerable Follicle"
is unobstructed. Read the page entitled
"The Changing Face of Adolescence,"
then "A Vulnerable Follicle." Put down the
first acetate overlay and read the text
accompanying it. Continue the sequence
until you have examined each overlay and
read the text material. Finally, examine
and read "The Scarred Aftermath."

The Changing Face of Adolescence

During puberty, when the body begins to mature, the sex glands become active and trigger changes that cause acne, a skin disease that often lasts from 3 to 10 years. The skin on a young person's face, chest, and upper back is especially vulnerable to the disease.

The illustration below shows a section of skin about 1/30-inch thick from the upper cheek of a 13-year-old. It is magnified about 50 times. The section is composed of two main layers—the *epidermis* (outer layer) and the *dermis* (inner layer). Specialized cells of the epidermis perform different functions. Some cells produce hair, some oil, and some sweat. Most epidermal cells produce the surface of the skin. When these cells die, they become a hard, paper-thin membrane called the horny layer. The dermis contains the skin's blood vessels, nerves, and connective tissue.

The epidermis folds deeply into the dermis to form the many hollow structures that hold hairs, sweat glands, and *sebaceous* (oil) glands. Each hair grows in a sac called a hair follicle. Extending from the follicle wall is the sebaceous gland, which manufactures *sebum*, a fatty substance, and delivers it to the skin's surface through the hair follicle.

During adolescence, male sex hormones—produced by both sexes—stimulate skin growth. The epidermis thickens and the sebaceous glands nearly double in size. Soon, the glands are producing much more sebum.

The small, weak hairs on the cheeks, forehead, chest, and back do not grow proportionately. Medical scientists speculate that the small hairs sometimes cannot stir or draw out the sebum as fast as it is secreted into the follicles. As a result, these follicle openings can become blocked. This sets the stage for acne. The Trans-Vision® that follows shows changes in a part of this section of skin as acne attacks.

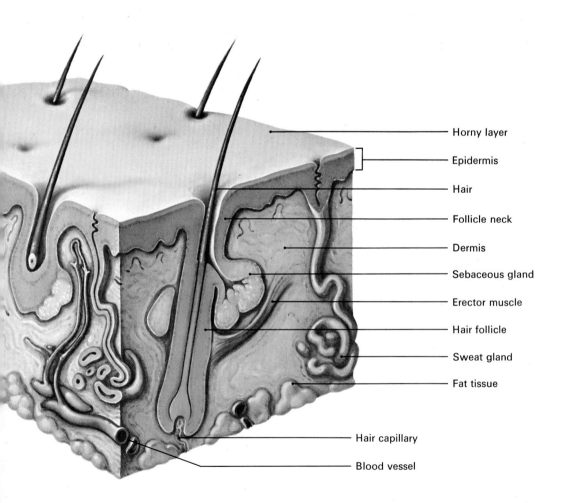

- Horny layer
- Epidermis
- Hair
- Follicle neck
- Dermis
- Sebaceous gland
- Erector muscle
- Hair follicle
- Sweat gland
- Fat tissue
- Hair capillary
- Blood vessel

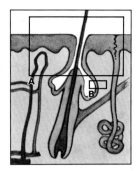

A Vulnerable Follicle

A. Sebum reaches the skin's surface through the follicle neck. The fine hair's growth, movements, and wicklike action draw sebum up slowly. Microscopic bacteria in the follicle feed on sebum, producing fatty acids. These acids and dead cells are pushed out by the sebum. This view is magnified 100 times. **B.** Cells at the lining of the sebaceous gland multiply rapidly. They are pushed away (to the left) from the lining by new generations of cells. Later, as the maturing cells approach the follicle, they will disintegrate and release sebum. This view is magnified 300 times.

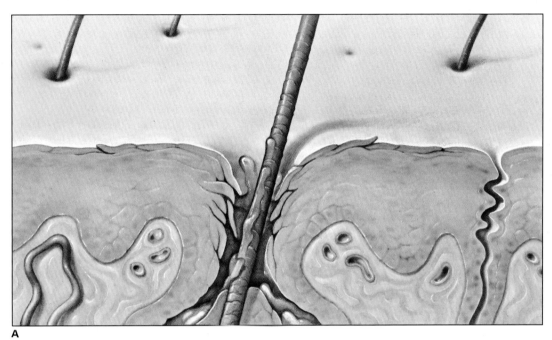

A

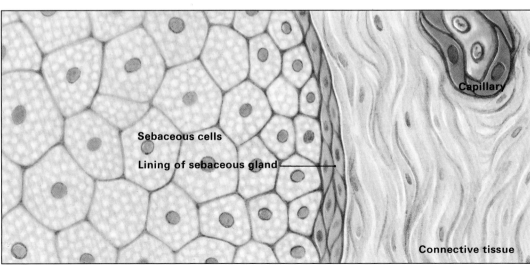

Capillary

Sebaceous cells

Lining of sebaceous gland

Connective tissue

B

worthy that the most severe forms of acne—with cysts, scarring, and numerous lesions—were more than twice as common in males.

Many dermatologists have since observed that race and heredity are related to acne. Examinations of Caucasian, Negro, and Oriental adolescents revealed that acne, especially the severe form, is much more common in Caucasians. It has also been found that severe acne tends to run in certain families, and that identical twins usually have similar kinds and degrees.

During the last hundred years, a great variety of acne treatments have been developed. Most are directed against the overproduction of horny cells and sebum, and the plugging of the hair follicles and sebaceous glands. Formerly, one of the mainstays of treatment was small local doses of X rays, which slow down the rapidly growing sebaceous cells. Thus the amount of sebum delivered to the surface through the follicles is reduced. Although the X rays reduce sebum production, the effect lasts only a few months. Because of the potentially harmful effects of X rays, the treatment cannot be given often, and it is never used now except in very severe cases that are resistant to all other measures.

Drugs that counteract the male sex hormones also reduce sebum production and perhaps also the production of horny cells. These drugs, called anti-androgens, include female sex hormones. Many dermatologists, including Dr. Victor H. Witten of Miami University and I, have given a form of estrogen, the female sex hormone, to selected acne patients for the past 15 years with very good results. Experience has shown, however, that estrogens reduce sebum production effectively in males only when given internally in such large doses that they feminize the patient. Therefore, the treatment usually can be given only to women. Understandably, researchers are looking for new and better anti-androgens.

The effectiveness of acne treatments has been studied by collecting and measuring the sebum on the skin surface before and after treatment. Some researchers remove the surface material with fat solvents. In another method, a site on the forehead is firmly covered for three hours by a 1-inch square of absorbent paper. The oily material is then collected from the paper, weighed, and analyzed.

Other treatments attack the plug in the follicle. For decades, dermatologists have removed plugs with a blackhead remover, a tiny metal instrument shaped somewhat like a box wrench. They place the instrument directly over the blackhead and press against the skin. This squeezes out the black plug.

Some of my colleagues believe that the plug cannot be removed by less drastic external treatment because it is wedged in too firmly and too deeply. However, I am convinced that the deeper materials can be cleared out if the horny material from the follicle neck and opening is steadily peeled away. This is the basis for treatment with peeling lotions and ointments, which usually contain sulfur and resorcinol.

Text continued on page 172

The Scarred Aftermath

After two to three weeks, the pus has drained from the pustule, leaving a wide cavity in the skin's surface. The hair, hair follicle, sebaceous gland, and erector muscle have been destroyed.

Blood vessels in the fat tissues have extended new capillaries into the ravaged area. These bring food, oxygen, and special blood cells, called fibrocytes, to restore skin tissue. The fibrocytes (not shown) have helped form a temporary scar by excreting collagen, a connective-tissue protein. Collagen fibers link together, forming a framework that supports blood vessels, fibrocytes, and watery tissues. This temporary repair consists of about 10 per cent solid matter.

During the next two to six months, the temporary scar gradually loses about 20 per cent of its fluids, and becomes drier and denser. As it dries, it contracts, forming a pit in the skin surface. The fibrous scar tissue is less flexible than the original epidermis and connective tissue it re-placed. The inset shows the interface between the scar tissue, permeated with capillaries, and normal connective tissue.

Not all acne eruptions go through the papule and pustule stages. If a pimple is squeezed or lanced, for example, it immediately fills with blood that coagulates to form a temporary plug. Within hours, new capillaries and fibrocytes begin the repair process. The same process occurs more gradually when white blood cells are able to stop the inflammation at an early stage.

Dermatologists do not know how to avert the onset of acne, or why its blemishes usually become less frequent when a person reaches his 20s. Until medical research can prevent the outbreak or scarring of acne, dermatologists can only work to reduce the infection and the disfigurement acne causes. They do this by prescribing treatments that can purge excess dead cells, sebum, and plugs from vulnerable adolescent skin.

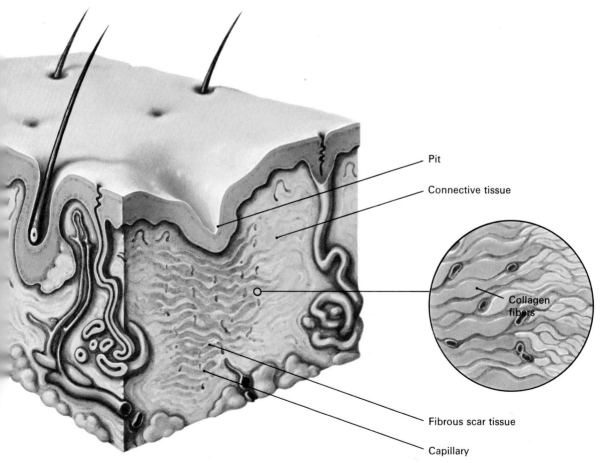

Pit

Connective tissue

Collagen fibers

Fibrous scar tissue

Capillary

Prepared by the editors of *Science Year*, The World Book Science Annual. Artist: Lou Bory Associates
Consultants: Hermann K. Pinkus, M.D., professor and chairman, Department of Dermatology,
and Earl J. Rudner, M.D., assistant professor of dermatology, Wayne State University School of Medicine
Printed in U.S.A. by the Trans-Vision® Division, Milprint Incorporated.

The skin can also be peeled physically by using abrasives, scrubs, hot compresses containing sulfur compounds, by "sunburning" with ultraviolet light, and by applying a slush of dry ice and acetone.

A treatment aimed at clearing the follicles was reported by Kligman and his co-workers in 1967. They applied vitamin A acid, which is also called retinoic acid, to the faces of volunteers with acne. This caused inflammation and peeling similar to that produced by the usual peeling remedies. But Kligman believes that the main benefit of vitamin A acid is in freeing enzymes in the cells, and that these enzymes help soften and dissolve small plugs at the openings of the follicles. He also believes that vitamin A acid speeds up the formation of horny cells. Those in the neck of the follicle would then push out the softened plugs from below.

To demonstrate this effect, the researchers injected test sites with radioactive isotopes, which are absorbed only by reproducing cells. The results indicated that the horny cells were produced faster when the follicles were treated with vitamin A acid. This new acne therapy had not been generally confirmed or approved by the U.S. Food and Drug Administration by mid-1971, but it was being intensively studied by many groups of researchers.

Treatments have also been directed against the bacteria involved in acne. Ever since 1950, when Dr. Lawrence C. Goldberg of the University of Cincinnati Medical School first reported the effectiveness of tetracycline, dermatologists have used it as well as other antibiotics. These drugs, given orally over long periods of time to patients with severe acne, are very successful. But investigators do not agree on how the drugs work. Many think that antibiotics kill large numbers of the *Corynebacteria acnes*. Others have shown that antibiotics also inhibit the bacteria and cellular enzymes that produce such acids.

Dr. Alan R. Shalita at New York University has studied the role of certain staphylococcal bacteria in acne. He believes that they, like *C. acnes*, liberate the fatty acids and are important in producing acne. Observations by Strauss and Pochi in 1971 also tend to implicate a factor other than the enzymes of *C. acnes* in producing fatty acids.

In an entirely different approach, dermatologist Edward C. Gomez and biochemist Sung Lan Hsia of the University of Miami Medical School showed in 1968 that the androgen testosterone is changed in the skin to a much more active form called di-dehydroepitestosterone. In February, 1971, biochemist Gail Sansone and dermatologist Ronald M. Reisner of the University of California, Los Angeles, Medical School published preliminary data indicating that acne-prone skin makes this conversion to active androgen more rapidly than normal skin does. Perhaps further studies will show that more di-dehydroepitestosterone is formed or retained in acne-prone follicles than in nonsusceptible ones.

Perhaps different sebaceous glands have different thresholds of response. This idea was supported by Strauss and Pochi in 1964. They

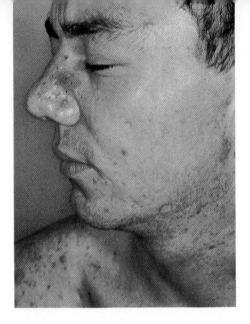

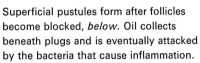

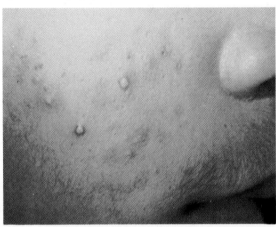

This acne patient's oily skin has large blackheads, the lesions formed by plugs in large follicle openings. Inflamed papules are clearly visible on his shoulder.

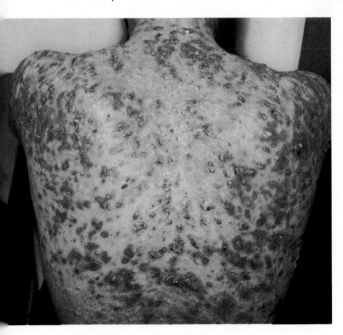

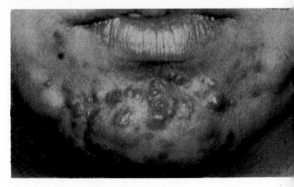

Inflammation can spread deeply into skin, *above,* causing interconnecting tunnels typical of acne conglobata.

The severe acne conglobata covering a young man's back, *above,* causes extreme discomfort and makes it difficult to wear a shirt. After acne subsided, *right,* it left deep, pitted scars and a few blackheads on a man's face. Such severe scarring is becoming rare as researchers learn to control acne.

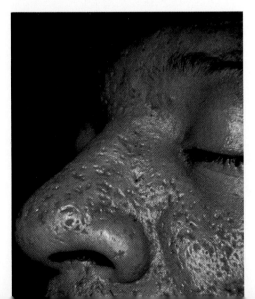

The Author Advises His Readers

If you are in your teens or early 20s, it is normal for you to have a few blackheads, whiteheads, and pimples from time to time. These are the symptoms of acne. Because individual skins differ greatly the causes of acne and the methods of treating it vary. You will have to experiment to find the best way to care for your skin.

There are dozens of lotions, creams, ointments, soaps, foams, abrasives, sprays, and other acne preparations on the market. Do some testing to find one that helps you; then use it regularly. They all work in the same ways, mainly by causing the skin's outer layer of dead cells to peel. This helps prevent the cells from accumulating with the excessive oil common to adolescent skin. If the dead cells and oil are not removed, they can clog the follicles, the openings in your skin containing the hairs and the oil glands. This clogged material can irritate your skin.

You may be able to produce the peeling without medication. Simply wash regularly and thoroughly with hot water and soap. Wash often enough to keep your skin feeling dry and slightly taut. Some skins require only one scrubbing a day, while others must be scrubbed with a rough cloth or complexion brush before they become dry and tight enough to peel.

You must use just the right preparations and the right amount of scrubbing. Otherwise, you may make your skin red, swollen, itchy, or too flaky. Once you have found an effective treatment, continue it without interruption. Keep in mind that your skin will usually tolerate more vigorous treatment in hot, humid weather. The drying procedures you use may have to be cut back in dry, windy, or cold weather because of the natural "chapping" effect of such weather.

The same stimuli that cause some of the follicles to clog cause others to pour out more oil and grease. Skin and hair that are excessively oily commonly accompany acne, and oily hair and dandruff seem to aggravate some cases. Thus you may need to wash your hair and scalp frequently. You may also find it helps to wear your hair off your face and forehead.

In temperate climates, acne tends to clear up in the summertime. Sunbathing often helps, probably because mild peeling follows exposure to sunlight. To find out whether the sun helps clear up your acne, sit in the sun for a brief time. Do not use a sun lamp unless your doctor recommends it and instructs you on how to use it.

Many dermatologists believe that diet has little effect on acne. Others, including me, have observed that certain foods can make some acnes worse, perhaps because of an allergy or hypersensitivity.

Test yourself. Go without chocolate, shellfish, milk, nuts, pork, or any other food you suspect for a couple of weeks. If it seems to lead to an improvement, then deliberately eat that particular food again – go on a binge with it for a few days – and see if your acne gets worse.

While certain bacteria play a role in producing acne lesions, acne does not begin as an infection. Furthermore, it is not contagious. You cannot transmit it or catch it from anyone. If you squeeze a pimple, blackhead, or whitehead, you do not risk blood poisoning. But if you squeeze a pimple incorrectly, you may force its contents deeper into the tissues and thereby increase the local inflammation. You may also leave behind a larger follicle opening or a visible scar. Keep your hands off until your doctor shows you how to remove blackheads and open pus-filled pimples.

Most acne lesions do not leave a mark, but sometimes they do cause pitlike scars. If you notice that your skin is scarring, don't take advice from the barber or beautician, your gym teacher, or even your grandmother. Go immediately to your doctor or a skin specialist. Modern dermatology has developed excellent means to keep acne under control and to keep you presentable while nature is curing it.

Above all, don't let your acne make you feel dirty, abnormal, or unwanted. Acne is affected little by hygiene or diet, and it has nothing whatsoever to do with sex habits. Whether severe or mild, acne is a nuisance. But it is also a sign that you are coming into physical maturity. [M.B.S.]

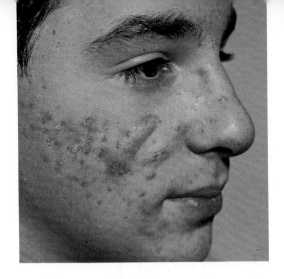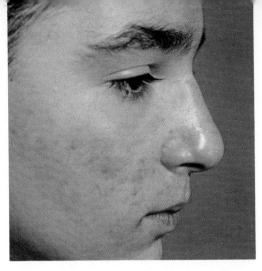

The face of a patient before and after medical treatment shows that modern research is beginning to bring acne under control.

found that greater amounts of sebum reached the foreheads of acne patients than those of persons without acne, even though both groups had the same quantity of circulating androgens.

A particularly encouraging development in acne research was the formation of a United States organization called the National Acne Association, in December, 1969. This group of more than 50 teams of dermatologists and research workers joined to intensify and coordinate their efforts and to get more attention and funds for their work.

There are many unanswered questions about acne and many problems in acne research. The greatest are probably the lack of properly trained investigators and financial support to encourage young people to enter the field. There is a need for more biochemists, dermatologists, endocrinologists, immunologists, microbiologists, and morphologists.

Teams of these specialists could pursue a multidisciplinary research program to better understand the events that lead to acne, how sebum is synthesized, and how androgens affect the glands. Working together, they could investigate ways to combat the bacteria that may be concerned and to neutralize the irritating fatty acids. In addition, the mechanics of how the horny layer thickens in the follicle neck must be defined. This research could develop much better treatments for acne during the 1970s, if the National Acne Association's program were provided sufficient funds.

Meanwhile, dermatologists and researchers continue their efforts against a seemingly simple but vastly complex disease of the skin. Although the significance of this common problem often escapes notice, blighted faces do not. Only their owners know the depth of acne's psychic as well as physical scars.

For further reading:

Brauer, Earle W., *Your Skin and Hair,* Macmillan, 1970.

Englebardt, Stanley L., "If Your Child Has Acne," *Parents' Magazine,* October, 1970.

Marples, Mary J., "Life on the Human Skin," *Scientific American,* January, 1969.

Sternberg, Thomas H., *More Than Skin Deep,* Doubleday, 1970.

New Life Line For the Heart

By Arthur J. Snider

**A new surgical technique, using a vein from the leg
as a by-pass around critically narrowed arteries,
gives hope to victims of coronary artery disease**

The pain was frightening and familiar to Joseph Montalto. It felt as
if a band was relentlessly tightening across his chest. Sometimes the
pain radiated to his left shoulder and down his arm. Brief—usually lasting for only a few minutes—but incredibly intense, the attacks often
seized the 40-year-old union business agent when he climbed stairs,
when he walked against the wind, or even when he became agitated.
His physician described the agony as angina pectoris.

Angina pectoris is the heart's cry for more oxygen. It is a signal that
one or both coronary arteries, which carry oxygen-rich blood to the
heart's own tissue, have narrowed dangerously. As a result, they cannot meet the heart's demand for increased oxygen during periods of
physical exertion and emotional stress. To ease Montalto's pain, his
physician prescribed small white pellets of nitroglycerin. This remark-
ably effective drug, placed under the tongue whenever an attack occurs, dilates an angina victim's coronary arteries, allowing them to
carry more blood to the oxygen-starved pump. Usually, it begins to
work in less than a minute.

During a 12-week strike by his union in the spring and early summer
of 1970, however, Montalto found himself under severe tension and increasing anginal anguish. He took more and more nitroglycerin pills

Surgeon holds back patient's heart to show a completed double by-pass.
The Y-shaped vein detours clogged sections in two coronary arteries.

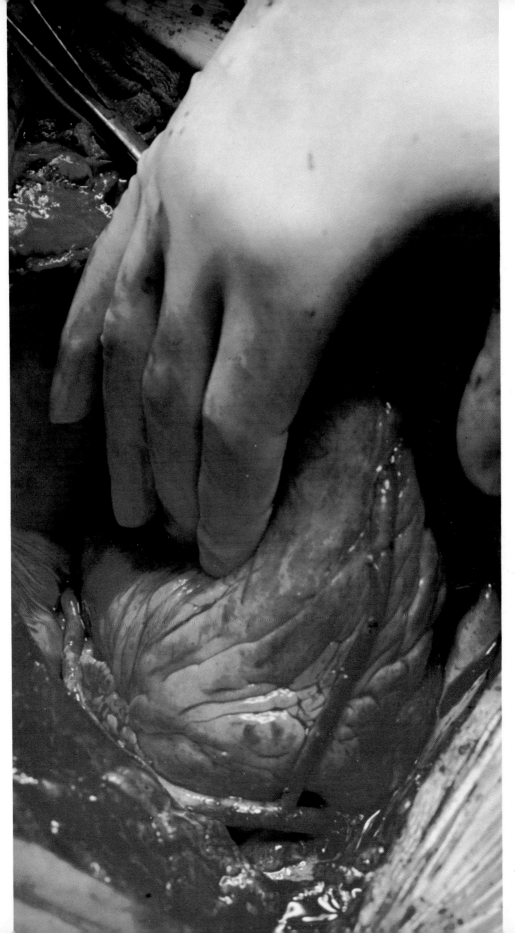

A New Vein for Old Arteries

In this typical by-pass graft, the Y-shaped saphenous vein is used as a detour around blocked portions in the circumflex and anterior descending branches of the left coronary artery.

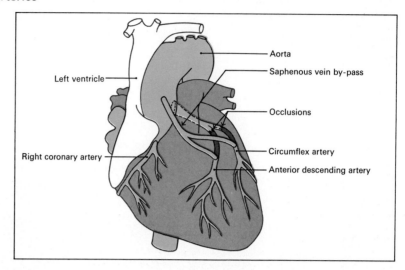

Aorta

Saphenous vein by-pass

Left ventricle

Occlusions

Right coronary artery

Circumflex artery

Anterior descending artery

The author:

Arthur J. Snider, science editor of the *Chicago Daily News,* received an American Medical Association award for medical journalism in 1971.

but finally, when no quantity could relieve his pain, his doctor referred him to Rush-Presbyterian-St. Luke's medical center in Chicago.

To determine the condition of Montalto's coronary arteries, cardiologists at the hospital used an X-ray technique called angiography, or coronary arteriography. In angiography, which is carried out under local anesthesia, a cardiologist inserts a long, thin tube into the femoral artery in the leg or the brachial artery in the arm. He guides the tube into the aorta—the major blood vessel leading from the heart —from which the coronary arteries originate. A small quantity of dye injected into the tube enters the blood stream and is observed on a television monitor and photographed as it passes through the coronary blood vessels. This allows the cardiologists to read the path of the dye through the vessels like a road map.

Montalto's angiograms revealed extreme and severe obstructions of his two coronary arteries. The flow of oxygenated blood to his heart had approached the absolute minimum necessary to maintain the action of the heart muscle. Because of the arteries' condition, the decision was made to operate, using a new surgical procedure called coronary artery by-pass grafting. The technique promised not only to relieve Montalto's pain but also to reduce the likelihood of a heart attack triggered by complete blockage of the blood flow.

In a four-hour operation, Dr. Hassan Najafi, a cardiovascular surgeon, opened Montalto's chest to expose the heart. At the same time, another surgeon made a small incision in the patient's thigh and snipped off a 10-inch segment of the saphenous vein. This vein is expendable, because the blood it carries can be returned to the heart by adjacent vessels in the leg.

The surgeons divided the vein in half. Each section was to serve as a 5-inch detour, or by-pass, around the blocked portion of the coronary channels. The surgical team completed preparation for placing Montalto on the heart-lung machine. Before activating the machine, however, Dr. Najafi applied a special clamp to reduce blood circulation through the aorta. Then he cut two small openings in the aorta and connected the end of a vein section to each opening. Dr. Najafi was careful to reverse the vein segments when he inserted them. This prevents blood from being blocked by the vessel's tiny flap valves, which normally stop the backflow of venous blood returning to the heart. The doctors activated the heart-lung machine and stopped Montalto's heart. Najafi then inserted and secured the other end of each vein segment into an opening created in each coronary artery at a point just beyond its obstruction.

When the heart-lung machine was disconnected and Montalto's heart started beating again, the results were immediately evident. The heart, now receiving a rich supply of blood, worked vigorously, and the patient's formerly pale features took on normal color. In 10 days, Montalto left the hospital, free of pain. In three months, he was ready to return to his job.

In angiography, surgeon injects dye, *below left,* through a previously inserted catheter that leads from the patient's major leg artery to his heart. The dye's progress through the coronary arteries can be followed on television screens, background, and is filmed for review and closer analysis, *below.*

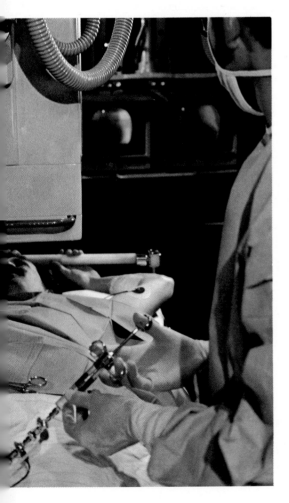

Angiogram, *below,* shows that anterior descending artery is blocked. Blockage appears at top center as almost complete gap in thick horizontal line that represents the artery.

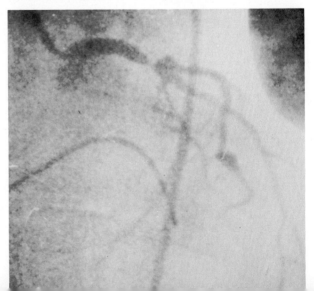

A surgical team makes a chest incision, *left.* Bone edges are packed with wax, and ends of small veins surrounding the heart are cauterized to prevent bleeding. Doctor in foreground holds a small sternal retractor, used to hold back ribs. At the same time, a separate surgical team works on patient's leg, *above,* to locate the saphenous vein and also the femoral artery, in which a catheter will be inserted for heart-lung machine connection. A long section of the saphenous vein is located and snipped off, *below left.* The vein is prepared, *below,* by trimming away fat and by cutting to length.

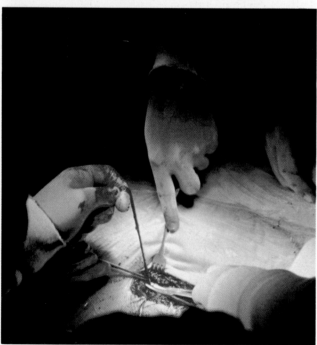

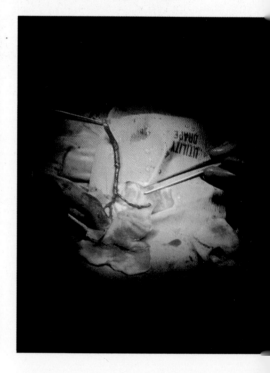

Montalto's dramatic recovery underscores the resiliency of the human heart. Beating 70 to 80 times a minute, the 9- to 12-ounce, pear-shaped organ pumps 5 quarts of blood through the body every minute. About a pint is siphoned off by the right and left coronary arteries connected to the aorta. Some cardiologists refer to three coronary arteries because the left one divides into two branches. Each artery subdivides into smaller and smaller branches, until a network of vessels covers the entire heart muscle. This network terminates in capillaries so small that the red blood cells must pass through them in single file.

If a blood clot forms where one of the main arteries has narrowed it creates a sudden and complete blockage (coronary thrombosis). The portion of the heart muscle served by this artery becomes deprived of oxygen and dies. This process is called a myocardial infarction, or, more commonly, a heart attack. The likelihood of it occurring depends on the condition of the arteries. Montalto's arteriosclerosis had reached a stage where a clot could have formed at any time.

Arteriosclerosis is a disease of the walls of large arteries. It causes nearly 90 per cent of all *ischemic* (obstruction of flow of arterial blood) heart disease. First, yellowish, fatty streaks appear in the internal lining of the arteries. These streaks slowly develop into deposits, particu-

The heart is completely exposed with a large sternal retractor as the surgeons prepare the femoral artery for insertion of the tube that will connect it to the heart-lung machine.

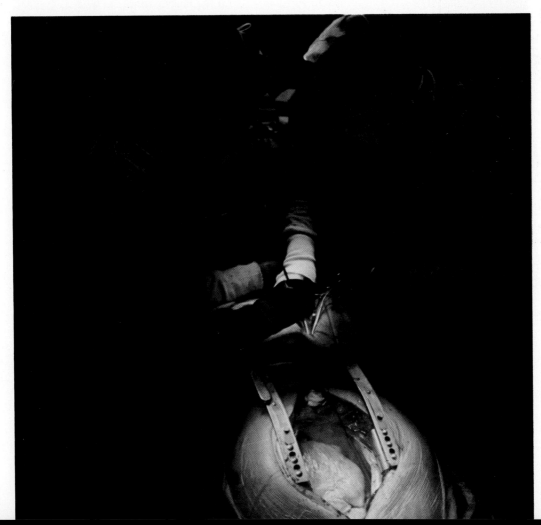

larly where the arteries divide or curve. The deposits build up and are soon covered by a gradually thickening fibrous shell, forming an arteriosclerotic plaque. As the plaque becomes larger, the channel narrows. Most of the plaque builds up in those portions of the coronary arteries that lie on the outer surface of the heart muscle. When a sub-branch of an artery burrows into the muscle itself, it seems to acquire an "immunity" from disease. In fact, most blockages occur only a short distance downstream from the mouth of the coronary arteries— 50 per cent within an inch and 70 per cent within 2 inches.

Since 1915, heart attacks have been the leading cause of death among adults in the United States and much of the rest of the indus-trialized world. Ever since coronary thrombosis was first described in 1912, physicians have sought a cure. But the toll has continued to mount. According to a report issued in December, 1970, by the Inter-Society Commission for Heart Disease Resources, more than a million persons a year in the United States experience heart attacks and some 600,000 die each year of coronary artery disease.

For decades, physicians considered the heart itself untouchable. The famed surgical pioneer, Theodore Billroth, had warned in the late 1800s: "The doctor who attempts to operate on the heart cannot wish to preserve the respect of his colleagues." Physicians treated arte-rial conditions mainly with drugs, diet, and rest, with limited success.

A number of early indirect surgical procedures were developed in the 1920s. For example, in 1922, surgeons tried cutting bundles of nerve cells, called ganglia, that go to the heart. The operation, called cervical ganglionectomy, stopped the pain, but it did nothing to over-come arterial disease.

About 10 years later, surgeons tried a surgical technique called thyroidectomy. The thyroid gland was removed to slow down the body's metabolic processes and ease the strain on the heart. Later, the same result was achieved by giving the patient radioactive iodine, which slowed down the action of the thyroid gland. Ironically, how-ever, both techniques caused a build-up of more fat deposits in the coronary arteries.

The era of "revascularization"—the introduction of new blood ves-sels into the heart muscle from another source in the body—began in 1932. Dr. Claude S. Beck of the Cleveland Clinic sutured a portion of the chest muscle to the heart's surface, thereby obtaining more blood through newly developed capillaries between the two tissues. During the next 15 years, surgeons amended this technique, grafting flaps of tissue from other nearby organs—intestine, liver, spleen—to obtain auxiliary sources of blood. Beck later tried to stimulate the flow of blood within the heart itself by irritating the surface of the heart muscle with abrasives. He and others who adopted the technique used such irri-tants as talc, sand, asbestos, powdered beef bone, iron filings, and carbolic acid. This work fell into disfavor, not only because of its inelegance, but also because many physicians were convinced that pa-

Surgeons, *far left,* prepare to connect patient to heart-lung machine. Signal on screen of heart monitor, *left,* shows that the exposed heart is beating satisfactorily.

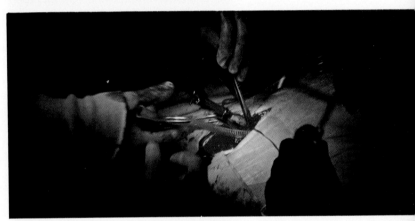

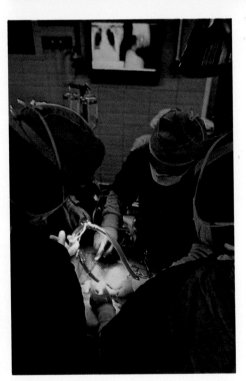

Outflow tube of the heart-lung machine is inserted in the femoral artery, *above.* Surgeon taps inflow tube to clear it of air, *left,* before it is inserted into the patient's heart. Lighted X-ray pictures were used as guides in making initial incision.

tients who benefited did so only psychologically. Because anginal pain is often a result of fear of the next attack, some patients may have improved simply because they had undergone chest surgery that they believed helped their condition.

The next milestone in the surgical assault on coronary disease came just after World War II. Canadian surgeon Arthur M. Vineberg conceived of cutting an artery that transports blood to some other parts of the body and implanting it into the heart muscle. He chose the left internal mammary artery because it is rarely afflicted with disease, is not essential in the artery-rich breast, and is located near the heart.

After four years of laboratory trials on dogs, Vineberg performed the first such implant in a human being on April 28, 1950. Vineberg cut the internal mammary artery of a 44-year-old man from its bed along the chest wall and inserted the open end into a tunnel previously prepared in the heart muscle. Three years later, Vineberg reported that the patient had progressed "from a condition of complete disability" to "walking 10 miles through the bush." Vineberg was convinced that the implanted mammary artery stimulated a network of dormant heart blood vessels that increased in size and linked up to feed a substantial portion of the organ. For patients with particularly severe artery disease, he added another source of blood. He brought up a section of the omentum, a lacy membrane that surrounds internal organs and is rich in blood vessels, and wrapped it around the heart.

In 1959, Dr. William H. Sewell, now of the Guthrie Clinic in Sayre, Pa., modified this operation by moving a strip of chest muscle and con-

Saphenous vein section is placed over the heart to check for fit, *below. Top right,* surgeon lifts the heart, exposing earlier incision in the circumflex artery (just left of his hand). *Bottom right,* a segment of vein section is held with forceps after being sutured into the artery.

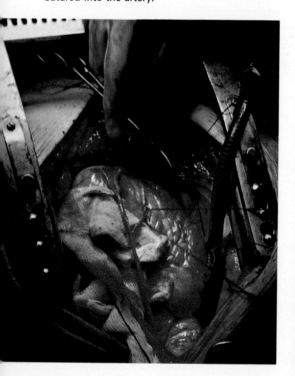

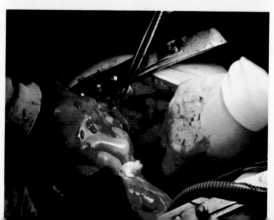

nective tissue along with the mammary artery to add support to the vessel and reduce the risk of damage. Soon, both the left and right mammary arteries were being used.

Skeptics were unconvinced that the Vineberg operation was effectively bringing new blood to the heart muscle, but the doubt was eliminated in 1958 when Dr. F. Mason Sones, Jr., and Dr. Earl Shirey of the Cleveland Clinic used the newly developed coronary angiography to examine the vascular network inside the heart muscle. The angiograms confirmed Vineberg's belief that tiny dormant vessels within the heart called arterioles gradually expand to receive and distribute the new blood supply. Experiments in 1971 on pigs and dogs, reported by Dr. Wolfgang Schaper of the Janssen Research Foundation in Beerse, Belgium, confirmed and revealed more about this remarkable process of collateral circulation. In some cases, Schaper found, small vessels grew to 10 times their previous diameter, enabling them to carry 45 times as much blood to the heart.

Before this discovery, the Cleveland surgeons had been using a surgical technique called endarterectomy to remove localized obstructions in the coronary arteries. They opened the artery at one end of the obstruction and reamed out accumulated fatty deposits through a second incision at the other end. Vascular surgeons had been reaming out plugs in the larger arteries of the leg for 15 years. But applying such surgery to coronary arteries, which are usually 1 to 4 millimeters in diameter, proved a more formidable task. Reaming tended to pile up fatty debris in the coronary arteries in front of the artery's tiny

Working on the front of the heart, surgeon makes an incision in the anterior descending artery. Another segment of the saphenous vein is in forceps above the surgeon's hand, ready to be sutured in place.

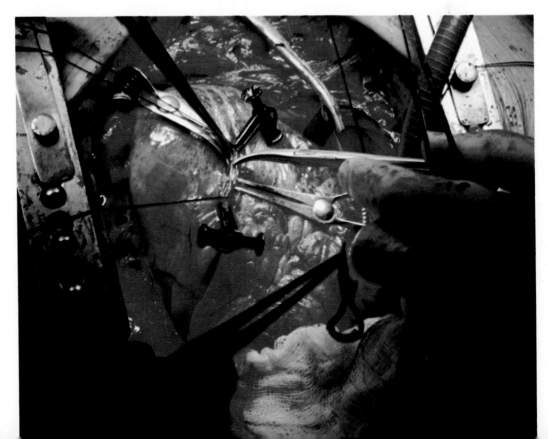

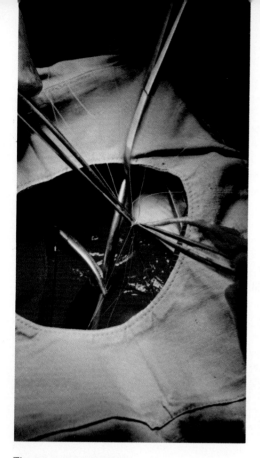

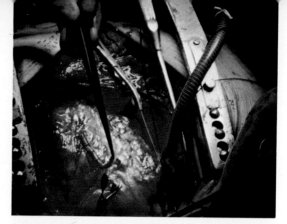

The surgeons suture the saphenous vein segment to incision in anterior descending artery, *above.* Free end of vein segment is then lifted from heart, *top right,* and joined, *right bottom,* to vein segment from the circumflex artery, forming a Y shape.

side branches, much as a snowplow blocks driveways while clearing city streets. This caused many deaths from clots.

The surgeons had also tried a tactic used by seamstresses to widen a sleeve or let out a pair of pants. They increased the diameter of the blood vessel by inserting an extra bit of tissue at the narrowed area. This technique, however, was helpful only to patients with obstructions confined to a single site. Multiple patch grafts were too risky to be practical.

The Vineberg operation also proved to have a major shortcoming. The heart muscle needed three to six months for collateral circulation to become active, and many coronary patients could not survive that long. An operation was needed to bring about an increased blood flow instantly. This had been achieved in the leg, where vascular surgeons simply cut out an entire segment of the obstructed artery and replaced it with a healthy vein segment. But the much narrower coronary arteries were more difficult to work with.

Angiography provided the breakthrough that opened the door to saphenous vein by-pass grafting. In the early 1960s, a resident physician from Argentina, Dr. René G. Favaloro, was reviewing films of angiograms in the research laboratories of the Cleveland Clinic. He became convinced that a by-pass graft of the coronary arteries was feasible. After months of painstakingly suturing vein grafts to the coronary vessels of animals, Favaloro unexpectedly had to use the technique on a human patient in May, 1967.

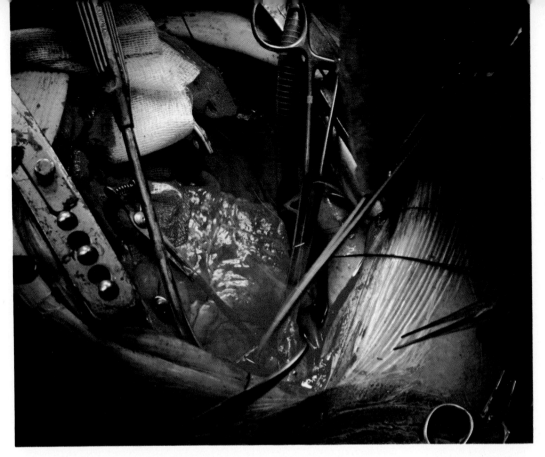

A 55-year-old woman who had been suffering anginal pain for three years lay on an operating table, scheduled to receive a patch graft. But Favaloro and the other operating surgeons realized the disease was too extensive for such a simple procedure. So they removed a segment of vein from her groin and performed a by-pass graft. The success of the operation encouraged Favaloro to try others. Soon after, he discovered that the saphenous vein in the leg was best able to withstand the high arterial pressures without breaking or ballooning and was least likely to deteriorate or cause blood clotting. Between May, 1967, and October, 1970, Favaloro and fellow staff members performed 1,310 by-pass grafts. Nearly 90 per cent of the patients who received by-pass grafts resumed their normal activities, totally free of debilitating anginal pain.

Angiograms have also ruled out many unnecessary operations. Of the first 1,000 patients undergoing angiography at the Cleveland Clinic for suspected coronary disease, only 650 were found to need surgery. "Most of the others went back to leading normal lives without any treatment other than the reassurance that was derived from the negative angiograms," said Donald B. Effler, head of thoracic and cardiovascular surgery.

In January, 1968, a few months after Favaloro's successful operation at the Cleveland Clinic, Dr. W. Dudley Johnson and his associates at the Marquette University School of Medicine in Milwaukee performed their first by-pass graft. Since then, the Milwaukee sur-

The saphenous vein is connected to an opening previously made in the aorta, near the lower-right edge of the exposed heart cavity. This completes the new life line for the heart.

geons and others throughout the United States have extended the operation to patients with obstructions in smaller and smaller coronary artery branches, some as small as 1 millimeter (less than 1/6 inch) in diameter. After three years' experience, Johnson announced that double and triple by-passes now make it possible to operate on 90 per cent of referred patients, including some with coronary disease so extensive they previously had been considered inoperable.

In supporting the decision to use the by-pass surgery for more patients, Johnson told physicians at a medical meeting: "It is obvious that patients most in need of help have been denied reconstructive surgery or, on occasion, have been placed in the unwarranted category of heart-transplant candidates."

Donald Ross, the London surgeon who performed Great Britain's first three heart transplants said, after visiting the United States in 1970, that he plans no further transplants because the saphenous vein by-pass will make most of them unnecessary.

While many cardiologists are referring more of their patients for by-pass surgery, some remain unconvinced. Two New York cardiologists, Arthur M. Masters and Charles K. Friedberg, for example, acknowledge angina is relieved in a significant number of patients, but they point out that only time will tell if the patients will live longer.

Most cardiologists and particularly cardiovascular surgeons, however, insist the saphenous vein by-pass operation is here to stay. Not

Surgeons simultaneously close the incisions in the chest and the leg.

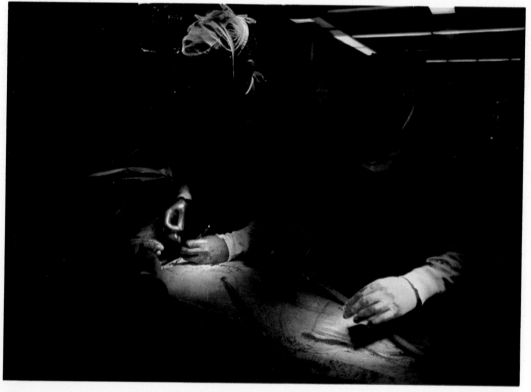

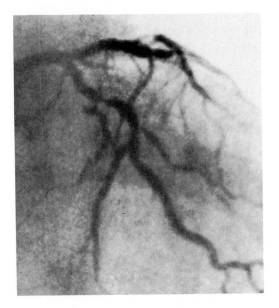 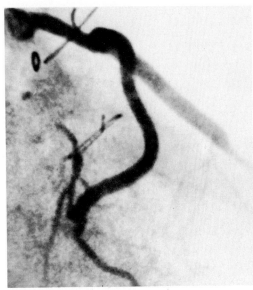

Angiograms before, *above left,* and after, *above,* the surgery reveal a new surge of blood to the heart. The Y graft has clearly by-passed blockages in both the anterior descending and the circumflex arteries.

only is the use of the operation growing at a dramatic rate, it is also now being combined with other forms of heart surgery. In 1971, cardiologists were performing by-pass grafts with heart-valve replacements. At the University of Texas Southwestern Medical School in Dallas, Dr. Harold C. Urschel launches a double attack on coronary artery obstructions. He performs by-pass grafts with gas endarterectomy—injecting carbon dioxide gas through a fine-gauge needle to clear out deposits and reopen circulation. Single and multiple by-pass operations have also been performed on patients whose diseased coronary arteries have led to heart conditions more severe than angina. A number of surgeons reported on their work at the annual meeting of the American Association for Thoracic Surgery, in Atlanta, Ga., in April, 1971. For example, Dr. Frank C. Spencer of New York City told of 21 vein by-pass operations performed to alleviate congestive heart failure. Dr. Ben F. Mitchell of Dallas reported 42 emergency by-pass operations to prevent impending heart attacks.

Dr. Effler foresees such widespread demand for the by-pass operation that surgeons at community hospitals far removed from key medical centers will ultimately develop the skill to perform it. "The only development that will replace the operation," says Effler, "is a medical cure for hardening of the arteries."

For further reading:
Effler, Donald B., and Sones, F. Mason, Jr., "Revascularizing the Myocardium," *Hospital Practice,* September, 1967.
Favaloro, René G., *Surgical Treatment of Arteriosclerosis,* Williams & Wilkins, 1971.
Sewell, William H., *Surgery for Acquired Coronary Disease,* Charles C. Thomas, 1967.

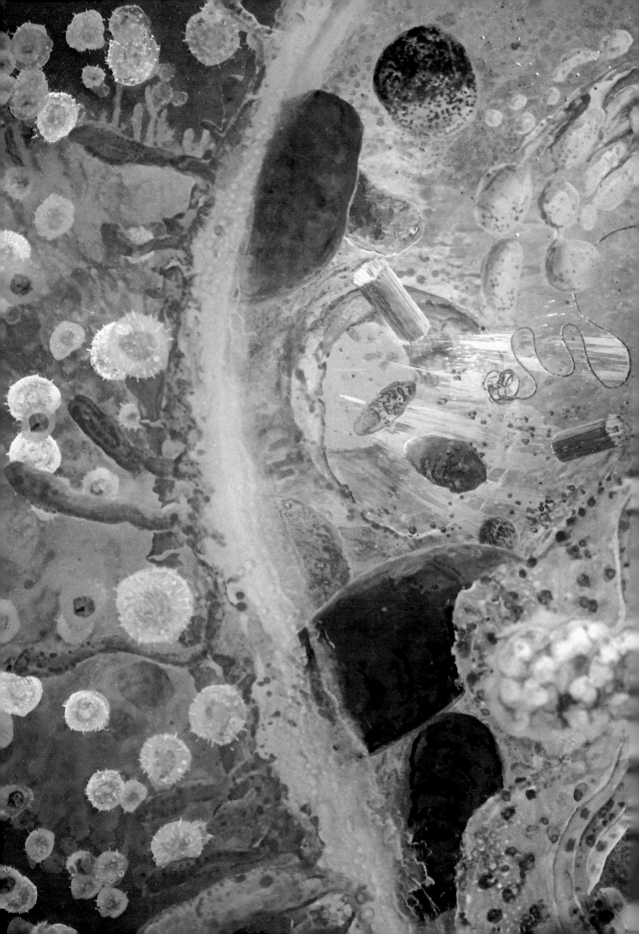

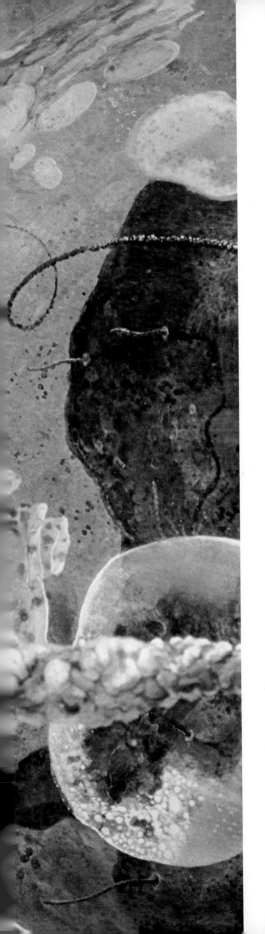

Cell Fusion And the New Genetics

By Joshua Lederberg

The union of cells from dissimilar organs or animal species may help to explain their great diversity

A reliable technique for fusing cells from man and other mammals has ignited a worldwide explosion in genetic research. Scientists have been combining specialized cells such as those of the liver with those of the blood. Even species lines are no barrier—hybrids of human and mouse cells were among the first produced. Human cells have also been fused with those from hamsters, rats, chickens, and other animals.

Classical genetic experiments, the kind that most people learned about in school, have revealed virtually all that we know of genetics. But, these experiments focus only on the offspring that develop from the fusion of a male and a female sex cell, such as the egg and sperm of higher organisms. Scientists discovered long ago, however, that all cells in the body contain the basic units that determine an organism's characteristics. These units, called genes,

The contents of one cell explode into another in an early stage of induced cell fusion.

are biochemical chains so small that they cannot be seen even through the most powerful microscopes. In most organisms, the genes are found on structures called chromosomes. The chromosomes, which are large enough to be seen through a simple light microscope, are located in the nucleus of the cell.

Classical experiments are useless for any direct study of genetic changes in the new organism's body cells after the egg is fertilized. Yet, the body cells comprise nearly all of the living creature, and they are the center of the most important and scientifically challenging processes of life. One such process is differentiation, through which embryonic cells develop into specialized cells such as those of the brain, skin, and blood. By fusing differentiated cells and examining the activities of the hybrid cells, scientists are beginning to understand better why the parent cells differ from one another. Ultimately, this research should also reveal much about how the body's organs perform their vital processes. Learning, which is possible only because of brain differentiation, and immunity, an extension of lymphoid tissue differentiation, are only two examples. Furthermore, some diseases, such as cancer, are examples of differentiation gone awry, and can be studied through cell-fusion experiments.

Cell-fusion has also proved to be a powerful tool in one place where classical genetics has uncovered information very slowly. Elaborate experiments with many generations of test animals have been used for gene mapping—determining the chromosomes on which specific genes are located. Scientists only rarely found human beings with family histories that could reveal what experiments reveal in animals. And controlled crossbreeding of people was, of course, unthinkable. But through cell fusion, a scientist can crossbreed human cells, and follow a single chromosome and its genes through the generations of cell division that follow.

The most important factor in fusing cells is an inactivated virus called inactivated Sendai virus (ISV). The ISV causes fusion, ordinarily a rare and random event, to become common and predictable. Pathologist Henry Harris of the University of Oxford in England developed the technique of adding ISV to mixtures of cells. He had read of research that seemed to indicate that cells fused in tissues that were infected with various viruses. Harris selected the Sendai strain after testing many viruses and grading their ability to promote fusion. He also discovered that ultraviolet light inactivates the virus, destroying its ability to infect but not its ability to induce cell fusion. Thus, experimenters have been able to use the virus without worrying about possible confusing side effects from virus infection of the cells.

In most experiments, the cells to be fused are from tissue cultures in which all cells are genetically identical and have been reproducing for many generations. Cells from such strains are the easiest to work with because they have adapted to growth under controlled conditions, yet they still maintain many of the characteristics of the tissue from which

The author:
Joshua Lederberg is professor and chairman of the department of genetics, and director of the Kennedy Laboratories for Molecular Medicine at Stanford University. He shared the 1958 Nobel prize in physiology and medicine for his studies of the genetic material in bacteria.

they came. However, cells taken directly from the organs of laboratory animals or human beings can also be fused.

Fusion begins when each inactivated virus attaches itself to the surface of two cells. Anywhere from two to dozens of cells may be clumped together. Then, the viruses open the walls of adjacent cells, forming channels between them. Next, the channels widen and the cells fuse, forming one large hybrid cell. At this point, the hybrid cell is multinucleate—that is, it has one nucleus from each of the cells that formed it.

After the fused cells incubate for a few days, some of them begin reproducing through mitosis, a process in which a cell divides and forms two new cells. Each of these daughter hybrid cells has a single nucleus containing chromosomes from the several nuclei in their parent cell. Some of the new, single-nucleated hybrid cells may then reproduce. After several generations, some of the daughter cells form strains of identical cells that can be kept alive and reproducing in cultures.

The simplest and most useful hybrid strains are those produced after only two cells are fused. But the nature of even these hybrids is often not as predictable as one might expect. For example, the cells of a strain originating from the fusion of a mouse cell and a human cell would be expected to have 86 chromosomes, because each mouse cell has 40 and each human cell has 46. However, stable strains of mouse-human hybrid cells rarely have 86. This is because the original hybrid and its offspring tend to lose chromosomes in the mitotic divisions that precede the emergence of a stable strain. In mouse-human hybrid cells, more human than mouse chromosomes are lost. This may be because the human chromosomes are geared to a slower mitotic rate than are those of the mouse.

Pathologists were the first scientists to suspect that it would be possible to fuse cells. They had seen and reported multinucleated cells, particularly in diseased tissue, as early as 1875. However, genetics was not then a full-fledged science, and, in the years that followed, pathologists and geneticists exchanged little information. As a result, cell fusion was virtually unknown to geneticists, the scientists who could best use it.

It was not until the 1950s that geneticists took a real interest in cell fusion. At that time, genetics was in the midst of a revolution triggered by the recruitment of bacteria and viruses as experimental organisms.

Investigators believed that only very closely related body cells, such as those from the same organ of two individuals of the same species, might fuse. But, because such cells are so much alike, it would be hard to tell the fused from the unfused cells. Researchers needed strains of cells that were closely related but that had distinguishable features that might show up in a fused cell. Such features are known as genetic markers because they are usually caused by a mutated gene. Scientists could use such cells to prove fusion by locating a single cell with the effects of the mutant genes from more than one cell strain. However, almost no genetic markers were then known for closely related cells. Nevertheless, in 1956 and 1957 some scientists reported experiments in which they used markers to detect genetic exchange between cells. These markers, however, were technically difficult to detect with the tools then on hand. Although the researchers were on the right track, their results were inconclusive.

Then, in 1960, a group of French researchers approached the problem from another angle. Certain strains of mouse cells grown in culture for many years had developed many visible changes in the size and shape of some of their chromosomes. The French researchers reported that when cells from the strains were mixed and cultured together, single cells appeared that had mixtures of the unusual chromosomes from the different strains. This showed that cells from different cell strains had fused. To many scientists, this seemed an all too easy solution to such a long-standing problem—to prove that cell fusion could occur. But several other investigators soon corroborated the findings.

The studies created optimism and renewed interest in cell fusion. Some of the researchers affected were pathologists who focused their attention on the influence of certain viruses. Following this line of investigation, Harris and a co-worker J. F. Watkins fused human and mouse cells with the aid of ISV in 1965. It was this that opened the floodgates of cell-fusion research.

Fusing cells from species so widely separated by evolution completely eliminated the problem of detecting fused cells. The chromosomes are perhaps the best markers in mouse-human cells, because mouse chromosomes are shaped like Vs and human

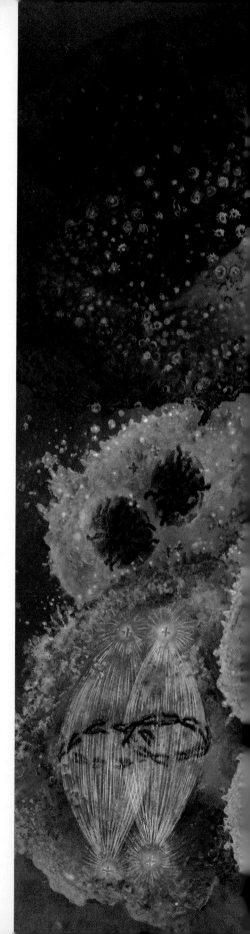

Two cells, top background, are connected by a channel that was induced by one of the small sphere-shaped viruses. The hybrid cell produced, left center, has two nuclei, each with chromosomes forming in preparation for reproduction through mitosis. When mitosis begins, bottom left, the chromosomes align on mitotic spindles. (Mouse chromosomes look like Vs; human chromosomes like Xs.) Spindle fibers pull chromosome halves to opposite ends of the cell, bottom center. But some human chromosomes do not split, causing a deficiency of human chromosomes in the upper of the two daughter cells about to form, right foreground.

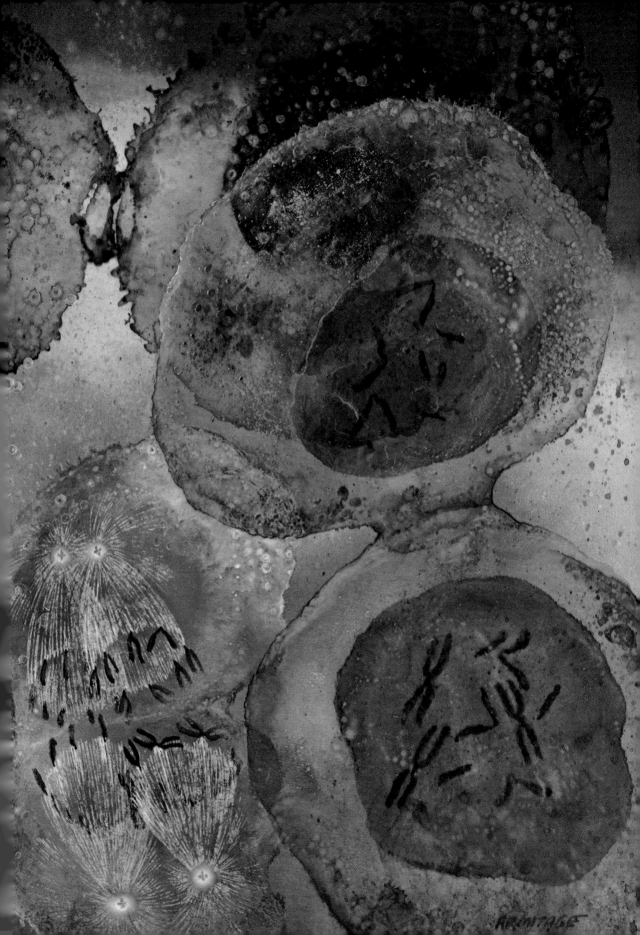

chromosomes are shaped like Xs. Genes that can act as markers are also abundant in such a fusion.

In 1970, biophysicists Theodore T. Puck and Fa-Ten Kao of the University of Colorado reported a study in which they fused human cells with Chinese hamster cells. Chinese hamster cells contain fewer and much larger chromosomes than do mouse cells. The large chromosomes are particularly easy to distinguish from human chromosomes, which are about the same size as those of a mouse.

One of the most exciting areas in cell-fusion research is fusing cells from two different, fully differentiated strains. One of the first such experiments was performed by geneticist Boris Ephrussi and co-workers at Case Western Reserve University in Cleveland. The scientists fused deeply pigmented hamster cells with unpigmented mouse cells. The mouse cells were known to have the genes for pigmentation, because some of the mouse's skin cells were pigmented. The cell strains produced by cell fusion would answer some questions about how the pigmentation genes are regulated in differentiated cells. For example, would the pigmented cell turn on the genes for pigmentation in the unpigmented cell?

The hybrid cells formed no new pigment and soon were totally unpigmented. However, they produced many other hamster gene products. This indicated that although the genes of the unpigmented mouse cells had inhibited the pigmentation genes of the hamster cells, the mouse genes did not inhibit all the hamster genes.

This experiment raises many questions. The more obvious ones include: Are the hamster pigment-forming genes irreversibly repressed? Will these pigmentation genes reassert themselves if one particular mouse chromosome is removed? That is, can researchers identify a particular mouse chromosome as source of the pigmentation-gene inhibitor and, therefore, the gene that produces it? Is the absence of pigment caused by a specific gene or by a general imbalance of the whole hybrid chromosome set?

In another experiment, Harris used cell fusion to study differentiation. In birds, red blood cells retain a nucleus throughout their lifetime in the circulation. However, this nucleus is apparently genetically dormant—that is, its genes are not operating. Harris set out to learn if the nucleus keeps the full repertory of normal genetic information. He chose the bird red blood cell because its nucleus can be separated from its cytoplasm and, with the aid of ISV, fused with a cell from any of a number of species. This eliminates any factors in the blood cell's cytoplasm that might influence the dormant genes.

Harris fused the nuclei of chicken red blood cells with mouse cells. The first thing he saw indicating change in the dormant nucleus was

A photomicrograph of the edges of two cells about to fuse shows inactivated viruses (arrows), each of which can start a fusion channel.

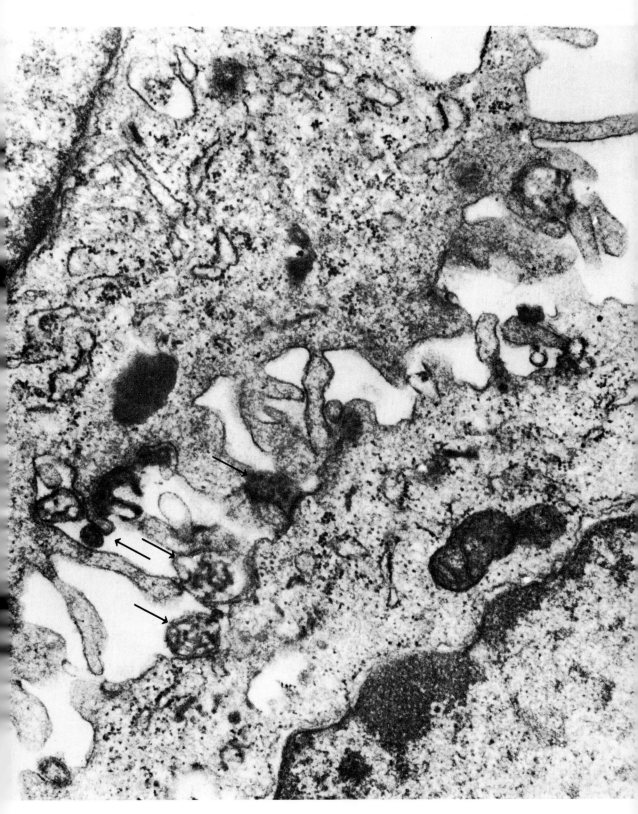

In differentiation study, two cells, extreme left, contain pigmentation genes, but lower one is unpigmented. Cells fuse, center, forming binuclear hybrid cell, far right. The pigment begins to deteriorate, and no new pigment is produced. The original unpigmented cell probably contained a gene that inhibits pigment-producing genes.

a 30- to 40-fold increase in its size. He then found bird gene products, including certain antigens and enzymes, in the hybrid cells. This proved that the dormant nucleus was activated by the nucleus and cytoplasm of the cell with which it was fused. It is particularly interesting that the genes in the dormant cell nucleus of a bird can be activated by the cells of such a biologically distant organism as a mouse.

To study cancer, researchers, primarily in England, France, Israel, Sweden, and the United States, have fused cancer cells and normal cells. One aim is to determine if malignancy is dominant or recessive. In classical genetics, these terms are used to indicate which of two different genes for the same characteristic will be expressed when they are present in a single organism. For example, a child who inherits a

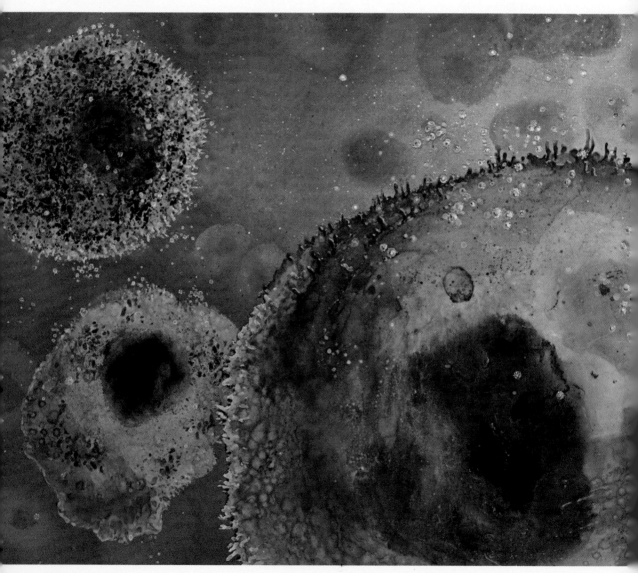

gene for blue eyes from one parent and a gene for brown eyes from the other, will have brown eyes. The gene for brown eyes is dominant; the one for blue eyes, recessive.

By mid-1971, the results of the cancer-normal hybrid cell experiments appeared inconsistent. In some of the studies, the hybrids have been normal—that is, they have not caused cancer when injected into test animals. This seems to indicate that genes on some of the chromosomes from the normal cells were dominant to those that caused cancer. Further support for this view arose from experiments in which some of the cancer-normal hybrid cells produced cancerous daughter cells after losing some of the chromosomes that came from the normal cell. In addition, the nonmalignant, cancer-normal hybrid cells proved

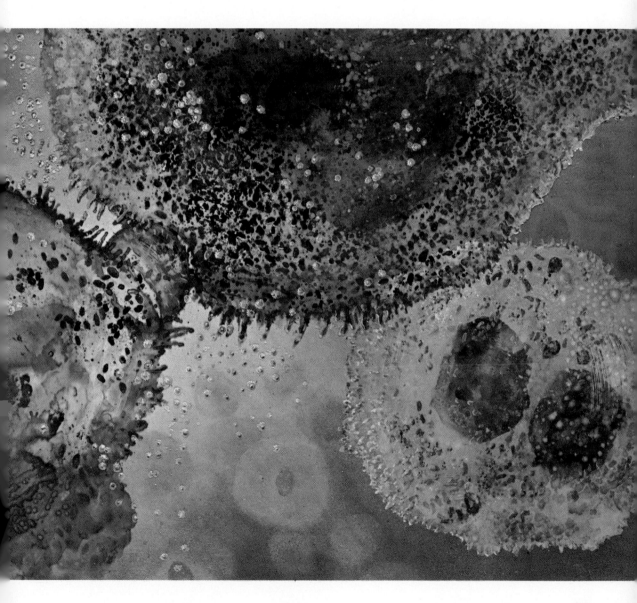

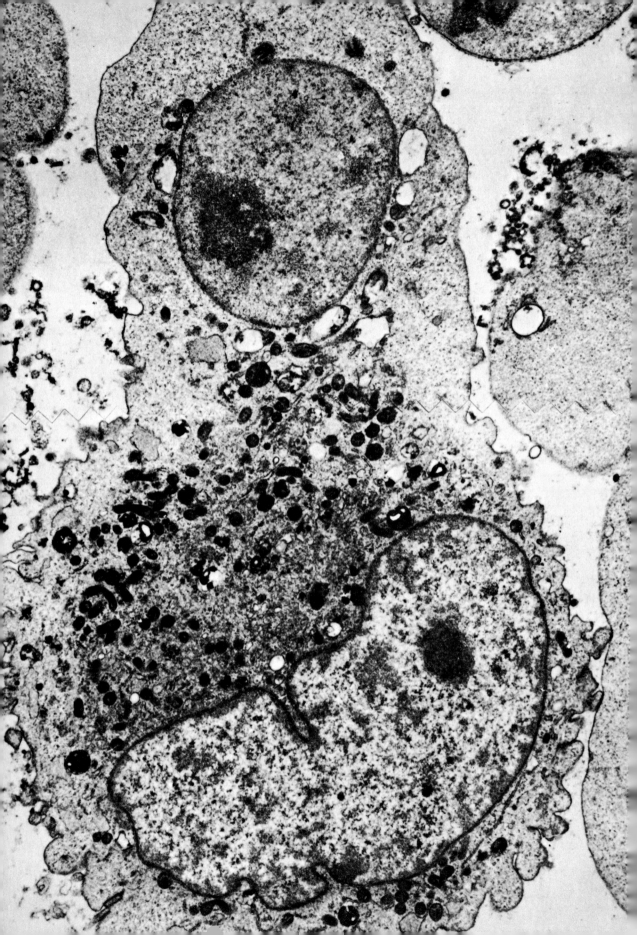

to be a vaccine against cells from the original cancer. Evidently, the nonmalignant hybrid cells retained the cancer-cell antigens, causing the animals injected with them to manufacture cancer antibodies.

In 1969 and 1970, however, Harris and co-workers reported studies that yielded hybrids of some kinds of normal and malignant cells that acted quite differently when injected into test animals. Even without the loss of chromosomes, these cells had some tendency to multiply and invade normal tissue, as do malignant cells.

Thus, although the cancer studies provide much to think about, they have not answered the original question—is malignancy dominant or recessive? But the fault is with the question. Genes are properly classified as dominant or recessive only when they are in pairs on two chromosomes, as they are in normal cells. Because the genetic arrangements in hybrid cells are much more complex, this terminology is inapplicable. The data from the fusion work, however, should prove useful to those who will formulate more meaningful questions.

Several other groups of scientists have used cell fusion to study another aspect of cancer. The virus called SV40 is one of many that can transform normal cells into cancer cells by injecting its genes into them. Human cells made cancerous by SV40 were used to start cultures of human cancer cells. Such cultures continue to produce cancer cells generation after generation without further use of the SV40 virus. Presumably this is because SV40 genes attach to human chromosomes and are reproduced along with them. When the scientists fused the human cancer cells with normal mouse cells, the initial hybrids were malignant. However, those hybrids that had lost all but three or fewer human chromosomes after several mitotic divisions were not malignant. It did not seem to matter which of the human chromosomes were lost. This indicates that the cancer-causing viral genes were incorporated into not just one, but many of the chromosomes of the originally infected cells. This also supports other work indicating that the transformed cells carried many copies of the viral genes.

Further study showed that some hybrids of normal cells and human cancer cells from SV40-infected strains suddenly liberate complete SV40 viruses fully capable of infecting new cells. As far as researchers could determine by examining the human cancer-cell cultures, the original viruses that made the parent cells malignant had disappeared many generations before. Their reappearance indicated that copies of the entire genetic content of SV40 were incorporated into the chromosomes of the original transformed cells, and were reproduced and passed on from generation to generation. Other experiments, however, showed that not all of the SV40 viral genes need be incorporated in the chromosomes to make some cells malignant. In these cancer cells,

**Fused Cells
Stop a Cancer**

The malignant cells of a mouse with abdominal cancer caused cancer in a normal mouse. However, malignant cells fused with normal cells did not cause cancer in a normal mouse, and made it immune to cancer cells.

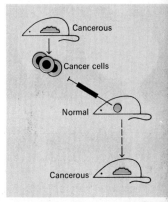

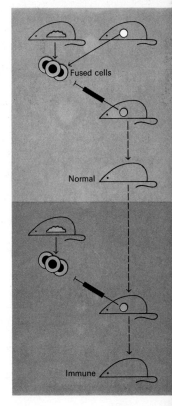

A photomicrograph, *opposite page*, shows the final stages of fusion of a chicken embryo's cell, top, with the cell of an adult mouse.

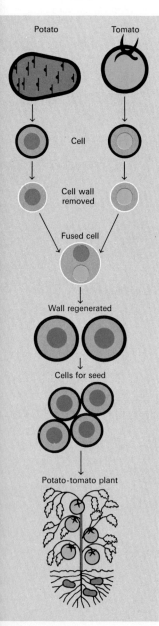

Potato   Tomato

Cell

Cell wall removed

Fused cell

Wall regenerated

Cells for seed

Potato-tomato plant

though, complete SV40 viruses can be liberated after massive infection with other kinds of viruses that probably provide substitute genes for those that are missing.

The implications of all this go far beyond what we may learn about cancer. These experiments prove that a virus can add some or all of its genes to a human or animal chromosome. They also indicate that a bottomless reservoir of new viruses may exist in the genes in cells of men and animals—viruses that we will not know about until either cell fusion or some other stimuli activate them. Such dormant viruses known to exist in some bacteria can be awakened by radiation and many components of air and water pollution.

The special value of cell fusion for gene mapping is based upon the chromosome loss that occurs as the hybrid cells reproduce. As early as 1967, researchers used this technique to find out which human chromosome contains the gene that produces the enzyme thymidine kinase. They fused human cells with mouse cells of a strain that cannot produce the enzyme. Then, they put the hybrid cell in a specially prepared nutrient solution that would kill cells unable to produce thymidine kinase—those that during mitosis had lost the human chromosome containing the gene for the enzyme. The scientists then found surviving cells that contained only one human chromosome. This chromosome had to be the one that carries the thymidine kinase gene.

A more generalized gene-mapping method similar to this technique would be to use cells from several hybrid strains that have only a few human chromosomes. A microscopic examination of cells from each strain would reveal which human chromosomes they contain. Scientists then could merely examine cells in each line through a microscope and analyze them for various chemicals. It is then relatively simple to correlate the retention or loss of a trait or chemical from the human cell with the retention or loss of a specific human chromosome.

In 1970, two teams of scientists used a related technique to prove that the genes for two specific enzymes are linked—that is, on the same chromosome. One team was led by geneticist Walter F. Bodmer, then at Stanford University, the other by F. H. Ruddle of Yale University. The scientists fused human and mouse cells and analyzed the hybrid cells from a series of resulting strains. Each team looked for about 15 different human enzymes. The scientists reasoned that enzymes produced by linked genes would always be present in those cells that retained the genes' common chromosome and absent in those cells that had lost the chromosome. The enzymes peptidase B and the B subunit of lactate dehydrogenase, which were among the enzymes that both teams were checking, were consistently absent or present together. This indicated that the genes for their production are linked.

Ultimately, the knowledge derived from cell fusion promises to be of incalculable use to man. For example, much of what scientists foresee for genetic engineering, the purposeful manipulation of chromosomes and genes, will require knowing at least on what chromosome a gene is located. The proof that viral genes can be incorporated into a human chromosome strengthens the hope of scientists that, ultimately, they may use harmless viruses to introduce useful new genes into human cells. The first genes selected for such treatment will probably be those that will repair single-gene defects like phenylketonuria, which causes brain damage, and hemophilia, in which the blood does not clot normally.

Also, many products of differentiated cells, such as specific enzymes and antibodies, could become important in medicine if we could produce them in larger, predictable quantities. Cell fusion should enable scientists to learn to increase the rate at which such substances are produced by cells in culture. This has already been achieved with bacterial cells. Researchers have developed strains of bacteria that commonly produce up to 100 times their normal amount of a specific biochemical. For example, several such strains were used to produce the enzymes that were added to detergents.

The manipulative possibilities of cell fusion may see their first dramatic use in the most ancient arena of genetic engineering—agriculture. A single plant cell can grow into an entire plant that is genetically identical to the plant from which the cell was taken. Perhaps the cells of two desirable species can be fused to produce a form that cannot be produced by cross-pollination, the customary method of hybridization in plant breeding.

The cellulose wall in which most plant cells are generally encased are barriers to fusion. These must be removed without injuring the cells. In some plants, this can be easily achieved with an enzyme that destroys the cell wall. In 1970, botanist Edward C. Cocking of the University of Nottingham in England reported that he had fused plant cells for the first time. Cocking used the enzyme to remove the walls of corn and wheat cells, and suspended the cells in solution. He triggered the fusion by adding sodium nitrate to the solution.

It is not far-fetched to predict that similar fusions will produce some extraordinary results, perhaps sooner than a responsible scientist dare guess. Cocking already envisions a new plant that would grow tomatoes above ground and potatoes below. If this seems frivolous, consider other crops that cell fusion might produce. For example, rice is both high-yielding and a very popular food in many overpopulated countries. Unfortunately, however, its nutritive value is low. Through cell fusion, we might someday produce a ricelike plant that has as much nutritional value as, say, soybeans. This would clearly establish the practical value of the new genetics, for cell fusion would then have a tremendous impact on our crowded and poorly fed world.

For further reading:

"Accelerating Somatic Cell Genetics," *Nature*, Oct. 24, 1970.
Harris, Henry, *Cell Fusion*, Harvard University Press, 1970.
Klebe, Robert J., and others, "Controlled Production of Proliferating
  Somatic Cell Hybrids," *The Journal of Cell Biology*, 1970, Vol. 45, No. 1.

The NAL: Accelerating Particles and People

By Allan Davidson

The National Accelerator Laboratory looks outward toward society as it probes inward for the secrets of inner space

Sprawling over a 6,800-acre site 35 miles west of Chicago, near Batavia, Ill., is the new home of the National Accelerator Laboratory (NAL). Operated for the Atomic Energy Commission (AEC) by Universities Research Association—a group of 51 major universities in the United States and one in Canada—the NAL houses the first in a new generation of superaccelerators designed to explore the nature and behavior of nuclear and subnuclear particles. Technically, an alternating-gradient proton synchrotron, the NAL particle accelerator, which will be used by scientists from many nations, is the world's largest single instrument for basic research. It is the most powerful, costly, and useful tool ever developed for high-energy particle physics.

The NAL also is unique in other ways: It is perhaps the first multi-million-dollar government-financed project that is running ahead of schedule (by one year), below its cost estimates (about one-fifth less than the $250 million authorized), and in excess of its specifications

Aerial view shows the NAL's huge main ring and grounds in their early stages. Once the village of Weston, the spacious site was donated to the AEC by the state of Illinois.

This Cockcroft-Walton preaccelerator forms protons—the NAL's projectile particles—from hydrogen gas molecules. The protons are then accelerated in stages to final energy.

(two and one-half times its authorized energy of 200 billion electron volts). For once, the taxpayer is clearly getting his money's worth.

Ever since the Greek philosopher Democritus (460?-370? B.C.) described the atom as a solid indivisible particle of matter, scientists, not fully believing him, have pondered its basic structure. Probing deeper and deeper into the nuclei of atoms, they have uncovered an astonishing array of new and strange particles—photons and electrons; muons, mesons, and baryons; pions, kaons, and nucleons; positrons, hyperons, and neutrinos. These elusive little bits and pieces of matter can be positive, negative, or neutral in electrical charge. They range in mass from 0 to more than 3,000 times the mass of one electron. They have lifetimes ranging from infinity to less than one hundred trillionth

of a second, and their average diameter is about one ten thousand billionth of an inch.

Although these and other properties of elementary particles have been measured, particle physicists realize that there is much they still do not know. The major challenge facing them is to find some apparent order—some underlying basis—in this array. Why are there so many particles? Why do they have such widely varying masses, life spans, and modes of decay? What is the relationship among the binding forces that hold them all together? And, are there even smaller, more "elementary" particles still to be discovered?

The key to probing this complex inner space is energy—the deeper the nuclear penetration, the greater the energy required. The type of energy that is used for producing unstable fundamental particles is kinetic energy, or the energy of motion of a stable particle, such as a proton, traveling at great speed. The method used to impart this energy to a proton or other stable projectile particle is electrical force.

In general, the greater the energy of the projectile particle, the more powerful its impact, and the greater its effect on the target nuclei. This is why particle accelerators have had to be made increasingly bigger and more powerful.

Deep investigation of nuclear structure became possible only when there was an available energy source of millions of electron volts (eV). The earliest cyclotron used for experiments, developed in 1932 by American physicist Ernest O. Lawrence, produced 1.2 million electron volts (MeV) and had a diameter of 11 inches. The NAL accelerator will reach 500 billion electron volts (GeV) and has a diameter of 1.24 miles. Essentially, however, all particle accelerators, whatever their size, do the same thing. They accelerate subatomic stable particles (usually electrons or protons) to extremely high energy, and fire them into tiny targets. The resulting impact shatters target nuclei into unstable fragments, or particles, and the various properties of these subatomic bits and pieces—their mass, spin, half-life, and so on—are recorded by highly sensitive instruments.

The protons begin the first stage of their 4-second, 900,000-mile journey at nearly the speed of light in the tunnellike linear accelerator, center. They are then injected into a circular booster accelerator, lower right, and finally, to the main ring, large circle at upper right.

Basically, the NAL proton synchrotron works like this: Protons, which are positively charged, are formed from hydrogen gas by stripping away negatively charged electrons in a preaccelerator. The protons are then given a 750,000 electron-volt (750 KeV) jolt of energy and are injected into a 500-foot tube, a linear accelerator, where they are accelerated to 200 MeV. At 200 MeV, they are injected into a booster accelerator that speeds them up to 8 GeV. The booster is a synchrotron, 500 feet in diameter, which makes the protons race in a circle, giving them a jolt of energy at each pass. When they reach 8 GeV, their real journey begins. The protons are injected into the main ring, more than 20 feet below the ground, where they are given a 2.8-MeV jolt during each of 190,000 passes around the ring's 4-mile circumference. A series of 774 bending magnets, each weighing 10 tons, and 240 focusing magnets keeps the protons—in a beam about a quarter of an inch in diameter—on course through the 2- by 5-inch vacuum chambers that thread through the magnets.

When the beam reaches its terminal energy of 500 GeV, a switching magnet bends the beam, which hurls down a corridor. At the end of the corridor is the target. The proton beam, now traveling more than 99.999 per cent of the speed of light, smashes the target nuclei into a shower of elementary particles, and instruments record the action.

The main ring tunnel, shown during construction before being covered, *left,* is 4 miles in circumference and is buried more than 20 feet underground. It houses the 1,014 huge magnets that bend and focus the protons, keeping them in a quarter-inch beam, *above.* The beam is accelerated through a 2-by-5-inch vacuum chamber that threads through the magnets, *below.* When the beam reaches its maximum energy, it smashes into a target, or it can be transported to experimental areas.

The entire acceleration takes only about 5 seconds, but during that time the proton beam travels nearly 900,000 miles, about twice the distance to the moon and back.

The beam can also be extracted and transported to any of three target stations in the experimental area, each of which is equipped to investigate specific particle properties. One of these target areas will house the largest bubble chamber in the world. Here, the particles will be slowed down in liquid hydrogen and their tracks, which appear very much like the lines of bubbles in a glass of ginger ale, will be photographed. In another, secondary beams produced by the main beam's collision with the target will provide particles for a wide range of investigations. Perhaps the most glamorous of these searches will be for some theoretical and possibly even more fundamental particles called quarks.

"Our knowledge of the behavior of particles at the very highest energies attained in the laboratory," summarized Robert G. Sachs, formerly asso-

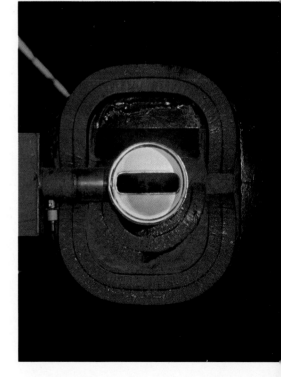

The NAL planners have done more than develop the world's most powerful research tool. They have also made the NAL an instrument to help fill some of the human and ecological needs of the community. For example, power-line poles, *top,* were designed for beauty as well as function. Buffalo, *above,* have found a haven here in an attempt to rescue a nearly extinct species of American wildlife. A sense of the past has been maintained by restoring some old houses, *below,* that once dominated the area.

ciate director for high-energy physics at the AEC's Argonne National Laboratory, near Chicago, "is exceedingly confusing and the problem seems to be exceedingly complex. Yet some underlying rhythm and meaning appear to be emerging, and many physicists feel that we are on the verge of uncovering the basic laws of subnuclear structure . . . The reason for going to even higher energy is the conviction that, for energies greater than the masses of known particles, an ultimate simplicity will set in, and the truly simple underlying laws may be revealed."

The role that the new superaccelerator will play in this search for the underlying structure of nuclear matter is but a part of the NAL story. Reflecting the feeling of many high-energy physicists, the scientists of the NAL have become increasingly uncomfortable at the thought that perhaps they have been too isolated in that miniature universe of particles locked inside the nucleus of the atom. The interrelation of those particles with people and society has occupied more and more of their attention. Thus, the NAL planners were not content with just developing a bigger and better research tool for science. They also saw the growing need to relate science to the social and ecological realities of the world about them.

In an attempt to show that technology can live with nature, the spacious Batavia site has been designed to keep pollution to a minimum and to restore some of the native species that were driven from the plains of northern Illinois by the relentless progress of city builders. For example, buffalo that once roamed throughout the state can now be seen only in zoos. Soon, however, visitors to the NAL will be able to observe herds of these majestic animals. In addition to the obvious benefits of preserving a dying species, the buffalo also will serve a more pragmatic function—their grazing will eliminate the need for costly, air-polluting power lawnmowers. Similarly, Canada geese, once indigenous to the Midwest, but now nearly extinct, have been imported in the hopes that they will nest in the parklike NAL grounds. A series of large cooling ponds, designed to reduce thermal pollution, is being developed on the site. Already, migrating birds have used a few existing ponds as resting spots.

With a concern for history, the NAL is restoring a number of stately old homes, built by early German, Swedish, and English settlers, for use by visiting physicists and their families. A few farmhouses will be converted into museums as a reminder of what the land and its people were once like. And, with an eye for esthetic beauty, even the utility poles that bring in electricity from nearby power plants have been specially redesigned.

It is in the area of human values, however, that the NAL scientists have made their most dramatic gesture; their concern is not only in accelerating protons, but people as well. The NAL planners have pledged themselves to help the underprivileged, the undereducated, and the unemployed. "In any conflict between technical expediency and human rights, we shall stand firmly on the side of human rights," reads an early NAL policy statement on the laboratory's goals.

What this means, in practice, is aggressive employment practices to reach underprivileged youth, largely in the Chicago area. An office has been set up on the city's South Side. Unemployed teen-agers and high school drop-outs are recruited from the ghetto streets and sent to an apprentice-training program at Oak Ridge, Tenn.

Motivation gets major emphasis in selecting trainees. No penalties are imposed for previous school or job problems or even police or prison records. At the Oak Ridge school, they are trained as drafts-men, mechanical and electronic technicians, data processors, and in other technical areas. Their progress is checked closely; in some cases their future supervisors fly to Oak Ridge to check on the trainees. After completing the three-month course, the trainees are returned to the NAL to start work in their new specialties. To date, there have been few drop-outs among the nearly three dozen young men selected.

A special effort has also been made to hire minority contractors for construction work at the site. Black contractors won contracts for a wide range of projects, including a $750,000 contract for magnets for the main ring. A contract for cutting trees and landscaping on the site went to a group of American Indians, whom the NAL helped organize as an all-Indian tree-cutting company.

We have come a long way since Democritus in our understanding of the nature of matter. Nevertheless, during the 20 or so years that particle physics has been recognized as a scientific specialty, it has been accused of being not only costly, but also irrelevant—seemingly remote from practical, everyday life. Clearly, the NAL program represents an attempt to make this difficult but exciting science relevant—to bring people and particles together. NAL director Robert Rathbun Wilson probably said it best:

". . . I am thoroughly confused about the distinction between particles, accelerators, and society. If I have not been able to resolve society into its elemental particles, I have at least found out that it is easier to accelerate particles than it is to accelerate societies. But in the course of giving a very large acceleration to our particles, let us hope that we can contribute at least a small acceleration to society."

Curbing
The Energy
Crisis

By Alvin M. Weinberg

**In the face of his skyrocketing energy
needs, man must continue his search
for sufficient and safe atomic power**

On some of the hottest summer days of 1970, thousands of New Yorkers struggled up endless stairs to sweltering offices and apartments. Under dimmed lights, they squinted at their newspapers to read about the cause of their plight—the power shortage that slowed subways, starved air conditioners, and halted elevators.

Scientists have long known of the crucial role that power plays in today's world, and they have seen a crisis coming. But brownouts and other relatively minor discomforts are not the threat they fear. They worry about the catastrophe that could result if the world's exploding population uses up the world's irreplaceable natural resources.

As these resources dwindle, the need for energy will grow. For energy is the ultimate "raw material." With an essentially inexhaustible supply of energy, we can convert salt water into fresh water, atmospheric nitrogen and water into fertilizer, and limestone and water into methane and other organic chemicals that are useful in industry. Indeed, all we need is energy and the elements that are, for all practical purposes, infinitely abundant. From these elements—hydrogen, oxygen, carbon, nitrogen, sodium, potassium, iron, aluminum, silicon, phosphorus, titanium, and perhaps a few others—we can manufacture whatever scarce materials we need.

Our traditional energy sources—coal, oil, and natural gas—are diminishing as power demands multiply. Man could so severely deplete the world's minable coal, by far the most abundant of the three, in 100 years that it would become too expensive for most uses. This could virtually end our production of iron and steel by coke-reduction. The impending shortage of the fossil fuels and their growing importance as a source of basic materials make the need for substitute energy sources one of man's most pressing problems.

The most appealing new energy source is the atom. The atoms in a single gram of uranium 235 can release as much heat as 3 tons of coal. But for decades, scientists have found atoms reluctant to release the energy easily and safely. They have tried two methods, atomic fission and atomic fusion.

The first has been surprisingly successful. By 1971, more than 100 fission power plants were either in operation, being built, or on order in the United States alone. In fission, uncharged atomic particles called neutrons strike and split the nuclei of uranium atoms. Energy, called the energy of fission, is released as heat that can drive steam-powered generators. But in today's fission plants, only uranium 235 can be split in this way. Natural uranium contains less than 1 per cent uranium 235, so the fuel for these plants is relatively rare and is costly to extract. Furthermore, the radioactive waste produced by these plants is very dangerous. This has made its disposal a difficult problem.

In atomic fusion, energy comes from combining the nuclei of atoms rather than splitting them. Atomic nuclei, however, have like charges and repel each other as we try to force them together. Physicists knew that this would make controlled fusion harder to achieve than fission. Most efforts have centered around a single basic technique. A confined gas is heated to a temperature at which the atoms move extremely fast. They then collide with sufficient force to separate their electrons from their nuclei. The ionized gas, called a plasma, that results is heated still further, until the nuclei move fast enough to overcome the force of their like charges. They strike each other and fuse.

I recall vividly the announcement, in 1967, of a major step toward controlled fusion. It was in the magnificent Hofburg Palace in Vienna, Austria, at the special 10th anniversary session of the International Atomic Energy Agency. I had just completed a talk on the outlook

The author:
Alvin M. Weinberg is the director of the Oak Ridge National Laboratory in Oak Ridge, Tenn. He is a member of the Science Year editorial advisory board.

for energy from fission. Lev A. Artsimovich, Russia's foremost expert on controlled fusion spoke next. In his dry, matter-of-fact manner, he announced that he and his co-workers at Moscow's Kurchatov Atomic Energy Institute had come 10 times closer to controlling fusion than had anyone before. With an experimental device called Tokamak T-3, they had heated a plasma of 2×10^{13} (20 trillion) hydrogen nuclei per cubic centimeter (cm^3) to 2 million°C. and were able to keep the nuclei together for 10 milliseconds.

Understandably, to Artsimovich's audience, mainly diplomats and scientists without expert knowledge of fusion, these figures meant little, and these men missed the full implication of his report. Scientists had tried with little success since 1944 to extract energy from fusion in a high-temperature plasma. By the time of the Vienna conference, progress had been stalled for years. Artsimovich's data indicated it had begun again.

Some scientists and engineers who did understand the Russian results questioned them because they were so spectacular. In 1969, however, a team of British researchers took their equipment to Artsimovich's laboratory and verified his claims. Artsimovich visited the United States in September, 1970, to report his latest results. By then, the Russian device had heated 4×10^{13} hydrogen nuclei per cm^3 to nearly 6 million°C. and confined the plasma for 20 milliseconds.

In 1971, experimental devices based on the Tokamak design are being built everywhere. There are five in the United States alone—at Princeton, N. J.; Oak Ridge, Tenn.; Cambridge, Mass.; San Diego, Calif.; and Austin, Tex. The significance of these efforts to fill our energy needs is best understood against the background of what these needs may soon be.

Since the source of all the energy we use is heat, it is convenient to discuss our needs in kilowatt-hours of heat. In the United States, energy of all kinds is consumed at a rate of about 50 billion kilowatt-hours of heat per day. Even if the population remained unchanged, this energy demand would about double when our scarce natural resources run out and we have to manufacture them. But, suppose everyone in the world reached a living standard comparable to that of Americans, and made the same energy demand. And suppose that, as some demographers believe possible, the earth's population reaches 20 billion by the year 2100. It would take almost 10 trillion kilowatt-hours of energy per day to meet the need. Where could we find it?

Only three possibilities seem practical—fossil fuels, the sun, and the atom. To be sure, there are others—hydro, tidal, and geothermal sources, for example. But none of these could be developed to meet even a fraction of our estimated energy budget.

Of the fossil fuels, the world's coal—about 7.6 trillion tons—would last only 15 years with a 10-trillion-kilowatt-hour daily consumption of heat. The reserves of oil and natural gas would add less than 18

A dramatic rise in world energy needs coupled with the limitations of fossil fuels will require ever greater dependence upon nuclear power.

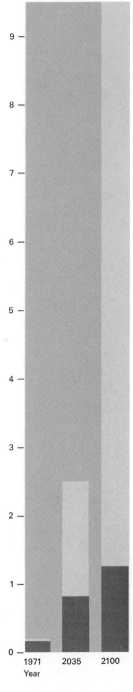

Future Sources Of World Energy

Nuclear fuels

Fossil fuels

Trillions of kilowatt-hours per day

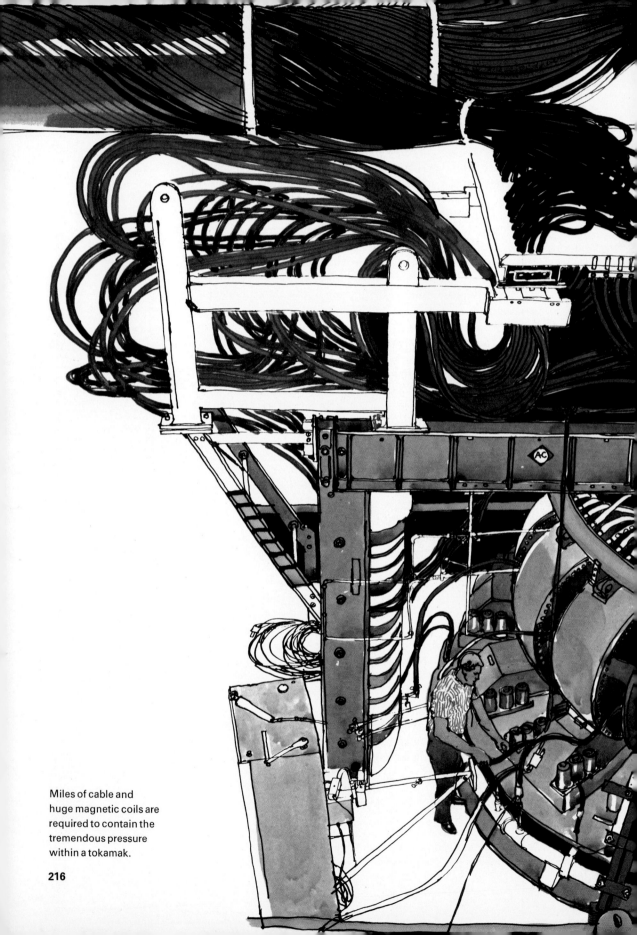

Miles of cable and huge magnetic coils are required to contain the tremendous pressure within a tokamak.

216

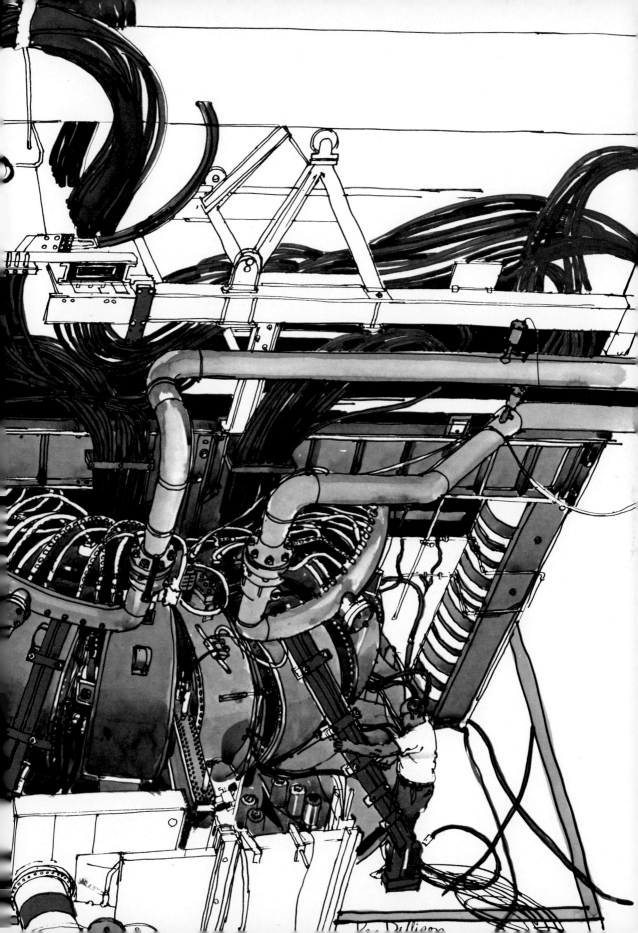

months to that total. In any case, burning these fuels at that rate would probably create intolerable air pollution.

Solar energy absorbed by the earth is hundreds of times our estimated need. It is essentially inexhaustible and obviously nonpolluting. But it is spread over such a huge area that about 14 square miles of collector surface would be needed to gather enough energy for a 24-billion-kilowatt-hour per day electrical generating station operating at 10 per cent efficiency. And solar energy would require a storage system because it is available directly only when the sun shines.

Atomic energy offers no collector problem and would be continuously available. In addition, nuclear fuels for processes now being developed are virtually inexhaustible. While today's fission plants depend on rare uranium 235, tomorrow's will convert one of two plentiful substances, uranium 238 or thorium 232, into fissionable fuel before consuming it. The top kilometer (about 3,000 feet) of the earth's crust contains enough of these substances to meet the requirement of 10-trillion-kilowatt-hours of heat for 20 million years. Fusion power plants will probably use lithium and two forms of hydrogen—deuterium and tritium—as fuel. Seawater has enough deuterium to meet the need for 3 billion years and the lithium in the upper kilometer of the crust could meet it for 15 million years.

The coming generation of fission reactors are called breeders. As they consume fissionable fuel—elements whose atoms are relatively easy to split—they breed more from fertile materials. Fertile materials are those that are easily converted into fissionable fuel. Despite the success of Tokamak T-3, breeders are much closer to practical reality than are fusion devices. They use one of two fuels, uranium 233, which is made from the fertile thorium 232, or plutonium 239, which is made from the fertile uranium 238. In either case, about 2.5 neutrons are released by each fuel atom split. One neutron is used to split another fuel atom, maintaining the chain reaction. This leaves about 1.5 available to change the fertile material into new fuel. In a typical fission device, the fertile material is in a jacket surrounding slender pins that contain the reacting fuel. The fertile material catches the extra neutrons that leak from the pins. Both the fission and conversion reactions generate heat that can drive conventional steam-powered generators.

The goal of most breeder reactor researchers is the liquid metal fast breeder reactor. It will use a stream of liquid sodium to carry the heat to the boiler. Liquid sodium has excellent heat-transfer characteristics, yet operates at pressures much lower than if water were used.

Nuclear fusion plants would use deuterium or deuterium and tritium as fuel. Both are forms of hydrogen. But in addition to the proton that is in every hydrogen nucleus, deuterium has one neutron and tritium has two. The gases are heated, causing their atoms to move faster and faster. Their electrons are knocked off as they collide with one another. The resulting plasmas are mixtures of positively charged nuclei and negatively charged electrons.

Nuclear Fusion in Plasma

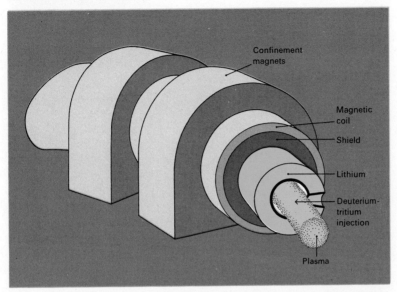

Confinement magnets

Magnetic coil

Shield

Lithium

Deuterium-tritium injection

Plasma

A tokamak experimental fusion device uses huge magnets to confine the hot, pulsing plasma of deuterium and tritium, *above*. These two forms of hydrogen are heated until their nuclei collide and fuse. This creates energy, helium, and neutrons. Some of the neutrons enter a jacket of lithium 6. In fusing with the lithium atoms the neutrons create more of the tritium fuel.

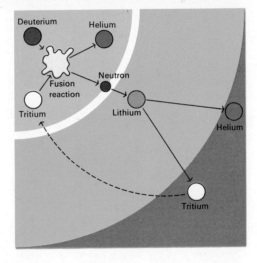

Deuterium

Helium

Neutron

Fusion reaction

Tritium

Lithium

Helium

Tritium

Huge amounts of additional heat impart so much speed to the nuclei that some overcome the repulsion caused by their like charges, strike each other and fuse. When they fuse, deuterium nuclei release 30 to 40 times as much heat as is needed to fuse them at an appreciable rate. The deuterium-tritium reaction gives up about 1,750 times the heat necessary to make it take place in a practical reactor.

The deuterium-tritium reaction is the focus of most fusion work today. Not only does it release more energy, but it also requires only about 10 per cent of the temperature needed for the deuterium-deuterium reaction. But, while deuterium is relatively common in seawater, only 1 in about 10^{18} of the hydrogen atoms in seawater is tritium. Thus, scientists must find some other tritium source. The plan

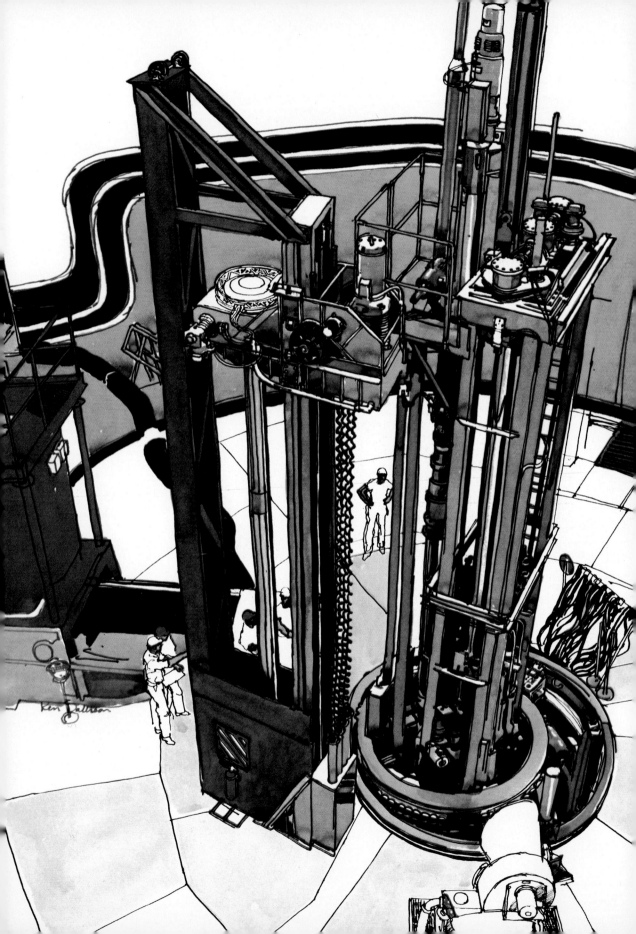

is to make it in the reactor from lithium. The deuterium-tritium reaction gives off free neutrons that can be caught by lithium, converting it to tritium. The conversion releases heat that adds another 25 per cent to the reactor's energy output.

A successful fusion device will require more than the heat needed to begin its reaction. It will have to hold enough nuclei close together so they can strike each other and fuse often enough to create sufficient heat to sustain the reaction. These requirements led physicist J. D. Lawson of England to devise a general rule, now called the Lawson criterion, that establishes the point at which a fusion reaction becomes self-sustaining. The rule says that the number of ions per cubic centimeter (n) multiplied by the confinement time in seconds (τ) must be greater than 10^{15} for the deuterium-deuterium reaction. For the deuterium-tritium reaction, $n\tau$ must be greater than 10^{14}. Scientists find it convenient to measure fusion progress by the $n\tau$ value achieved.

In 1955 in Geneva, Switzerland, when controlled fusion was first publicly discussed by Homi J. Bhabha, president of the first International Conference on Peaceful Uses of Atomic Energy, the most successful experiments had reached $n\tau$ values of about 10^7. The highest temperature achieved was around 500,000° C. Bhabha then predicted that controlled fusion would be a reality by 1975. I bet him $20 in 1962 that his schedule would not be met.

Now I am not so sure his timetable was wrong. Artsimovich's latest results are an $n\tau$ of about 8 x 10^{11} at almost 6 million° C. United States experimenters have also done well. For example, the Scylla 4 reactor at Los Alamos, N. Mex., has achieved an $n\tau$ of about 2 x 10^{11} and perhaps 50 million° C., and the Livermore, Calif., device called 2X has reached an $n\tau$ of about 7 x 10^{10} at about 80 million° C.

Yet, after two decades of very hard work by some of the world's most capable physicists, we stand, in 1971, at only about 1/100 of the Lawson criterion. This is because the search for controlled fusion has uncovered problems far more difficult than anybody expected.

The major problem we face in fusion research is containing the pulsing, twisting, high-pressure plas-

An experimental breeder reactor is "fed" fuel rods by complex automatic equipment. Most of the reactor is in a huge tank of liquid sodium beneath the floor.

ma. It has been compared to holding a quivering mass of jelly with a few rubber bands. Granted, the plasma is very thin. Even with 10^{15} particles per cubic centimeter, it is only 1/10,000 as dense as air. Yet, at the extreme temperatures required, it reaches a pressure of about 300 pounds per square inch—some 20 times that of air. Very strong steel vessels normally would be used to contain it, but no wall made of any material could withstand a temperature so high that it literally tears atoms apart.

Scientists use the force that a magnetic field exerts on the charged nuclei to contain the plasma. This is the principle of the "magnetic bottle." The plasma is wrapped in an intense magnetic field that causes the nuclei to turn in tight little orbits around the lines of force. But the nuclei can move unrestrained along the length of the field's lines of force. So, if the lines enter the bottle as they enter the ends of an ordinary bar magnet, the nuclei easily drift out along them. Thus, the bottle's ends are its weak points. Many scientists prefer an endless bottle—a torus, or doughnut shape, such as Artsimovich's Tokamak T-3.

Unfortunately, no magnetic bottle can totally prevent the escape of the nuclei. The escape routes are devilishly hard to anticipate. Most plasma research has been aimed at identifying the various routes, then plugging the leaks. Thus far, the nuclei always seem to have been able to find new ways out.

One escape mechanism is simple cross-field diffusion, also called classical diffusion, in which two particles collide and displace their respective centers at random. If the particle escape was due only to classical diffusion, containment would be easy since this process is very slow. However an actual plasma can exhibit instabilities that enhance the leakage. The simplest are the so-called hydrodynamic instabilities, which allow the plasma to escape confinement much as water sloshes out of a moving pail. Experiments by Russian physicist M. S. Ioffe in 1961 demonstrated what American mathematician Harold Grad predicted six years earlier. A magnetic distribution with the absolute field strongest outside the bottle and getting progressively weaker toward the interior, but never reaching zero, would control hydrodynamic instabilities.

But, alas, physicists found that even after hydrodynamic instabilities were cured, the plasma escaped much faster than it would from cross-field diffusion. There were a host of microinstabilities, wild and bizarre oscillations within the plasma, that could grow without bound and destroy the confinement. Scientists could not predict these oscillations by thinking of the plasma as a fluid. They had to deal with it as a collection of particles that began oscillating due to the confining magnetic field and the attraction or repulsion of neighboring particles. More than a hundred kinds of microinstabilities have been identified, and fusion researchers have been trying to find magnetic-bottle configurations that will be able to reduce the microinstabilities to acceptable proportions.

Tokamak is the most successful configuration thus far. It has two primary magnetic fields. One surrounds the major axis of the torus, the axis that passes through the hole in the doughnut. The other field surrounds the minor axis, which encircles the hole. The two interact, forming a field whose lines of force cross each other and seem to suppress both the hydrodynamic instabilities and the microinstabilities—at least up to the particle densities and temperatures achieved so far.

Other magnetic configurations have also shown promise. The 2X machine at Livermore and a device called IMP at Oak Ridge are examples of magnetic mirror devices—long, magnetic bottles with very strong fields to cap their ends.

James Tuck's Scylla 4 at Los Alamos represents quite a different approach. It creates a sort of microthermonuclear explosion. The device is a long tube that is subjected to a suddenly rising magnetic field. The resulting shock wave compresses and heats the plasma. Tuck makes no attempt to control the pesky instabilities. Instead, he tries to heat and compress the plasma enough to complete the reaction and generate a net energy gain before instabilities can destroy the confinement.

A related but still more extreme approach is being tried in several places, notably by University of Rochester physicist Moshe J. Lubin. He simply blasts a pellet of frozen deuterium and tritium with a high-powered laser. This scheme uses no magnetic field. Rather, it attempts to add energy to the pellet so rapidly that a significant gain is generated before confinement is destroyed by the blowing apart of the pellet. Its immediate obstacle is the lack of a powerful enough laser, although Lubin believes he might achieve fusion if he were able to use 100 of the best lasers now available.

One nasty problem besets both fusion and fission devices—radiation damage. Each square centimeter of the vacuum wall of the fusion device will be bombarded with 2×10^{20} neutrons every day. Such punishment will pepper the metal—perhaps niobium and vanadium, which resist corrosion by lithium. Tiny holes will form and the metal will become brittle and swell. The wall, which will become intensely radioactive, will have to be replaced regularly—perhaps every year or two.

Fission in a Breeder Reactor

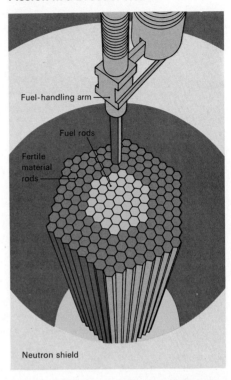

Fuel-handling arm

Fuel rods

Fertile material rods

Neutron shield

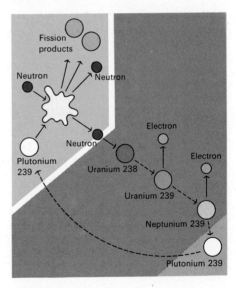

Fission products

Neutron

Neutron

Electron

Neutron

Electron

Plutonium 239

Uranium 238

Uranium 239

Neptunium 239

Plutonium 239

Special, remote-controlled equipment inserts a long rod of fertile material—U 238—in a breeder fission reactor, *top*. The reacting fuel atoms give off a number of neutrons, *above*, some of which begin a chain of reactions that ultimately convert the U 238 into additional fuel—plutonium 239.

Of even more concern to energy scientists is the impact of nuclear power on the environment and on the safety of people. "All power pollutes," according to the *Environmental Handbook*, published for the 1970 environmental teach-in. In a sense, this is correct. And since a power supply is necessary for man's survival and nuclear power plants are the sources of the future, the logical question is how do these plants affect the environment.

Two environmental hazards exist—waste heat and radioactivity. Concern over heat from new nuclear plants is becoming widespread. And at the needed levels of energy production, heat pollution could, indeed, become a serious local problem, but certainly not a worldwide one. Local heat problems could be avoided by clustering large power reactors near the sea or even offshore. Locating them some distance from the cities that will be their major market may require improved transmission methods. But the needed improvements are certainly within the reach of modern technology.

Radioactivity is of particular concern in fission devices. A fusion reactor poses roughly $1/10,000$ to $1/1,000,000$ as much potential radioactivity risk as a fission breeder of similar size. Tritium is one of the least toxic radioactive isotopes and is relatively inoffensive compared to the strontium 90 or plutonium produced from fission.

Notice that I said potential radioactivity risk. There is no danger unless radioactivity escapes into the environment. But even the best-run nuclear reactor will leak tiny amounts. Controversy in 1971, sparked largely by two scientists at the Lawrence Radiation Laboratory, at Livermore, centers on whether such amounts present an unacceptable risk. Biophysicist Arthur R. Tamplin and John W. Gofman, a physician and nuclear chemist, argue that they do.

Until the middle of 1971, Federal regulations said that nuclear plant discharges are allowable if they expose "suitable samples" of the general population to an average of no more than 170 millirems of radiation per person per year. For comparison, the average background radiation from natural sources of cosmic rays, gamma rays, and other radiation in the United States is about 100 millirems per year. In addition, we get 50 to 60 millirems per year from medical X rays. In Ceylon, 10 million people receive about 200 millirems per year from the uranium in the rocks near their homes.

Gofman and Tamplin say that we must assume radiation damage will be caused in proportion to the dose, even when the dose is very small. And they charge that the old standards allow all 200 million Americans to receive the additional 170-millirem dose per year. Excessive radiation is known to cause cancer and genetic damage that can result in birth defects. Therefore, say Gofman and Tamplin, present standards can result in many thousands of new cases of cancer and defective children per year.

Although standards-setting groups, such as the Federal Radiation Council, also assume that danger may continue even with very small radiation doses, it cannot be scientifically proved. To prove a 1 per cent genetic effect in mice, for example, would require experiments with several billion mice. Also, while a suitable sample is difficult to define, it is intended to indicate the maximum radiation allowed. Thus, this dosage limit would have to apply to the population exposed to the greatest risk. Since the effect of radioactivity diminishes as it leaves its source, this population is usually made up of people who live closest to the plant. The amount received by all other people would be far less.

The argument about radiation emission standards was dramatically ended, for all practical purposes, in June 1971, when the Atomic Energy Commission reduced the allowable emission from nuclear power plants to about 1 per cent of the previous figure.

However, the question of a catastrophic release from a fission nuclear reactor still remains. If a high-powered nuclear reactor lost all of its cooling during operation, or if its nuclear reaction went out of control, large amounts of radioactivity might be released. For this reason, nuclear reactors are among the most carefully designed and built devices man has ever created. They have many backup systems that make a serious accident that could destroy the reactor and release radioactivity to the public very unlikely.

A fusion reactor cannot, in principle, "run away." However, it too can melt if cooling is not supplied to it, and it would also pose a radiation hazard in the very unlikely event of such an accident.

Thus, fusion is safer than fission, but neither is totally safe. Under these circumstances, mankind faces a difficult decision. We will need all but unlimited energy for survival; yet, the only practical sources to satisfy these needs are fission and possibly fusion. Both methods pose some, though an extremely small, risk of accident.

What, then, is the proper course for mankind? This is a question that goes beyond science. To me, it seems that the only prudent course is the one we are now following: Develop the safest and cleanest possible fission breeders while we vigorously try to solve the remaining problems of controlled fusion. In a battle for survival and ultimate victory over material want, dare we do any less?

For further reading:

Boffey, Philip M., "Radiation Standards: Are the Right People Making Decisions?" *Science*, Feb. 21, 1971.

Gillette, Robert, "Nuclear Reactor Safety: A Skeleton at the Feast?" *Science*, May 28, 1971.

Gough, William C., and Eastlund, Bernard J., "The Prospects of Fusion Power," *Scientific American*, February, 1971.

Lubin, Moshe J., and Fraas, Arthur P., "Fusion by Laser," *Scientific American*, June, 1971.

Rose, David J., "Controlled Nuclear Fusion: Status and Outlook," *Science*, May 21, 1971.

Weinberg, Alvin M., and Hammond, R. Philip, "Limits to the Use of Energy," *American Scientist*, July-August, 1970.

A New Look at Prehistoric Europe

By Robert C. Cowen

The Europeans of 2,000 years before Christ and earlier lived in a state of near-savagery, or so thought European prehistorians. These scholars—archaeologists who work with the eras before recorded history—believed that Europe was overshadowed by the rich and flourishing civilizations of the Egyptians, the Sumerians, and the Mycenaeans. Recent evidence from a number of sources now suggests, however, that the early inhabitants of northern Europe were actually highly accomplished people, whose culture included sophisticated features that were unknown to the more "civilized" cultures of the East.

Most archaeologists were schooled in a theory called diffusionism. This notion holds that civilization and advanced technology arose in Egypt and the Middle East and then, starting about 2500 B.C., diffused, or flowed, northward into Europe. In the late 1800s and early 1900s, Swedish archaeologist Oskar Montelius suggested that certain cultural influences, such as the design of tools and weapons, may have flowed into prehistoric Europe from the Middle East. Building on this idea, British archaeologist V. Gordon Childe, from 1925 until his death in 1957, developed the larger and grander framework of diffusionism to explain European prehistory.

Childe's ideas and methods were to influence an archaeological generation. "The irradiation of European barbarism by oriental civilization," in Childe's words, was clear because of at least three

Converging on the past from several directions, scientists have found an advanced culture for ancient northern Europeans

important links between Europe and the civilizations of the Aegean and the Middle East. These links were: (1) the spread of collective burial tombs to Spain and France; (2) the diffusion of metallurgy to Europe through the Balkans; and (3) the use of Mycenaean methods in the construction of certain megaliths—large, standing stone structures—in Britain.

Now, however, Childe's dominant belief is being challenged, and prehistorians must reconsider their whole approach to the ancient northern Europeans. Three lines of research are converging to bring a new view into sharp focus. The first is from the archaeologists themselves. In 1966, a British archaeologist, Colin Renfrew of Sheffield University in Yorkshire, England, challenged the archaeologists' neat diffusionist framework. In articles published in the late 1960s, Renfrew led a host of younger prehistorians in the United States and Europe who questioned the validity of diffusionism as the only guiding principle in European prehistory. Renfrew went a step beyond his colleagues by developing in considerable detail how social and cultural changes could have been due to local factors.

The second development is based on a refinement during the past five years in the technique of radiocarbon dating. New age determinations show that prehistoric objects in Europe are much older than was thought. This offers a stunning confirmation of the skepticism of Renfrew and the other archaeologists. Although the carbon "clock" of ancient chronology is still being reset, experts generally agree, for example, that the beginning of bronze metallurgy in eastern Europe had been underestimated by as much as 1,000 years. This date has now been shifted from about 2000 B.C. to about 3000 B.C.

The third revolutionary development is largely the work of a non-archaeologist who has been able to relate prehistoric European culture to his own field. Alexander Thom, professor emeritus of engineering science at the University of Oxford in England, has demonstrated that high intellectual achievement existed some 4,000 years ago in what was thought to be the "darkest," least advanced, part of Europe. By tramping the hills and fields of Great Britain for decades, Thom has painstakingly surveyed hundreds of ancient megalithic sites. These include *dolmens* (tombs), *cromlechs* or *henges* (circles of standing stones), *menhirs* (single upright stones), *alignments* (parallel rows of stones), and *rings* (circles of small flat stones). In articles published in British journals in the late 1960s and in books published in 1967 and 1971, Thom shows that the ancient megalithic sites were laid out with a precision that would tax a modern surveyor.

This three-pronged evidence of highly accomplished and creative people inhabiting Bronze Age Europe cuts right across the diffusionists' assumption that civilization and virtually all advanced technology spread to Europe from the East. Childe had recognized that he was working with untested assumptions. If these should prove false, he said, Europe's prehistorians would have to start again. And so they

The author:
Robert C. Cowen is natural science editor for the *Christian Science Monitor*. He has spent the past two years in London, England, covering European science.

must. As Renfrew puts it, they now have to find "an explanation in European terms for what took place in prehistoric Europe."

Archaeologists have long been able to date the major accomplishments of the Eastern civilizations. These dates are determined basically from written records—the civil calendar of the ancient Egyptians and, to a lesser extent, the record of Middle Eastern kings. Archaeologists also have confidence in the extension of these dates to other civilizations that traded extensively with the East, such as the Minoans on the Mediterranean island of Crete. Dating the early civilizations in Europe and Great Britain, however, proved more difficult because there were no written records to guide prehistorians. Generally, they had to infer dates from parallel developments in the Near East and the Aegean civilizations. Much of this dating was done in the late 1800s and the early 1900s by British archaeologist and Egyptologist Sir Flinders Petrie, who correlated the discovery of Egyptian objects on Crete with Minoan objects in Egypt.

Renfrew's early doubts centered around the diffusionists' explanation of the striking similarities between the ancient Aegean and Spanish monumental stone burial tombs, particularly features of their vaults. Later archaeologists thought Crete might have been a center of influence. For instance, in studies during the early 1960s, British archaeologist Beatrice Blance suggested that colonists from the Cyclades, a group of more than 200 islands lying off the southeast coast of Greece in the Aegean Sea, might have landed in Spain. There, a settlement at Los Millares had round *bastions* (fortified strongholds) similar to those on the island of Siros in the Cyclades. Blance dated Los Millares at about 2500 B.C., which was supported by radiocarbon dating of organic remains found at the site.

"OK," Renfrew allows, "she had a point. There are these similarities. But there are two ways to interpret them. You can say that one culture came from another. This view . . . was superficial. Having worked in the Cycladic Islands, I don't feel the sites were that similar. What similarities there were seemed due to parallel development to meet similar needs. After all, round bastions are a kind of fortification that can arise independently. There were similarities between Spain and the Aegean because similar things were happening in both places. It was an era of proto-urbanization (city-building). In the Aegean, this went on to become Minoan civilization. In Spain, for some reason, it didn't flower into such glory."

Renfrew published this assessment in 1966, before the readjustment in radiocarbon dating. With recalibration, however, the date now is about 3000 B.C., at least as early as collective burial began in the Aegean. "The recalibration is the clincher," Renfrew says, "and is the key to the whole problem."

Chemist Willard F. Libby discovered carbon 14 in 1947, and he and his colleagues developed radiocarbon dating as an archaeological tool. Radiocarbon dating, for which Libby won a Nobel prize in chemistry

New Chronology for Prehistoric Europe

Western Europe

▓▓▓ Introduction of agriculture

■■■ First megalithic tombs

▓▓▓ Bronze metallurgy

▓▓▓ Megalithic observatories

Revised dates for major accomplishments of the ancient Europeans show a shift backward in time, ranging from about 600 years for development of bronze metallurgy to 2,000 years for the beginning of agriculture.

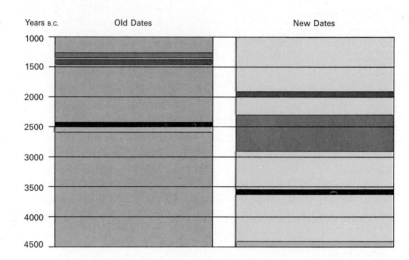

in 1960, is based on the cosmic ray production of new atoms. It depends on a complex chain of nuclear, geophysical, and biological processes. Cosmic rays plunging into the atmosphere transform nitrogen atoms, the most abundant atoms in the air, into carbon 14 (C^{14}), a radioactive form of carbon. Within perhaps a few days—the actual time is not established—these C^{14} atoms react with oxygen in the air to form radioactive carbon dioxide. This, in turn, mixes thoroughly with ordinary carbon dioxide and forms a concentration of about .03 per cent in air. The "hot" carbon dioxide spreads throughout the world, mixing into the sea, in plants through photosynthesis, and in the living tissues of the animals that eat the plants.

Meanwhile, the hot carbon atoms decay at a steady rate. One by one, they give off a burst of radiation and change back into nitrogen. At any time, the rate of decay of C^{14} atoms in the atmosphere and in organic matter equals the rate at which new C^{14} atoms are created by cosmic rays. The whole cycle is in equilibrium. As long as plants and animals remain alive, they maintain a C^{14} concentration that is equal to that of their environment. When they die, however, the equilibrium ceases; the C^{14} begins to decay at a steady rate and the radiocarbon clock starts ticking. After 5,730 years—the half-life of the radioactive material—50 per cent of the original C^{14} concentration will remain.

In order to date any carbon-containing specimen, such as animal bones or charred wood, scientists have only to determine how much of its original C^{14} has disappeared. Then, using the decay rate, they calculate the time this disappearance required. In practice, the process requires elaborate equipment, skilled handling, and a series of steps that contain many possibilities for error. One of the greatest uncertainties lies at the very heart of Europe's archaeological revolution: How do we know a specimen's original C^{14} equilibrium concentration? We cannot assume that the environment has always had the same level of C^{14} that is found today. If the original C^{14} concentration is underestimated, the dating will be too young, and vice versa.

The environmental level depends on the rate of C^{14} creation in air. That level depends, in turn, on the rate of cosmic ray influx, which itself depends on the strength of the earth's magnetic field. The magnetic field deflects more or fewer of the incoming charged particles of cosmic radiation away from the earth, depending on its strength at a given time. A period of relatively weak magnetic shielding would intensify the cosmic ray bombardment, creating a higher C^{14} concentration in the environment. Conversely, relatively strong shielding would lower the C^{14} level.

Geophysicist Vaclav Bucha, working at the Czechoslovak Academy of Sciences and later at the Institute of Geophysics and Planetary Physics at the University of California, Los Angeles, has traced the earth's magnetism back 20,000 years. Bucha works with magnetism found in rocks and ceramic materials, such as pottery fragments, that have been positively dated by some other technique. He has found substantial variations in their magnetism. In papers published in 1969 and 1970, Bucha has shown that the earth's magnetic field over the past 8,500 years was weakest about 4000 B.C. and strongest between 400 and 100 B.C. It swings between values of .6 and 1.6 times the present magnetic intensity. This alone would be reason enough to take a fresh look at carbon dating accuracy.

Fortunately, tree-ring research, or dendrochronology, has given investigators a unique opportunity to determine the extent of deviations in the radiocarbon clock. In the late 1960s, dendrochronologist Charles W. Fergusson and his colleagues at the Tree-Ring Laboratory

Bristlecone pine, *above*, in east-central Nevada is over 5,000 years old.

Comparing Radiocarbon and Tree-Ring Dates

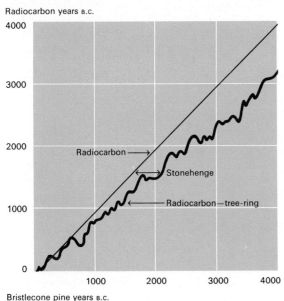

Radiocarbon years B.C.

Radiocarbon dates, calibrated against the bristlecone pine tree-ring dates (wavy line), show a gradual departure with time from the previously assumed radiocarbon dates (straight line).

Bristlecone pine years B.C.

at the University of Arizona in Tucson, established a chronology for the oldest trees known, the bristlecone pines. Specimens taken near Ely, Nev., date back some 5,000 years. Other specimens found in California's White Mountain area go back 7,500 years.

By using Fergusson's bristlecone pine tree-ring chronology, chemist Hans E. Suess and co-workers at the University of California, San Diego, could determine the extent of deviations in radiocarbon dates from 5300 B.C. to the present. They radiocarbon dated ancient samples of the pine wood and compared the radiocarbon ages with the tree-ring ages. In 1969, Suess published a calibration curve that shows considerable variation from conventional radiocarbon dates. Interestingly, the ups and downs of environmental C^{14}, as indicated by the bristlecone pine studies, parallel Bucha's curves for magnetic variations. When the earth's magnetic guard is down, C^{14} concentrations are up, and vice versa.

Suess' recalibration is based only upon specimens taken from one species of tree in a restricted geographical area. Data from new studies might upset it. But at present, the recalibration is widely accepted. Renfrew claims that he feels "quite comfortable" with the recalibration. "The real issue is the big shift of dates," he explains. "I feel absolutely confident on this point. I think the broad picture that emerges is OK."

The resetting of the carbon clock does nothing to upset the relative chronology of prehistory within Europe. It moves all the European dates backward together. But, in doing this, it cuts through Childe's links between European prehistory and the Middle East. New carbon dates show that the custom of burial in monumental tombs came not from the East by way of Spain, as Childe had believed, but developed long before in Brittany (in northwestern France) and in Iberia (now Spain and Portugal). In fact, these European tombs are older than the 4,500-year-old Great Pyramid of Khufu in Egypt and may be the earliest of man's surviving monuments. Similarly, revised dating indicates that the complex art of bronze metallurgy, Childe's second link, arose semi-independently in Europe.

New radiocarbon dates cut, especially, through Childe's third link— the presumed influence of Mycenaean civilization on prehistoric Britain. Some archaeologists assumed, from certain carved markings and signs, that Mycenaean craftsmen from ancient southern Greece might have supplied their expertise between 2000 and 1500 B.C. to construct Stonehenge, the most impressive of the huge stone megaliths in Britain. Original carbon dating from organic material (a piece of burnt wood) taken from the Stonehenge site on the Salisbury Plain in

Among the hundreds of megaliths in Europe are, *clockwise from upper left,* a tomb in Brittany, France; the standing stones of Callanish, a Scottish Stonehenge; and stone showing cup and ring marks at Cumberland in England.

Stonehenge, *above*, stands alone
and full of mystery on England's Salisbury
Plain. Its designers, the Windmill Hill
Folk of about 2400 B.C., *left,* may have
used a megalithic yardstick to work
out the positions of the heel stone and
initial uprights. About 2000 B.C., the
Bell Beaker Folk, *right,* raised the
remaining uprights in place, using an
elaborate system of levers, ropes, and logs
to hoist the huge stones, which weigh up
to 50 tons, into predug foundation pits.

Wiltshire in 1952 dated the monument to 1848 B.C. ($\pm$275 years). Revised radiocarbon dating now moves it back to earlier than 2200 B.C., which means that Stonehenge is older than the Mycenaean civilization itself.

Certainly, the ancient Europeans must have been highly intelligent people, and this has been confirmed by Thom's investigations. His data indicate that megalithic structures were precisely laid out in integral multiples of a universal length, much as we use the foot, yard, or meter. Thom calls this length—exactly 2.72 $\pm$ .003 feet—the megalithic yard. In addition, he finds that the designs of many of these structures were based on properties of triangles. This indicates that these early builders had an extensive knowledge of practical geometry. But probably Thom's most dramatic discovery is that these large megaliths had well-defined sight lines for observing the rising or setting of the sun, the moon, and certain stars.

Thom was by no means the only person to suggest that these precisely aligned stones could have been used as astronomical observatories. In 1963, astronomer Gerald S. Hawkins of Boston University dramatically demonstrated such possibilities for Stonehenge. But, Thom has shown how megalithic man could have used his observatories to maintain an accurate calendar and to tell time at night by the rising or setting of prominent stars. They even had the capability, he claims, to study motions of the moon with the incredible precision of about one minute of arc, or about 1/30 of the moon's apparent diameter. Thom has found these features in stone alignments throughout Britain, and is now discovering them in Continental Europe.

As in Britain, Thom found that alignments and rings on the continent were set out in units of the megalithic yard. The builders used a unit of 2.5 megalithic yards, which Thom calls the megalithic rod. And again, he finds the ancients laid out their patterns with astonishing accuracy—to within 1 part in 2,000. To set out rows or rings of stones with such accuracy today, a modern surveyor would have to take into account such minor effects as the expansion of a steel measuring tape caused by the heat of the sun.

When Thom talks about precise layouts, he means patterns that use a universal length unit and well-defined rules of geometrical construction. Neither of these features would be apparent to a modern-day observer when he looks at an old ring or at rows of standing stones. Thom found them only after extensive statistical analysis of hundreds of such alignments. Starting with a collection of accurately surveyed dimensions—say a list of ring diameters—various statistical techniques can be used to determine if there was a standard unit of length and what the size of this unit might be. Thom worked in this way to discover the megalithic yard. The length units he found are either the megalithic yard or derivatives of it, such as .5 or 1.5 megalithic yards.

No one knows how megalithic man developed his designs. There are no surviving records or "blueprints." Perhaps they drew some of them

The stone alignments
at Carnac, France, *below,*
are one of the prehistoric
wonders of Europe. More
than 2,700 accurately
placed stones extend 10
to 13 lines abreast for
more than 2 miles. *Left,*
a Bell Beaker priest
makes an offering to
the midsummer sun as it
appears in line with
the great stone rows.

informally in beach sand, just as a modern designer would make preliminary drawings with pencil and paper. Thom thinks some of the designs were inscribed on stone before being laid out on the ground. He has analyzed scores of "cup and ring" marks—small depressions and incised circles cut into rocks and stones in Britain. From a statistical analysis of the marks, a length unit emerges which is exactly 40 times smaller than the megalithic yard—.816 inch. He believes the designers worked with fixed compasses fitted with quartz or flint scribing points. These compasses would be set accurately at, say, .5, 1, 1.5, 2, 2.5, and so on, megalithic yards apart.

Whatever method the ancient designers used to work out their structures, however, they may have embodied in them an advanced astronomical knowledge. Thom has identified many sighting lines set up on azimuths, or plotting angles, that are important in astronomy. Sometimes the lines are directly defined by two or more stones. Occasionally, stones mark exactly where to stand to view the astronomical body across a natural *foresight* (a reading in a forward direction) on the horizon. This might be a notch in a hill or an especially well-shaped peak that the setting sun just grazes during the summer solstice. Thom has shown that some layouts have sight lines that could have been used to maintain an accurate calendar, dividing the year into 16 more or less equal months. He has also shown that many sites could have been used for detailed studies of the motions of the moon and to predict eclipses with a precision not again attained in Europe until the Renaissance.

Thom is an example of a scholar competent in one field who has used his skills to shed unexpected light on another discipline. In fact, Thom's ideas were so novel, and his mathematical data and complicated statistics so difficult, that nonmathematically oriented archaeologists had difficulty understanding them. At first, he could not get his archaeological papers published, in spite of his standing as an Oxford professor. Now, he has won wide respect for his work, if not universal acceptance of all his conclusions. Alexander H. A. Hogg, secretary of the Royal Commission of Ancient Monuments in Wales, says he accepts Thom's astronomical arguments, but he has trouble seeing the case for the megalithic yard and Pythagorean geometry. Renfrew, on the other hand, thinks the megalithic yard is well established, but he has doubts about the validity of some of Thom's astronomical sighting lines. Archaeologist Glyn E. Daniel, of St. Johns College, Cambridge, England, one of the most distinguished megalithic specialists, summed up the general feeling among archaeologists by saying that "Trying to explain these phenomena [Thom's findings] as repeated accidents is even more improbable than accepting the admittedly unexpected conclusions."

Thom makes no pretense of being an archaeologist. He has confidence in his surveys and statistics and in what they reveal although he admits there are limitations to speculation. "We do not know the extent of megalithic man's knowledge of geometry and astronomy," Thom said in the introduction to his most recent book, *Megalithic Lunar Observatories*. "Perhaps we never shall," he adds. He offers any thoughts he has about the capabilities of megalithic man merely as a layman's speculations. With that reservation, he says he is impressed by what seems to have been a widespread "civilization" throughout Great Britain and parts of Europe. Many centuries, probably millenniums, of observation and study culminated in the observatories that flourished about 1800 B.C. This suggests a "scientific" interest that provided advanced training for specialists.

It was a culture that established and maintained a standard of length with a precision of 1 part in 2,000. How did they do it? If each community made its own "yardsticks," even if they copied a standard sent from settlement to settlement, detectable errors inevitably would have crept in. Thom thinks that standard rods, perhaps made of seasoned oak, were manufactured and sent out from a center, a sort of ancient "bureau of standards." But neither Thom nor anyone else has actually found one of these rods.

"What does all this imply about these people intellectually?" Renfrew asks. "It shows certainly that they were interested in studying the sun and the moon. . . . I think the point is these people wanted to record their knowledge. The monuments and circles were a means of recording. They were obviously fascinated by whole numbers, if you accept what Thom is saying. And there's a lot in what he's saying. You don't have to assume that they were up to all kinds of advanced things we don't know about."

The new discoveries from a wide range of disciplines present an exciting challenge to archaeologists to discover a more creative past than they had ever imagined Europe possessed. But what does it mean to the rest of us? What relevance does it have for mankind today?

Renfrew points out that archaeologists now must discover the social and economic factors that shaped cultural advances. Here were men as intelligent and as creative as we are. By studying their development, Renfrew says, we may shed a light on the basic dynamics of social and cultural change that will help us to understand better the challenges we face today. And Thom observes: "I've often said, if I've made any contribution, it is in finding out there was a civilization that had existed in these islands [the British Isles] and went bust. It's a warning not to take things for granted. Our own civilization could go bust, too."

For further reading:
Childe, V. Gordon, *The Dawn of European Civilisation,* Knopf, 6th rev. ed., 1958.
Hawkins, Gerald S., *Stonehenge Decoded,* Doubleday, 1965.
Renfrew, Colin, "Trade and Culture Process in European Prehistory," *Current Anthropology,* April-June, 1969.
Thom, Alexander, *Megalithic Sites in Britain,* Oxford, 1971.
Thom, Alexander, *Megalithic Lunar Observatories,* Oxford, 1971.

The World Series Of Science

By Philip M. Boffey

As its annual meeting continues to be the center of turmoil, the AAAS strives to broaden its impact on science and society

During the week between Christmas and New Year's in 1970, many newspapers in the United States carried tales of discord and disruption emanating from an unlikely source, a scientific convention in Chicago. The convention was the annual meeting of the staid and genteel American Association for the Advancement of Science (AAAS), the nation's largest general scientific organization. But some newspaper reports pictured the convention more like a gathering of rabble-rousers. There were stories of speakers being heckled by radical young scientists, of an eminent physicist needing the protection of bodyguards, and even a bizarre incident at one session in which the wife of a prominent biologist became so angry at a young radical that she poked him with her knitting needle, drawing blood.

By and large, the stories were true, but the antics of a relatively small group of hecklers and protestors tended to distract attention

The audience at AAAS sessions can expect to hear not only from the Establishment, but also from those who protest the uses of science.

from the main purpose of the meeting—to discuss significant advances in science and their importance to the improvement of society.

The AAAS, which has its headquarters in Washington, D.C., holds its annual meetings in a different city each year, and it tries to attract as large and diverse an audience as possible. Of all the hundreds of scientific meetings held in the United States annually, probably none is more accessible and understandable to the student, the teacher, and the ordinary citizen. Because of the great variety of topics discussed, this meeting has been dubbed "the World Series of science." Its sessions are open to the public. About one-fifth of those attending the 1970 convention registered as students.

The AAAS meetings have occasionally produced moments of high drama. In Chicago, excitement reigned during the appearance of Edward Teller, popularly known as the "father of the hydrogen bomb." Teller participated in a panel discussion that quickly turned into a debate between those who believe a scientist has a responsibility to help perfect his nation's armed might, and those who believe he should somehow resist allowing the government to use his work for military purposes.

The first speaker on the panel was Hungarian-born Albert Szent-Györgyi, a spry, 77-year-old Nobel laureate in physiology and medicine. Szent-Györgyi told the overflow crowd: "Because science is used for war, we have lost the respect of people and there is a revulsion against scientists."

Then Teller, also born in Hungary, rose to defend his work in helping the United States to develop thermonuclear weapons. From the start of his talk, Teller was heckled by radical students who stood on the speakers' platform holding placards denouncing him as a "war criminal." But Teller held his own against the hostile listeners. He likened the radical's attacks to the treatment he had received as a Jew in the "witch caldron" of Nazi Germany. And he declared that he had opposed dropping the atomic bomb on Hiroshima and Nagasaki in 1945. He preferred instead that it be used in a demonstration so that the Japanese could see its effects and recognize the hopelessness of continued fighting. Such a demonstration, he said, "could have proved to the world that science could have ended a terrible war, but instead we killed 100,000 persons."

The confrontation between Teller and the radicals had more drama than do most debates at scientific meetings. But it was not surprising that a sharp dispute over an important issue of public policy should break out at the AAAS convention. For in recent years the AAAS has attempted to develop broader interests and a greater concern for social issues than most other scientific societies.

When the AAAS was founded in 1848 by a few geologists and naturalists, it was about the only scientific society in the United States. It served a rather narrow professional purpose then: to promote the advancement of science by bringing researchers into stimulating con-

The author:
Philip M. Boffey is a science journalist and former writer for *Science* magazine, the official journal of the American Association for the Advancement of Science.

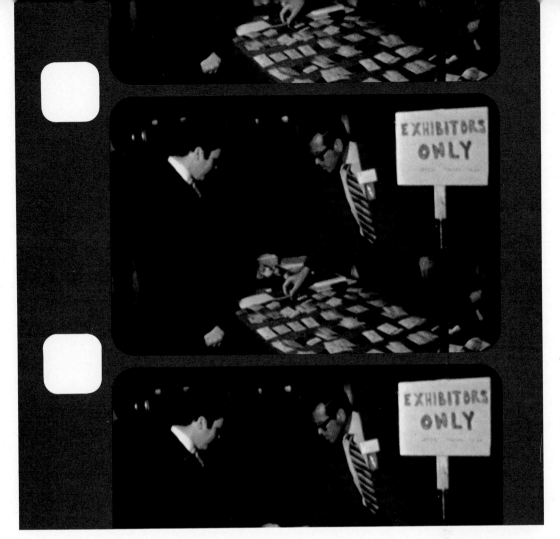

An exhibitor registers so that he can man a booth at the AAAS Annual Exposition of Science and Industry, where visitors can see the latest in scientific books and instruments.

tact at periodic meetings where they could hear colleagues discuss the results of their research. The AAAS also published *Proceedings*, which were volumes that contained the major papers presented at these meetings.

In those days it was relatively easy for the biologist to understand the research of the chemist or geologist. But over the years, as science became more complicated and scientists more specialized, the original functions of the AAAS were taken over by more specialized professional societies. By 1876, for example, chemists had become so numerous and had developed so many interests of their own that they left the AAAS and founded the American Chemical Society. Similarly, in 1899 the physicists founded the American Physical Society, and in 1948 the biologists founded the American Institute of Biological Sciences. These societies started to hold their own meetings and publish their own professional journals. In 1971, if a scientist wants to hear about the latest developments in his own field or wants to talk informally with his peers, he generally goes to the meetings of his own disciplinary society.

Meanwhile, the AAAS has developed into a broader organization that serves many branches of science. It attracts scientists who want to

Publications prepared
by the AAAS on a
variety of disciplines
are displayed at
the annual meeting.

discuss matters of interest to more than one discipline, to the scientific community as a whole, and to the general public. Also, its meetings are a handy place for scientists to meet colleagues in other fields.

The AAAS has some 133,000 members. About half of them are scientists at universities. The rest are industrial scientists, students, teachers, and others who are interested in science and its impact on society. The organization has four stated goals: (1) to further the work of scientists, (2) to facilitate cooperation among scientists, (3) to use science to promote human welfare, and (4) to increase the public understanding of science. It seeks to achieve these goals in many ways.

One of the ways is by distributing scientific information. The AAAS publishes *Science*, one of the world's leading scientific journals, which has a circulation of about 163,000. It also publishes many science bibliographies, books, and other documents. It awards prizes for outstanding work in science and science journalism. It distributes audiotapes and videotapes of significant segments of its annual meetings to high schools. And it issues reports on critical technical problems such as air pollution and the siting of nuclear power plants.

The AAAS also carries on several educational activities. The most important has been developing a science curriculum for elementary schools. Known as "Science: A Process Approach," the curriculum was used in the 1970-1971 school year by some 70,000 elementary school teachers to instruct more than 2 million pupils. Basically, this curriculum attempts to develop the children's skill in using scientific techniques such as observing and recording. It stresses the processes of science rather than its content.

The organization also stages the so-called Holiday Science Lectures, at which eminent scientists discuss developments in their fields with promising high school students. During the 1970-1971 school year, for example, R. H. Bing of the University of Wisconsin spoke on "The Inventive Side of Mathematics" in Fort Worth, Tex.; astrophysicist Kip S. Thorne of the California Institute of Technology discussed "The Violent Universe" in Phoenix, Ariz.; and biologist Carroll M. Williams of Harvard University described what it is like "Inside the Living Insect" in Portland, Ore.

Similar seminars are sponsored by the AAAS for congressmen, diplomats, and other government officials who must grapple with complex technical questions. In 1971, for example, congressmen heard talks on population control by Garrett J. Hardin, a biologist at the University of California, Santa Barbara; on waste management by Abel Wolman, a sanitary engineer at Johns Hopkins University; and on technology assessment by Michael Michaelis, an engineering consultant of Arthur D. Little, Inc., Washington, D.C.

Despite these activities, the AAAS has been attacked as a stodgy, do-nothing organization that defends the status quo and makes little effort to help solve such pressing problems as hunger, poverty, pollution, and overpopulation. Even most AAAS leaders would agree that the organization has failed to make as great an impact as it should. In an effort to stimulate the organization, the AAAS Council and the Board of Directors in 1969 approved a resolution calling for more intense involvement in public affairs and for a tenfold boost in membership by 1980. This goal, which would require the AAAS to corral more than a million new members within a decade, is unlikely to be met. But it reflects the desire of the organization to reach a greater number of people.

As the AAAS focus has changed from a preoccupation with the narrow professional needs of scientists to a broader interest in society at large, the organization has encountered complaints from a different direction. Its scientific journal and its annual meeting have been criticized for going overboard on social problems and for failing to attract enough reports on red-hot scientific advances. Indeed, the 1970 meeting contained an extraordinary amount of rehashing of old material. A survey indicated that 22 per cent of the papers presented had been reported previously, usually in a professional journal or at another scientific meeting. Still, if the chief goal of the AAAS meeting is to review the current state of science, it is not necessary to attract a lot of papers that report important breakthroughs.

For the most part, the AAAS simply tries to provide a forum in which important ideas can be discussed. However, one report at the 1970 meeting indicated that the AAAS intends to take an increasingly active role in proposing solutions for critical problems. Presented by the AAAS Herbicide Assessment Commission, the report constituted an indictment of herbicide use by U.S. military forces in Vietnam.

For roughly a decade, American military planes have been spraying chemicals on forests that are believed to hide the enemy and on croplands that are believed to feed the enemy. More than 8,500 square miles have been sprayed since the program began in 1961. Many persons had charged that the chemical sprays are causing great damage to the ecology of South Vietnam and to the health of the people, but there had been no definitive studies to prove this.

In December, 1969, the AAAS board of directors authorized Matthew S. Meselson, a biologist at Harvard University, to prepare a detailed plan for determining the impact of herbicides on the land and people of Vietnam. Meselson and three colleagues made a five-week inspection trip to Vietnam in August and September, 1970, consulted experts, and reviewed relevant scientific literature.

At the 1970 meeting, Meselson revealed that the spraying program has caused "extremely serious harm" to the land and to some of the people. In a preliminary report, he and his colleagues asserted that the crop-destruction program had been a near total "failure" because nearly all of the food destroyed would actually have been consumed by civilians, particularly the Montagnard tribes of the Central Highlands. The AAAS investigators also reported that from one-fifth to one-half of South Vietnam's mangrove forests had been "utterly destroyed," with no sign of new life coming back. Perhaps half the trees in valuable hardwood forests north and west of Saigon are dead, and worthless bamboo threatens to take over the area.

Not all scientists who heard the presentation were convinced that the herbicide program was as destructive as described. But the AAAS study seemed to achieve a political impact. The White House, which had previously been informed that Meselson would denounce the crop-destruction program, announced on Dec. 26, 1970, the day the AAAS convention opened, that authorities in Saigon were initiating "an orderly, yet rapid, phase-out of the herbicide operations."

Presentations at AAAS meetings usually do not have such immediate international impact. Most discussions try to focus on ideas of interest to a broad range of scientists. At the 1970 meeting this type of discussion was exemplified by a symposium on "The Chemistry of Learning and Memory." It was of concern to biologists, chemists, medical scientists, pharmacologists, and psychologists, among others.

At the symposium, Dr. Georges Ungar, professor of pharmacology at Baylor College of Medicine in Houston, caused quite a stir by

Anthropologist Margaret Mead, a familiar figure at AAAS meetings, fields a hard question at a press conference on social problems.

announcing that his laboratory had isolated a chemical substance from the brains of rats that appears to be crucial to the processes of learning and memory. Given the choice between a lighted or a dark enclosure, rats and mice normally prefer the dark. But Ungar's laboratory trained some 4,000 rats to fear the dark by subjecting them to electric shocks every time they tried to take refuge in a dark enclosure. Then he sacrificed the rats and injected extracts from their brains into untrained animals. The new rats promptly developed the same behavior pattern—they were afraid of the dark even though they had not been subjected to the electric shocks.

Ungar noted that this type of experiment had already been performed in at least 26 laboratories throughout the world. However, his laboratory is apparently the first to isolate the chemical material that contains the behavior-inducing information. Ungar was cautious about claiming applications for his work. But he did suggest that if man learns more about how the brain codes information, it may be possible to use man-made chemicals in treating mental retardation or the impaired memories of the aged. In the distant future, he said, it may even be possible to improve the intelligence of all people by chemical means. See THE ATTEMPT TO UNDERSTAND THE BRAIN.

Another type of discussion concerns matters of interest to the entire scientific community, such as federal financial support for scientific research. At a symposium during the 1970 meeting, Philip Handler, president of the National Academy of Sciences, warned that "in radio astronomy and particle physics our leadership is being lost to other nations." Glenn T. Seaborg, chairman of the Atomic Energy Commission, expressed concern about the climate of opinion in which science operates. He lamented "a negative reaction to science and technology on the part of a significant segment of the public." And some speakers complained that current Ph.D. degree programs turn out dull, socially unaware, science graduates.

Sometimes a more philosophical question sparks general interest, such as a controversial assessment of the future of science presented by biologist H. Bentley Glass, the retiring president of the AAAS in 1970. Glass suggested that science may no longer be exploring an "endless frontier," that the most significant laws of nature may have been discovered. He noted that telescopes have already plumbed space almost to the limits of the observable universe, and that research has revealed the universality of the genetic code, the evolutionary process, and other characteristics of living things.

"No rate of scientific development equal to that of the past half century can be long maintained," Glass said. "The laws of life, based on similarities, are finite in number and comprehensible to us in the main even now. We are like the explorers of a great continent who have penetrated to its margins in most points of the compass and have mapped the major mountain chains and rivers. There are still innumerable details to fill in, but the endless horizons no longer exist."

Not all scientists agreed with Glass, however. Indeed, some later pointed out in *Science* magazine that Lord Kelvin, the noted British physicist, had made a somewhat similar prediction in 1894 and was proved to be wrong. Kelvin had asserted: "The last half century has seen such enormous progress in science and technology that similar advances cannot be expected in the future. In my own field the possible changes can only affect the third or fourth decimal place." Yet only a few decades after Kelvin said this, the development of quantum mechanics began to change the whole realm of physics. Glass retorted that he was "aware of the immortal balderdash of Lord Kelvin," and said he had been careful not to say that science has discovered everything. Nevertheless, he insisted that, while our greatest advances are still to come in many fields, the pace of scientific advance is slackening.

As far as the general public is concerned, perhaps the most interesting part of a AAAS meeting are those sessions that feature sharp debate over such public issues as population growth, the arms race, environmental pollution, and the use of drugs. These debates seldom fail to attract widespread attention. At the 1970 meeting, for example, Barry Commoner, director of the Center for the Biology of Natural Systems at Washington University in St. Louis, was widely quoted on his assessment of the danger of mercury pollution. He warned that this "may well emerge as the most serious acute problem among the dismal array of problems which constitute the environmental crisis." Commoner particularly criticized the U.S. Food and Drug Administration for "minimizing the hazard" of mercury found in tuna and other foods as well as in the air, soil, and water.

Many of Commoner's sharpest comments were made in a large room that is generally off-limits to most of those attending the convention. This is the press room, almost certainly the busiest in the convention complex. Scores of reporters from newspapers and television stations gather there to attend press conferences and to interview scientists and write or film their stories. Throughout each day of the convention, scientists who will be delivering major papers are asked to visit the press room to summarize their papers and submit to intense, even hostile, questioning. The press room has by far the most complete collection of convention talks on file.

The range of material offered at a AAAS meeting and the size and complexity of the meetings often bewilders those attending for the first time. The 1970 program required 352 pages to describe the meeting's offerings. There were 15 major lectures and 120 symposiums. About 1,500 scientists presented papers. There were also tours and other special events. Five Chicago hotels were needed to accommodate the more than 5,700 people attending the convention.

The various sessions that make up a AAAS convention are held in a diversity of rooms and halls provided by the hotels. The talks vary widely in style and content. Some rely heavily on slides, and the lights flick on and off while those at the rear of the room strain to read the

Physicist Edward Teller
overcame harassment
at the lectern and
made protesting young
scientists listen to
what he had to say.

tiny print cast on the screens. If a talk is interesting and offers a fresh viewpoint, the audience usually peppers the speaker with questions and applause. If a talk is dull, as many of them are, people yawn, doze, whisper, squirm, and often get up and leave.

Those attending the 1970 meeting could hear talks on subjects ranging from "An Economic Analysis of Federal Air Quality Legislation," to "Theories of Interstellar Molecule Formation," to "The Vocal Repertoire of Chipmunks in California." About 1,600 persons attended an illustrated lecture on primitive life styles given by anthropologist Margaret Mead, while fewer than a dozen attended a discussion on the economics and politics of initiating and managing large-scale technological programs.

A glaring weakness of the AAAS meetings at present is that there is little centralized control over program content. Programs are arranged by the AAAS central meeting office, by AAAS disciplinary

groups such as the AAAS section on astronomy, and by the dozens of specialized societies that are loosely affiliated with the AAAS, such as the American Society of Zoologists, the Parapsychological Association, the American Mathematical Society, the History of Science Society, and the Potato Association of America.

There often are several sessions on the same topic. Environmental issues were particularly prevalent at the 1970 meeting. Fully 30 per cent of those who arranged symposiums at the meeting claimed that other programs overlapped theirs in content. Because of this duplication, anyone attending a AAAS convention is wise to study his program book carefully.

For many people, the highlight of a AAAS meeting is found not in the formal symposiums, but in hundreds of private encounters throughout the labyrinth of rooms in the convention hotels. Old friends reminisce over coffee, and new-found friends discuss their research in the corridors and elevators. Ambitious scientists swap gossip on job opportunities at various institutions, and young male researchers earnestly discuss a variety of esoteric subjects with their female counterparts. Cocktail parties abound, some sponsored by industrial concerns that hope to interest scientists in their products and others hosted by local dignitaries. Still others are informal parties hosted by individuals in their hotel rooms. And at night, when most of the day's work has ended, the more hardy conventioneers venture out into the surrounding city to sample the restaurants and night life, or to visit friends who live in the area.

Perhaps the best way to find out what the organization is trying to do is to attend one of the annual meetings in person. The next time the AAAS meets near your home, you, too, might want to stop in and sample its wares. Despite the many justified criticisms that have been leveled at AAAS meetings, the person who seeks a good general idea of where science stands at the moment, and who carefully chooses the sessions he will attend, can enjoy a valuable educational experience. A student, asked if he would recommend that other students attend the meeting, replied: "Definitely! I thought that if we . . . could just roam around to 10 meetings like this, it would be great—just a course in itself . . . I felt, as a lonely student from a small college in western Pennsylvania, I could have got up and said something [to the eminent scientists present] and they would have talked to me. This is really what grabbed me."

For further reading:

"AAAS: Conflict, Confrontation, Consideration," *Science News,* Jan. 9, 1971.

"A Time of Torment for Science," *Science News,* Jan. 2, 1971.

Boffey, Philip M., "AAAS," a series of three articles in *Science:* April 30, May 7, and May 14, 1971.

"Political Science, 1970," *Newsweek,* Jan. 11, 1971.

Sullivan, Walter, "Disputes That Were Not on the Agenda," *The New York Times,* Jan. 3, 1971.

Science File

Science Year contributors report on the year's major developments in their respective fields. The articles in this section are arranged alphabetically by subject matter.

Agriculture

Anthropology

Archaeology
Old World
New World

Astronomy
Planetary
Stellar
High Energy
Cosmology

Biochemistry

Books of Science

Botany

Chemical Technology

Chemistry
Dynamics
Structural
Synthesis

Communications

Computers

Drugs

Ecology

Education

Electronics

Energy

Genetics

Geochemistry

Geology

Geophysics

Medicine
Dentistry
Internal
Surgery

Meteorology

Microbiology

Oceanography

Physics
Atomic and Molecular
Elementary Particles
Nuclear
Plasma
Solid State

Psychology

Science Support

Space Exploration

Transportation

Zoology

Agriculture

An epidemic of southern corn leaf blight and the selection of an agriculturist for the Nobel Peace prize highlighted events in agriculture in 1970 and 1971. The blight caused the greatest production loss on a single crop in the history of U.S. agriculture. The Nobel prize winner, American crop expert Norman E. Borlaug, has probably done more to improve the physical welfare of people in a shorter period of time than anyone else in this generation.

Blight. Southern corn leaf blight (*Helminthosporium maydis*) was first observed in the United States near Belle Glade, Fla., in 1970. The fungus disease reached epidemic proportions there in late February and early March, then spread northward. By the first week in June, it was found throughout Florida, northern Georgia, and the coastal areas of Alabama, Mississippi, Louisiana, and Texas. By mid-June it was in Kentucky and Tennessee. By July 15, it was found as far north as Minnesota and Wisconsin and as far east as the coast of North Carolina. By September 1, the epidemic covered all of the major Corn Belt areas, including Wisconsin, Michigan, and the northeastern states.

According to Lloyd Allen Tatum, administrator for the Plant Science Research Division of the U.S. Department of Agriculture (USDA) Agricultural Research Service, Beltsville, Md., corn blight losses in 1970 totaled 710 million bushels. At the December, 1970, price of $1.50 per bushel, this amounted to about $1 billion.

The blight is composed of two strains, one of which is unusually virulent to corn containing "T male-sterile" cytoplasm. The other strain is a less serious yellow leaf blight. Corn with the T-type cytoplasm was used in about 90 per cent of all corn hybrids planted in 1970.

About 18 per cent of the seed corn farmers planted in 1971 was "N" resistant cytoplasm varieties, and approximately 30 per cent was of partly resistant blends. But the remaining 52 per cent was of the T type. However, almost all of the new seed produced in 1971 was expected to be other than the T type.

A blight watch, started in 1971, covered 80 per cent of the nation's Corn

Corn leaf, *left*, that did not have T-type cytoplasm, remained healthy in the midst of an epidemic of southern corn leaf blight that attacked leaves of other plants, *center*, and their ears, *right*.

Agriculture

Belt. It was designed to detect the spread of the blight so that farmers might apply fungicides to save their crops. The National Aeronautics and Space Administration produced infrared and natural-color aerial photographs with high altitude (60,000 feet) aircraft that were used to detect the progress of the disease.

Biweekly flights began in June over 210 sites, each of them 1 mile wide and 8 miles long, in Ohio, Illinois, Indiana, Missouri, Iowa, Minnesota, Michigan, and Nebraska. Ground observations were correlated with the aerial photographs. Purdue University's Laboratory for Applications of Remote Sensing and the University of Michigan Institute for Science and Technology, provided centers for the collection of data and the interpretation of photographs.

The Nobel prize was awarded to Borlaug on Dec. 10, 1970. The Iowa-born crop expert developed improved strains of wheat and rice while working for the International Maize and Wheat Improvement Center in Mexico. This research organization is financed by the Rockefeller and Ford foundations, and works in cooperation with the Mexican government. Variations of Borlaug's semidwarf Mexican wheat have been grown successfully throughout Asia in what has been called "the green revolution." See AWARDS AND PRIZES.

Crossbreeding was again widely practiced in beef cattle in 1971. About 30 per cent of all market cattle are now crossbred. Exotic breeds, when crossed with standard English breeds (Hereford and Angus, for example), grew faster on less feed and gave more milk than the standard breeds. More than 1,000 calves were expected in 1971 from crossbreed experiments at the USDA Meat Animal Research Center, Clay Center, Nebr. The most popular exotic breeds included Charolais, Limousin, and Main-Anjou, all from France, Simmental (which appear to have the fastest growing rate) from Switzerland, and Murray Grey from Australia.

New grain. Triticale, a new synthetic species of grain derived from both wheat and rye, was used on a worldwide scale for the first time in 1970. It is more productive, adaptive, and nutritious than either parent. Evangelina Villegas of the protein quality laboratories of the International Maize and Wheat Improve-

ment Center in Mexico reported that Triticale had greater potential for improved protein than any other cereal.

Chicken vaccine. Development of the first cancer vaccine for chickens was announced on March 1, 1971, by Ben R. Burmester of the USDA Regional Poultry Research Laboratory in East Lansing, Mich. The other members of the research team that developed the vaccine were Richard L. Witter, William Okazaki, and H. Graham Purchase.

The vaccine protects chickens from Marek's disease, a leukemialike virus that causes paralysis, blindness, and tumors. Marek's disease causes an estimated $200-million loss to poultry farmers each year. The vaccine reduces the incidence of tumor lesions by as much as 90 per cent. Licenses to manufacture the vaccine have been issued to three pharmaceutical houses.

Lunar soil tests. Seedling plants were germinated in lunar soil by Charles H. Walkinshaw, Haven C. Sweet, Subramaniaan Venketswaren, and Walter H. Horne at the Manned Spacecraft Center near Houston, Tex. The soil was brought back from the moon by the Apollo 11 and Apollo 12 astronauts. Nutritional deficiencies appeared in cabbage seedlings after several days' growth, but cultures of tobacco remained alive and green for several months. Root development was greatly stimulated in corn that was planted in the lunar soil.

Water management studies. Trickle, or drip, irrigation, a relatively new concept in agriculture, was described by E. Dale DeRemer, an agricultural consultant from Phoenix, Ariz., at the annual meeting of the American Society of Agronomy in August, 1970. The process involves the slow application of water at designated intervals by small-diameter hoses. Only the roots of plants become moist in this process, and one-third to one-half as much water is used as compared to other methods. There is also no salt build-up, according to DeRemer. Fertilizers may be added with the water. Another advantage of this type of irrigation is that the weeds grow sparsely.

The concept originated 20 years ago in the greenhouses of western Europe. Vast areas in the Negev Desert of Israel are now trickle-irrigated. The Israelis report high yield increases by this proc-

Agriculture

ess, and as little as 25 per cent of the normal amount of water is needed. Trickle irrigation is also used on over 9,000 acres of vineyards and orchards located near Melbourne, Australia.

Asphalt moisture barriers to be used to hold moisture in sandy soils incapable of retaining sufficient water for plant growth were produced in 1971 by the Amoco Moisture Barrier Company, a subsidiary of the American Oil Company. The material, in liquid form, is sprayed 2 feet below the surface of the soil with a special plow. It dries instantaneously to form a sheet of asphalt one-eighth of an inch thick.

Thermal pollution, a potential ecological problem, was turned to a useful purpose near Springfield, Ore. On a 170-acre demonstration project, seven farmers irrigated their crops with heated water from an electric power station. The warm water was pumped 2 miles through a buried 16-inch steel pipe.

The orchard and vegetable crops irrigated with thermal water were protected from early spring frosts. The experiment was under the direction of the Agricultural Division of Vitro, a subsidiary of Automation Industries.

Parallel experiments, conducted in 1970 by Larry Boersma at Oregon State University, used thermal water for warming the soil. Boersma's yield increases ranged from 15 to 64 per cent for corn, beans, and tomatoes. He also found that the growing season was lengthened by the use of thermal water.

Integrated pest control, the application of predators and other natural parasites, along with resistant varieties and crop management to farming, also made gains in 1970. According to USDA entomologists William H. Day and Robert W. Fuester, wasp parasites have reduced alfalfa weevil populations by 90 per cent in test areas in New Jersey and Pennsylvania. Predator mites have been unusually effective on the McDaniel spider mites and the European red mites that infest the apple orchards of British Columbia and Washington. John Garretson, a Yakima Valley fruit grower, reportedly reduced his insect control costs to one-half with these mites.   [SYLVAN WITTWER]

Anthropology

"The triumph of the average man" could well be the title of a study reported early in 1971 by Dr. Albert Damon of Harvard University. Between 1880 and 1912, anthropologists studied 2,450 Harvard students. More than 90 per cent of them had died by June 30, 1969, the date of Damon's study. Those of medium height (5 ft. 6 in. to 5 ft. 9 in.) and weight (132 to 152 lbs.) while in college lived the longest and had the most children, according to the study.

Neanderthal theses. One hundred years ago Neanderthal Man, then represented by remains from the Neander Valley in Germany, was diagnosed by an eminent pathologist of the day, Dr. Rudolf L. K. Virchow, as having suffered from rickets. The features that led Virchow to this conclusion, especially the heavy ridging above the eye sockets and the markedly bowed arm and leg bones, were later interpreted as normal in Neanderthals. However, the notion of Neanderthals with rickets did not completely die.

The updated thesis, stated in 1970 by Francis Ivanhoe of Cartwright Gardens, London, sees man as a tropical primate who began his existence with dark skin. The pigment in the skin protected him from overdoses of vitamin D and resulting kidney problems. When man moved out of his tropical homeland to Europe, where ultraviolet radiation from the sun tended to be more scarce, the dark skin became a liability. It filtered out so much of the vitamin D-producing radiation that Neanderthal Man's skeleton could not develop properly. The result, according to Ivanhoe, was rickets.

Meanwhile, D. J. M. Wright of Guy's Hospital Medical School, London, suggested in 1971 that Neanderthal Man may also have suffered from congenital syphilis. Wright arrived at this theory by observing the bulging of Neanderthal's skull, the curvature of his long bones, and his saddle-shaped nose. In addition, Wright noted that syphilis and rickets often go together in populations with poor nutrition.

Los Angeles Man. The search for the remains of the oldest human residents of the Americas goes on, but the results to date have been generally disappointing.

Anthropology

Continued

The latest entry, "Los Angeles Man," was found at the corner of La Tijera and La Cienega boulevards near Los Angeles International Airport. Although actually found in 1936, Los Angeles Man was not really "discovered" until 1971, when his remains were subjected to carbon 14 dating analysis by Rainer Berger, assistant professor of anthropology and geophysics, University of California, Los Angeles. If it proves to be accurate, the date obtained—more than 23,600 years ago—would make Los Angeles Man the earliest human skeletal remains yet found in the New World.

The disappointing part of the find is that it consists solely of the occipital bone, the bone at the back of the skull. This is much too little to allow scientists to determine anything significant regarding the appearance of this early American.

Early man in Kenya. In 1967, a Harvard paleontological expedition digging west of Lake Rudolf in Kenya discovered the lower jaw of an australopithecine with one tooth, a molar, still in place. This specimen is known as the "Lothagam Mandible," from the Lothagam deposits where it was found. According to Bryan Patterson, the leader of the expedition, it has now been dated by the potassium-argon technique at 5.5 million years old. This makes Lothagam about 1.5 million years older than the oldest hominid (manlike) fossils from the Omo Valley of southern Ethiopia, and thus the most ancient australopithecine remains yet discovered.

The 1970 expedition to the Lake Rudolf area of Kenya, led by Richard E. F. Leakey of the National Museum of Kenya, recovered 16 hominid specimens, more than twice as many as the 1968 and 1969 expeditions combined. The specimens came from two regions—Koobi Fora and Ileret—and included eight mandibles, portions of three craniums, two arm bones, and four leg bones.

On the basis of their robustness, Leakey tentatively assigned six of the mandibles to the genus *Australopithecus*. The other two mandibles are much more delicate, and Leakey suspects these may represent some early form of the genus

"Why don't we just have leftovers tonight?"

James Stevenson © 1971 The New Yorker Magazine, Inc.

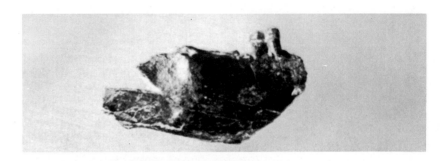

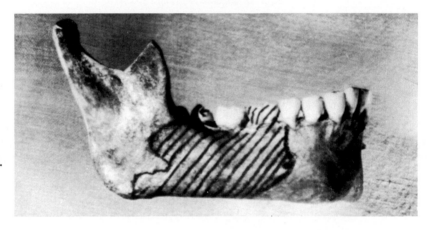

The Lothagam Mandible, *top*, a jawbone fragment found in Kenya, indicates that man lived as long as 5.5 million years ago. Shaded area of jawbone of modern man, *right*, shows where fragment fits.

Anthropology

Continued

Homo, to which modern man belongs. Two of the craniums, one with associated mandibular parts, are also thought to be australopithecine.

Medical anthropology. Americans may have their problems with smoking and lung cancer, but in the Far East it is betel nut chewing and cancer of the mouth. The betel nuts are chewed in a form called quids, which may also contain tobacco and other materials. Quids are popular in the Far East. However, they have been suspected for some time of being at least partly responsible for the high incidence of oral cancer there. Work with hamsters reported in 1971 by Kailash G. Suri, Henry M. Goldman, and Herbert Wells of the Boston University School of Graduate Dentistry, suggests that the tobacco in the quids, although not cancer-producing in itself, can enhance these qualities in the betel nut.

"Snowmobiler's back" is a new problem for doctors and for the patients who suffer from it. The condition is produced, according to Verne L. Roberts and Robert P. Hubbard of the University of Michigan Highway Safety Research Institute, when the vertebrae of the back are rammed together upon impact as the snowmobile driver "flies" his machine through the air.

Actually, the condition is not a new one. In 1971, anthropologists at the University of Chicago observed the characteristic fractures in Canadian Eskimo skeletons recovered from archaeological sites, some of which date back 1,000 years. About 45 per cent of the adult Eskimos studied had at least one compressed vertebra; and every vertebra below the neck in one middle-aged female showed some degree of fracture.

The cause of this plight in the Eskimos seems to be much the same as with snowmobilers—the jarring of a sled over rough ice, especially the compression ridges that build up near shore. With the introduction of snowmobiles to the Arctic, the problem will actually become more serious—perhaps not so much for the Eskimo driving the snowmobile, but certainly for the one on a springless sled towed behind, a practice common among Eskimos.   [CHARLES F. MERBS]

Archaeology

Discoveries in Thailand in 1971 add weight to the hypothesis that primitive men in various parts of the world independently discovered such cultural innovations as farming, pottery, and bronze working. Also in 1971, excavations for a pipeline in Alaska revealed many new archaeological sites, and a subway in Mexico City led to important archaeological discoveries.

Old World. Evidence for what may be the beginnings of plant cultivation as early as 10,000 B.C. was found at Spirit Cave in northern Thailand by Chester F. Gorman of the University of Hawaii. Gorman discovered carbonized seeds and skins of food plants including almonds, betel nuts, bottle gourds, water chestnuts, cucumbers, peas, and beans. Some of the seeds were larger than those from present-day wild varieties of the plants, so Gorman concluded that they must have been from cultivated plants. The material came from various levels in a stratum dated by radiocarbon dating techniques from about 10,000 to 6000 B.C. The results suggest that the cultivation of plants may have taken place in this area as early as, if not earlier than, in the Near East.

Stone tools found near the seeds were those of the Hoabinhian, a culture named after a site near Hoa Binh in North Vietnam where other artifacts were found. Partially polished tools and well-made pottery marked with cord impressions, dating from about 6600 B.C., were also found in Spirit Cave. These later tools were artifacts of a culture sometimes called the Late Hoabinhian. Its remains have been found as far west as Burma and are linked through pottery style with the earliest pottery-making groups in Formosa, northern China, and Japan.

Bronze casting may also have developed independently in Southeast Asia, possibly as early as about 3000 B.C. In 1965 and 1966, and continuously since 1968, archaeologists Wilhelm G. Solheim and Donn Bayard of the University of Hawaii, and Hamilton Parker of the University of Otago in New Zealand have made excavations at Non Nok Tha in northern Thailand. Among the items they have found is a socketed copper tool dated at about 3500 B.C.,

Two frescoes, of young boys and antelopes, were found on walls of building buried for 3,500 years under volcanic ash on Aegean island of Thera. The frescoes are now on display at the Athens Archaeological Museum.

Markings carved 32,000 years ago on a bone may represent a simple calendar, according to Alexander Marshack of Peabody Museum, Harvard. Drawing, *far right*, clarifies markings that indicate the lunar phases: new moon (black) is flanked by crescent moons, and full moon (white) is flanked by quarter moons. Arrows show the progression of symbols on the carving.

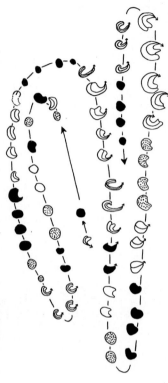

Archaeology

Continued

which makes it the earliest metal tool found in eastern Asia. They also found crucibles and double molds made of sandstone for casting axes—evidence of full-scale bronze working. These are dated at between 3000 B.C. and 2300 B.C., 500 to 1,000 years earlier than comparable evidence dates bronze working in India or China.

In Egypt, Manfred Bietak, leading a team of archaeologists from the University of Vienna, found an undisturbed rock tomb of the late Pharaonic period (about 1500 B.C.) at Asasif, near Luxor, in April, 1971. Since most of the ancient Egyptian tombs were robbed prior to their discovery by archaeologists, this find is unusual and important. The contents included small wooden statues, jewelry, and papyri bearing religious writings. The team also found a mummy in the innermost of three wooden *sarcophagi* (coffins).

Bietak also discovered another rock tomb close to the first. He believes the second tomb is that of a *vizier* (high government official) who lived in the 26th dynasty (from 663 to 525 B.C.). Stones

from the temple of Pharaoh Ramses II, who ruled in the 1200s B.C., were used in building the vizier's tomb, and one of these stones bears an effigy in relief of Queen Hatshepsut. Representations of this queen are rare, because the succeeding pharaoh ordered all representations of her destroyed.

In December, 1970, a group of scientists began to X ray some of the royal mummies in the Cairo (Egypt) Museum. The group was led by archaeologist James Harris from the University of Michigan and included Kent Weeks, an Egyptologist from the Metropolitan Museum of Art in New York City, and Arthur Storey, a physiologist and orthodontist from the University of Toronto. They discovered that gold ornaments, jewels, amulets, and statuettes had been hidden in the linen wrapped around the mummies. In some cases, such objects had been placed in the body cavity after the organs were removed. Harris said these objects were the first artifacts of undoubted royal association to be found since the discovery of King Tutankhamen's tomb in 1922.

Archaeology

Continued

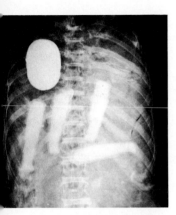

X ray of Queen Notmit mummy shows jewelry and small statues that were hidden inside her ribcage when she died in about 1080 B.C.

The mummies investigated included those of Pharaohs Seti I and Thutmose III and Queen Notmit. The Queen's body contains four figurines and a large scarab, said to be about the size of a saucer and covered with hieroglyphic inscriptions. The X rays also revealed bracelets and armbands thought to be made of gold, and sacred-eye amulets. Officials at the Cairo Museum have not yet decided whether to remove the jewels and ornaments, because removal would damage the mummies.

Iraqi paintings. A group of wall paintings, dating from the 800s B.C. were unearthed in April, 1971, among the remains of the ancient Assyrian capital of Nimrud. According to Fuad Safar, Iraq's Director-General of Antiquities, the paintings are still in excellent condition. Altogether, there are 20 stone panels, each measuring 13 by 10 feet. They were found in the ruins of the palace of King Ashur-Niza-Pal II. The red and black paintings were in a room that had served as the king's private bathroom, and they depict various religious ceremonies.   [JUDITH RODDEN]

New World. David Sanger of the National Museum of Canada excavated a large cemetery at Grand Lake, New Brunswick, which contained 60 graves of Red Paint Indians and about 400 ground and chipped-stone artifacts. The site, which resembles similar burial sites in Maine, has been dated about 1500 B.C. Sanger also directed excavations at 10 sites in the Passamaquoddy Bay region in New Brunswick that he was able to date by the indications of sea-level rise between 500 B.C. and A.D. 1000.

John P. Cook of the University of Alaska surveyed along the northern part of the proposed oil pipeline in Alaska, from Prudhoe Bay to Valdez. He located about 200 sites where archaeological remains might be excavated before the construction of the pipeline destroys the evidence. Some of the sites were scheduled for excavation, but others were not of sufficient interest.

Stuart Streuver of Northwestern University uncovered evidence that successive bands of prehistoric people had settled at the Koster site near Kampsville, Ill. At a depth of 28 feet, he found

Archaeologists digging at a site at Calico Hills in southern California found stone tools that may be 60,000 to 100,000 years old. However, these dates were hotly debated in the scientific community.

Archaeology

Continued

Markings on stone found in Bat Creek grave in Tennessee may be 3,000 years old and proof of pre-Columbian migration of Jews to New World.

Astronauts
See Space
Exploration

a grave containing the remains of a dog that had been buried by prehistoric men about 5200 B.C. The site was also occupied by hunters and gatherers from about 3000 to 2000 B.C. and about 1500 to 1000 B.C. Preliminary study indicates a close relationship between the occupants of this site and the Indians of the central Ohio Valley from about 3000 to 1000 B.C.

James L. Murphy of the Department of Geology, Case Western Reserve University, found the earliest evidence of maize agriculture in the eastern United States, in south-central Ohio. He recovered a carbonized 10-rowed ear of "Tropical Flint" corn (a cultivated variety) from the Daines Mound in Athens County, near the Ohio River. Radiocarbon dating established that the maize grew about 280 B.C. The earliest-dated cultivated maize from the southeastern United States was found at Fort Center, Fla., on the west side of Lake Okeechobee, by William H. Sears of Florida Atlantic University. This corn was grown about A.D. 300.

Richard A. Daugherty of Washington State University excavated six Indian dwellings at the Ozette site on the northern Washington coast. These six large wooden houses, not yet dated, were buried by a mud slide, and the mud preserved their planks, timbers, and much of their contents. Among the house furnishings were baskets, a loom, spindle whorls, sealing clubs, canoe paddles, art objects, bows and arrows, whaling and sealing harpoons, wooden boxes, and ceremonial objects probably associated with whaling.

Mexico. René F. Millon of the University of Rochester continued a study of the archaeological evidence for the geographic limits and population density of the urban center of Teotihuacán, Mex. The map he constructed indicates that Teotihuacán was similar in its social, economic, and religious configuration to the great Aztec capital of Tenochtitlán, which thrived during the early 1500's. Millon estimates that Teotihuacán must have had a population of 125,000 from A.D. 400 to 600, when the city covered some 20 square kilometers (8 square miles). The city contained a number of major religious centers, large one-story compounds where obsidian carving and other industries were con-

ducted. It also had centers where people who had emigrated to the city lived, including one of people from the Valley of Oaxaca, Mexico.

Great quantities of archaeological material were recovered, most of it from the Aztec period, during excavations for a new subway in Mexico City. A small temple and courtyard were found below Guatemala Street. The temple shows architectural details clearly linking it to the preceding Teotihuacán style, as well as to temple structures of the late Aztec period. In addition, mural paintings of religious scenes in red, yellow, orange, green, blue, and black were found. These were badly deteriorated, however, according to Jordi Gussinger of the National Institute of Anthropology in Mexico City.

South America. Excavations of rock shelters in the Bogotá highlands of Colombia provided evidence of human occupation some 12,000 years ago. The archaeological study was conducted in 1970 and 1971 by Thomas van der Hammen of the University of Amsterdam, G. Reichel Dolmatoff and Gonzalo Correal of the Colombian Institute of Anthropology, and Wesley R. Hurt of Indiana University. The scientists found many stone tools, including unifacial percussion flaked scrapers, blades, and scrapers with no projectile points.

In Ecuador, William J. Mayer-Oakes of the University of Manitoba found stone tools made of obsidian near San José. He believes the tools were used more than 12,000, and possibly 15,000, years ago.

In the central Peruvian highlands, Thomas Lynch of Cornell University found evidence of hunting camps that were occupied about 11,500 years ago in the Callejon de Huailas, a valley east of Chimbote. He also found evidence that a group of people lived at Guitarrero Cave about 12,500 years ago.

In Argentina, Marcelo Bórmida of the University of Buenos Aires reported that a group of excavated sites known as the Neuquense Chopper complex, along the terraces of the Rio Neuquén near the Andes Mountains, has now been dated at about 10,000 B.C. There, searchers found unifacial scrapers and flakes in the earliest level of a stratified cave, which was inhabited about 9,000 years ago. [JAMES B. GRIFFIN]

Astronomy

Planetary Astronomy. It has long been a goal of astronomers to photograph planets and galaxies from high altitudes, without earth's atmospheric interference. On March 26 and 27, 1970, Stratoscope II, a radio-controlled, balloon-borne, 36-inch telescope system, was launched from the National Center for Atmospheric Research (NCAR) Scientific Balloon Flight Station at Palestine, Tex. It obtained many such super-resolution pictures of Uranus and Jupiter, and of Jupiter's closest satellite, Io.

The Uranus pictures are remarkable in that they show a planet devoid of features. This finding was a mild surprise to visual observers who had occasionally reported seeing markings and cloud belts. The Princeton University experimenters—Robert Danielson, Martin Tomasko, and Martin Schwarzchild—presented a preliminary interpretation of the pictures at the third annual meeting of the Division for Planetary Sciences of the American Astronomical Society, in Tallahassee, Fla., in February, 1971. They reported that the observed limb darkening (the apparent fall off in brightness between the center and the edge of the planet) was very similar to what would be expected for a deep, cloudless molecular atmosphere.

This result confirms a conclusion reached from an examination of the spectrum of Uranus by Michael Price of the Illinois Institute of Technology Research Institute, Michael B. McElroy of Harvard University, and Michael J. S. Belton of the Kitt Peak National Observatory. Their report, in February, 1971, indicated a maximum absorption in certain spectral lines of molecular hydrogen. Since there is a limit to the apparent amount of hydrogen that can be observed, finding this maximum strength of hydrogen absorption on Uranus means that cloud particles, which would strongly reduce the hydrogen absorption, cannot be present.

The atmosphere of Uranus appears to be a sphere of pure gas—mainly molecular hydrogen (H_2) with a small admixture of methane (CH_4). Measurements at radio wave lengths reported in 1970 by Ivan Pauliny-Toth and Kenneth I. Kellerman of the National Radio As-

The launch balloon that lifted the 36-inch Stratoscope II telescope, left, off its Texas pad is inflated prior to flight. A larger main balloon, inflated at 10,000 feet altitude, carried the 4-ton telescope over 80,000 feet above the earth. Stratoscope II took high-resolution pictures of Uranus and Jupiter.

Astronomy

Continued

Russia's Venera 7 on Dec. 15, 1970, became the first spacecraft to land and operate on the surface of Venus. This replica of Venera 7 was on display at a Moscow exhibition.

tronomy Observatory indicate, however, that ammonia may be present in very deep levels of the atmosphere that cannot be probed by visible light.

Pictures of Io. Among the fascinating pictures returned by Stratoscope II telescope were some unique shots of Jupiter's inner satellite Io. Io has been of special interest in recent years because it has a special property of being able to "trigger" large bursts of radio energy from Jupiter's vast magnetosphere. The French observer Bernard F. Lyot noted many years ago that Io is a special object in the solar system because it possesses regions around the poles that are much darker than the rest of the satellite.

The super-resolution pictures obtained by Stratoscope II confirmed Lyot's observation. The Stratoscope experimenters also obtained a sequence of pictures of a total eclipse of Io as it moved into the shadow of its parent planet.

Venus landing. On Dec. 15, 1970, Venera 7, the fourth Soviet spacecraft to successfully penetrate the atmosphere of Venus, descended to the surface of the planet and operated in that inhospitable

environment for some 20 minutes. Since all of the previous Russian spacecraft ceased operating while still some considerable distance above the surface, this represents a "space first." The U.S. planetary spacecraft have all been non-landing, fly-by missions.

The difficulties of landing probes are underlined by the history of Venera spacecraft. In this latest effort, the Soviets added 220 pounds of thermal insulation to the descent probe and then had to remove all scientific experimental equipment–with the exception of atmospheric temperature and pressure sensors–in order to keep the spacecraft at its original weight.

The descent through the Venusian atmosphere was not without incident. At about the 400°F. level in the atmosphere, the parachute configuration was changed in order to slow the craft's descent through the dense regions of the atmosphere near the surface. The maneuver succeeded. Shortly afterward, however, there was an abrupt, and unexplained, speed-up followed by increasing oscillation of the falling craft. No

Astronomy

official explanation of this behavior has been made, but it would appear that there must have been some drastic failure in the parachute system. Nevertheless, the descent probe landed and operated on the surface of the planet, and the mission must be judged a success. The surface temperature was found to be about 900°F. and the pressure about 90 times that at sea level on the earth. See SPACE EXPLORATION.

Structure of Venus' clouds. The nature and structure of the Venus clouds was the subject of much discussion during the past year. A remarkable technical advance was made by a group of Harvard University and Smithsonian Astrophysical Observatory astronomers – Nathaniel Carleton, Wesley Traub, and Richard Wattson – who, for the first time, were able to measure the exact shapes of spectral lines formed in the atmosphere of Venus. They obtained extremely high spectral resolution with a specially designed spectrometer. Their early results have already indicated the existence of layering in the clouds.

James E. Hansen and Albert Arking of NASA's Institute for Space Studies in New York City made a detailed mathematical analysis of the large amount of data on the polarization of light from Venus that has accumulated over recent years. In February, they reported that the planet's cloud particles must be spherical, probably liquid, and have a refractive index between 1.43 and 1.47. The most remarkable fact is that no known material fits all three of these specifications simultaneously. Thus the problem of the composition of the Venusian clouds remains as enigmatic as ever.

The Phobos image. A remarkable accomplishment of the Mariner 7 spacecraft that flew by Mars in August, 1969, has been the discovery of the image of the satellite Phobos on one of the pictures of Mars as the satellite passed between the spacecraft and the planet. Two interesting facts have emerged from this photograph: (1) the satellite is noticeably distorted from a spherical shape, and (2) its *albedo* (ability to scatter and reflect light) is very low. In May, 1970, Bradford A. Smith of the New Mexico State University Observatory noted that the dimensions of the satellite are a meager 11 by 13 miles and its albedo is 0.065. [MICHAEL J. S. BELTON]

Stellar Astronomy. The remarkable spectrum of star HR 465 created considerable speculation during 1971. Nearly 10 years ago, William P. Bidelman, using the 120-inch telescope at the Lick Observatory in California, found that many of the star's spectral lines were caused by one or more of nine rare-earth elements, but he was unable to identify about 25 per cent of the lines.

Additional identification came in July, 1970, from the University of Michigan, where Margo F. Aller and Charles R. Cowley matched the wave lengths of some of the unidentified stellar lines with laboratory spectrograph lines from the radioactive element promethium.

This discovery offers rare evidence for a radioactive element in the stars. Because the half-life of the longest-lived isotope of promethium is only 18 years, promethium's presence in the atmosphere of HR 465 indicates it must have been created by some unknown process very recently.

Thin solar layer. A thin, previously unobserved layer in the sun's atmosphere was found by Harvard University astronomers who analyzed solar ultraviolet radiation. The measurements of emission lines of ionized carbon and neutral hydrogen atoms had been obtained by the sixth Orbiting Solar Observatory (OSO-6) satellite, launched in August, 1969.

Eugene H. Avrett of Harvard reported at a conference held Dec. 2, 1970, that the newly found layer, called an "isothermal plateau," lies about 1,200 miles above the visible surface of the sun. Both above and below this layer, the solar temperature rises steeply with height, but within the 90-mile-thick plateau, the temperature is nearly constant at 20,000°C.

Interstellar molecules. As radio astronomers search the sky with improved receivers, they are making new entries on the once short list of interstellar molecules. Hydrogen cyanide, which some scientists consider to have been a key ingredient in the early evolution of life, was detected in June, 1970, by Lewis E. Snyder and David Buhl, of the National Radio Astronomy Observatory (NRAO), using the observatory's 36-foot Kitt Peak, Ariz., antenna. In October, radiation from formic acid was discovered by Benjamin Zuckerman,

New Galaxies Through The Dust

The Maffei galaxies, obscured by the Milky Way dust, appear as small, faint, diffuse patches. Maffei 1, a giant elliptical galaxy, is the fuzzy image in center, *below*. Maffei 2 is a spiral galaxy. Its faint nucleus, *right*, is a weak radio source

Italian astrophysicist Paolo Maffei's article, "Infrared Object in the Region of IC 1805," which appeared in the October, 1968, *Publications of the Astronomical Society of the Pacific*, initially attracted little attention in the astronomy department at the Berkeley campus of the University of California, where I teach. I noticed it, but from what I understood, these Maffei objects were apparently nonstellar, quite diffuse, and very, very red. They were not obviously important.

Luckily, several graduate students in the department, especially Robert Landau, Gary Grasdalen, and Malcolm Raff, were interested in these new infrared objects. They began photographing Maffei 1, using the 30-inch telescope at the nearby Leuschner Observatory. Landau convinced me that we should use the more powerful 120-inch telescope of the Lick Observatory.

On a clear early autumn night in 1969 at Lick, we obtained photoelectric spectra of the nucleus of Maffei 1 in both the red and near-infrared regions. At no point could it be detected visually. We had to "home-in" on the infrared signal to guide the telescope. To my surprise, the data clearly indicated that we had a spectrum that could only arise from a heavily obscured galaxy. The object had a red band of titanium oxide typical of giant elliptical and spiral galaxies.

We were convinced that if Maffei 1 – which is within one-half degree of the dusty plane of the Milky Way – could be seen in an obscured location away from the galactic dust, it would appear very bright. My Berkeley colleague Ivan R. King compared photometric brightness profiles of Maffei 1 and normal elliptical galaxies, and found that the system would be near magnitude 6 (visible to the unaided eye), if it were located away from the intervening dusty clouds.

By this time, we realized something important was in the works. We called upon our colleagues at Berkeley, the California Institute of Technology (Caltech), and at the 200-inch Hale telescope on Mount Palomar to help us complete some of the preliminary work before the beginning of the rainy weather season.

James E. Gunn and Wallace L. W. Sargent of the Hale Observatories ob-

tained image tube spectra of Maffei 1. J. B. Oke of Hale scanned the entire optical spectrum, and Gordon Garmire of Caltech and Gerry Neugebauer of Hale obtained important infrared brightness of both Maffei 1 and 2 with several telescopes.

Luckily, rainy seasons need not stop the work of radio astronomers. Nannielou H. Dieter used the University of California's 85-foot radio antenna at Hat Creek. From her data and the photographic and spectroscopic data from the Leuschner, Lick, and Hale telescopes we deduced that Maffei 1 was a giant elliptical galaxy from 50,000 to 100,000 light-years in diameter. Our estimates indicated that it was one of our closest neighbors in space—merely 3.3 million light-years away.

In late 1970, John Bahcall of Caltech and I took long-exposure infrared photographs of Maffei 1 and 2 with Palomar's 48-inch Schmidt telescope. Gunn also obtained an image tube photo of Maffei 2. These pictures show more faint outer detail than we could see previously. Maffei 2, about the same distance as Maffei 1, is probably spiral, rather than elliptical.

If the two "new" galaxies are really close, they would be members of the association of galaxies called the local group. If the Maffeis are farther away, they could well dominate a small cluster of galaxies.

I believe we will learn much from more intensive and sophisticated study of Maffei 2. In fact, French radio astronomers have detected the 21-cm hydrogen line in Maffei 2, and their measurements prove that the galaxy has a spirallike rotation.

Clearly, Maffei's original discovery has led to a pleasant surprise. But there is much work yet to be done. The Maffei objects have provided one lesson for me—I will not ignore modest-sounding articles that do not seem relevant to my interests. [HYRON SPINRAD]

Astronomy

Continued

John A. Ball, and Carl A. Gottlieb observing with the NRAO 140-foot radio telescope at Green Bank, W. Va. Calling attention to the organic nature of this molecule, the researchers noted that it is usually found in "ants, bees, and nettle plants." See MOLECULES IN SPACE.

Gum Nebula. A team of astronomers including John C. Brandt and Theodore P. Stecher of the Goddard Space Flight Center and David L. Crawford of Kitt Peak National Observatory, reported in February, 1971, that the Gum Nebula, a very large and faintly glowing gaseous region in the southern sky, was larger and more distant than previously thought. Indeed, with its diameter of over 2,300 light-years, it is probably the largest nebula in the Milky Way.

Evaluating the number of free electrons in the nebula, the astronomers speculated that an enormous amount of energy must have been required to strip them from their parent hydrogen atoms. The ordinary energy-releasing processes in known types of nebulae are inadequate to produce an object this large. The only likely source of enough energy to ionize the atoms would be a supernova, or exploding star.

Dwarf galaxies. A normal galaxy consists of a vast assemblage of stars and star clusters, permeated by a thin medium of gas and dust. Occasionally, the gas and dust are concentrated into nebulae, in which new stars form. In December, 1970, the dwarf galaxies I Zw 0930+55 and II Zw 0553+03 were reported to be similar dense gas clouds, in which an unusual star formation process is at work.

Telescopic photographs revealed that II Zw 0553+03 consists of a relatively small core (estimated at 620 light-years in diameter) from which faint plumes extend far into space. I Zw 0930+55 is actually a pair of physically associated dwarf galaxies.

Measuring light from these objects with the 200-inch telescope on Mount Palomar in California, Wallace L. W. Sargent and Leonard Searle found a vital clue—the dwarf galaxies are much bluer than normal galactic systems, which indicates that the dwarfs contain a substantially greater proportion of the

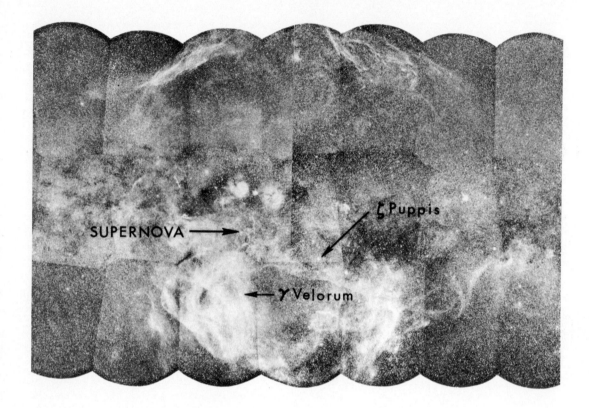

SUPERNOVA ⟶

ζ Puppis

← γ Velorum

Astronomy

Continued

The Gum Nebula, one of the largest objects in the Milky Way, has been identified as a new type of galactic object. The center region continues to glow today, 11,000 years after ionization by ultraviolet and X-ray emission from an exploding supernova.

young, massive blue stars than does a normal galaxy.

A radio telescope survey at the 21-centimeter wave length of neutral atomic hydrogen recorded emission from II Zw 0553+03, and their measurements led to a surprising interpretation by the California astronomers. According to their theory, which also applies to I Zw 0930+55, the volume of the dwarf galaxy is filled by a relatively dense neutral gas, averaging 500 hydrogen atoms per cubic centimeter, and the hot blue stars condense from this gas very rapidly.

Quasars. At the Hale Observatories, James E. Gunn studied the relatively nearby quasar PKS 2251+11. He showed that it is associated with a cluster of galaxies, whose brightest member has roughly the same red shift as the quasar. This is the first "classical" quasar – as distinguished from a quasarlike object – to be identified with a cluster of galaxies.

Since the cosmological interpretation of galaxy red shifts is well-established – they are caused by the general expansion of the universe – the implication of

Gunn's work is that quasar red shifts are also cosmological in origin.

A major contribution to the physical theory of quasars and radio galaxies was published in January, 1971, by Martin J. Rees at the University of Cambridge in England. Attributing the energy source of these powerful radiators to the rotational braking of collapsed magnetized stars in their central regions, he outlined the way in which low-frequency electromagnetic waves can accelerate electrons to velocities near the speed of light. The electrons then emit the observed radio waves. This can occur at great distances from the galaxy and thus may account for the enormous separation of the twin emission lobes that are typically found in the radio galaxies.

New X-ray stars. Scientists at American Science and Engineering, Inc., led by Minoru Oda, announced the discovery in April, 1971, of periodic X-ray pulsations from the galactic source Cygnus X-1. A similar effect was soon found in the source Centaurus X-3. As neither is a radio pulsar, a new class of X-ray stars may exist. [STEPHEN P. MARAN]

Astronomy

Continued

High Energy Astronomy. X-ray astronomy is more than 10 years old, but the observing time from rockets has totaled only a few hours. However, a new era in this branch of cosmic diagnostics began on Dec. 12, 1970, with the launching of the first orbiting observatory entirely devoted to the study of cosmic X-ray sources. It has been scanning the sky ever since.

The 300-pound observatory is the first in a series of small astronomy satellites of increased capability and complexity built by the National Aeronautics and Space Administration (NASA). It was put into space by a team of Italian engineers from the University of Rome Aerospace Center. The X-ray detectors had been developed by scientists from American Science & Engineering, Incorporated (ASE) who are among the world's leading X-ray astronomers.

The scientists had decided to put the satellite, Explorer 42, into a circular orbit more than 300 miles above the equator. Any other orbit would bring the satellite over the South Atlantic from time to time, where its sensitive X-ray "eyes" might be disturbed by the Van Allen radiation belt, which comes close to the earth in that region.

Because an equatorial orbit can be most easily achieved from a launch pad near the earth's equator, it was decided to launch the observatory from the coast of Kenya. From a launch pad erected on 20 steel legs in the Indian Ocean, an American Scout rocket carried the satellite aloft. Because Kenya, the host country for the satellite launch, was celebrating the seventh anniversary of its independence, the launching team christened the satellite Uhuru, the Swahili word for freedom.

Uhuru now circles the earth every 96 minutes and appears to be one of the most successful ventures in the scientific space program. Uhuru has already accumulated many times more data than have been obtained since the discovery of X-ray stars in 1962. Although only a small amount of the continuous stream of data that will ultimately become available from Uhuru has been analyzed, a new X-ray picture of the universe is beginning to emerge.

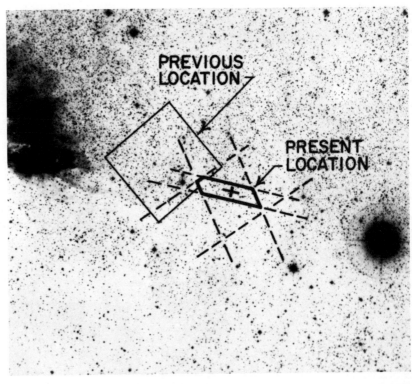

The general location of a pulsating X-ray star was determined more precisely (within the heavily outlined area) in 1971 by a new X-ray satellite, Explorer 42.

Astronomy

Continued

The 24 parabolic frames of India's first radio telescope provide a signal-collecting area of 93,000 square feet. The new telescope stands at Ootacamund in the Nilgiri Hills of southern India.

Before the Uhuru launch, some 40 sources of cosmic X rays had been spotted and astronomers had observed a uniform X-ray glow in the entire sky. They did not know the positions of many of the sources accurately, and this made identification with optical sources very difficult. Uhuru has found many new sources, and provided the astronomers with much more precise positions for the previously known X-ray sources.

Heavenly glow. A number of fascinating details began to emerge. Five physicists from ASE reported in February, 1971, that several odd galaxies are very copious X-ray emitters. In some cases, this confirmed earlier rocket observations by a group of X-ray astronomers at the Naval Research Laboratory.

Uhuru also observed that, in the X-ray energy range from 2,600 to 6,900 electron volts, the quasar 3C-273 emits 3×10^{38} watts of power. This fabulous flux of high-energy radiation is 750 billion times the total power output of the sun. If this is typical for quasars, then the X-ray glow of the heavens may be due solely to the quasars' emission.

Another fascinating observation was that some clusters of galaxies appear as extended sources of X-ray radiation. These could be a new type of cosmic X-ray source, and of great importance in understanding why galaxies cluster.

New cosmic object. Then, in May, 1971, seven ASE scientists announced that Uhuru had observed a completely new type of cosmic object in the X-ray source Cygnus X-1. This source lies in the Milky Way several thousand light-years from the earth. It radiates in the X-ray range at about a thousand times the energy of the sun. The scientists discovered that the X rays of Cygnus X-1 are emitted in pulses. And they concluded, on the basis of fluctuations in the signal, that one quarter of its X-ray radiation must come from a region smaller than the size of Texas. However, this object is quite different from the Crab Nebula pulsar. X-1 is not the remnant of a supernova and, unlike the Crab, it does not emit radio radiation. The nature of this new and fascinating compact high-energy object is still an enigma.　　[E. L. SCHÜCKING]

Cosmology. A group of nine scientists reported a fascinating observation in 1971 that adds to the mystery of quasars. The scientists are from the Massachusetts Institute of Technology (M.I.T.); the Goddard Space Flight Center, Greenbelt, Md.; the Jet Propulsion Laboratory, Pasadena, Calif.; and the University of Maryland. These radio astronomers used position measurements of the quasars 3C-279 and 3C-273 as a check on Einstein's theory of gravitation, i.e., general relativity.

On October 8 each year, 3C-279 disappears behind the sun. As it nears the sun's rim, the quasar's radio signal is deflected by solar gravity. The angle between the radio beams from 3C-279 and its neighbor 3C-273, which is not in the path of the sun, gives very precise measures of the light deflection.

These observations, published in April, 1971, did much more than confirm Einstein's theory. They led to a stunning discovery about 3C-279.

The radio astronomers' telescopes consisted of the Haystack antenna in Massachusetts, operated by M.I.T., and the Goldstone telescope in California, operated by the California Institute of Technology (Caltech). The two instruments, 2,400 miles apart, observed microwaves from the two quasars at a wave length of 3.8 centimeters and registered their new data on 600 miles of magnetic tape.

From this gigantic pile of information, astronomers used computers to synthesize the radio images of the two quasars. The result for 3C-279 indicated a double quasar with an angular distance between the components of 0.00155 ± 0.00005 second of arc. This angle is equal to that subtended by a pinhead 1,000 miles away. Never before had any object in the sky been looked at with such magnifying power.

Double structures had been observed in quasars before, but never on so small a scale and with such staggering accuracy. And it was the accuracy of the measurement that led to the great surprise. When the scientists remeasured the distance between the two components of the quasar in February, 1971, they found that this distance had increased by 10 per cent.

Quasar 3C-279 is believed to be about 3 billion light-years from the earth. A simple calculation using Einstein's theory of general relativity showed that at least one component of the quasar had been ejected with more than 99.5 per cent of the velocity of light.

According to the theory of special relativity, the mass of a body and also its energy become infinite if its speed approaches that of light. In this case, the energy of the explosion must have been staggering – an energy corresponding to at least 10 times the total energy $(E=mc^2)$ that the object would have if at rest.

Astronomers had previously found some indirect evidence that quasars eject objects with speeds comparable to the velocity of light. But the new observations of quasar 3C-279 show that extremely violent motions take place. Perhaps one should say "took place," because what we see now actually happened 3 billion years ago.

Quasars are old. They appear to be objects of the distant past, older and as extinct as the dinosaurs. This conclusion emerged from a careful survey and field count of quasars carried out by Maarten Schmidt of Caltech. Schmidt, who identified the first quasar in 1963, made a systematic study of quasars in a small region that may be representative of the whole sky. With the 200-inch Hale telescope on Mount Palomar in California, he identified 23 quasars in the small region. By carefully measuring their brightness and red shifts, Schmidt derived that their number increased with distance. Because looking out into space also means looking back into the past, he found how the number of quasars had changed during the evolution of the universe.

Extended to the entire sky, his study showed that quasars were most numerous about 8 billion years ago – about 2-billion years after the universe was created. Since then, their number has rapidly declined. Schmidt estimates there are 15 million quasars in the sky up to the limit of detection of the Hale telescope. These quasars were shining in the past and can all be seen, theoretically, today. However, an instantaneous inventory in the same region of the universe carried out now would yield only about 35,000.

One of the greatest puzzles about quasars is why there were practically

Astronomy

Continued

none, or at least very few, in the first 2-billion years of universal history.

Since the first quasars were discovered, a number of astrophysicists (among them Fred Hoyle, Geoffrey Burbidge, and William Fowler) have questioned whether these enigmatic objects were actually at cosmological distances. Their "local" theory of quasars assumed that these brilliant sources of radiation had been ejected by galaxies and were at least 100 times closer to home than assumed by the cosmological hypothesis put forward by Schmidt. While Schmidt believed that quasars are up to 100 times as bright as the most luminous galaxies, Hoyle predicted their luminosity might only reach one-hundredth that of a giant galaxy.

During recent years, astronomers have found evidence that favors the Schmidt theory. It was discovered that quasars have a number of properties in common with certain types of galaxies whose cosmological nature is not in doubt. These nuclear and Seyfert galaxies do not, however, share the most outstanding property of a quasar – that

of being the brightest object in the world's visible range. The conclusive proof for the cosmological nature of quasars was still missing.

The proof was supplied in the constellation Pegasus in March, 1971, by James Gunn of the Hale Observatories. Gunn found that the quasar PKS 2251+11 occurs in a cluster of galaxies with the same red shift as the quasar.

There is no doubt that the Gunn object is a bona fide quasar. It is 15 times brighter in the visual range than the brightest galaxy known, it is extremely blue, it has a starlike image under high resolution, and it is a strong radio source. Its red shift is 32.3 per cent, corresponding to a velocity of recession of 50,000 miles a second.

When Gunn photographed the quasar with the 200-inch Hale telescope, using an image converter to increase the light-gathering power of the huge mirror, he saw that the quasar was surrounded by a group of galaxies. The decisive check was now to discover whether these galaxies showed the same red shift as the quasar.

Observations by radio telescopes (yellow line) that are 2,400 miles apart indicated there are separate components in quasar 3C-279. Observations four months later (black line), *right*, made it appear as if the components were flying apart at 10 times the speed of light, *far right*.

Quasar Mystery: Faster than Light?

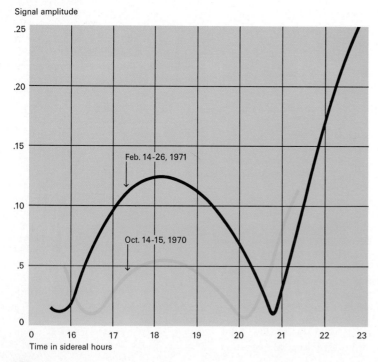

Signal amplitude

Feb. 14-26, 1971

Oct. 14-15, 1970

Time in sidereal hours

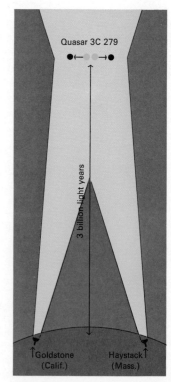

Quasar 3C 279

3 billion light years

Goldstone (Calif.)   Haystack (Mass.)

Astronomy

The brightest galaxy in the cluster did show a red shift of 33 ± 1 per cent. This clinched the case. There is, however, a puzzling feature in this galaxy–a large spiral some 300,000 light-years from the quasar. The spiral radiates a bright light of the type that originates in a hot hydrogen gas. However, as Gunn pointed out, this puzzle can be solved if one takes into account that the hydrogen gas in the galaxy is heated by the neighboring quasar. In the March, 1971, *Astrophysical Journal Letters* Gunn wrote: "We conclude that the object PKS 2251 + 11 satisfied the most stringent requirements for inclusion in the class of quasi-stellar objects and that it is almost certainly at the same distance as a galaxy of essentially the same red shift."

New energy sources of quasars and nuclei of galaxies were proposed by three physicists–John Bahcall; Curtis G. Callan, Jr.; and Roger Dashen–at the Princeton, N.J., Institute of Advanced Studies. The scientists took up a suggestion made earlier by a number of their colleagues that quark reactions might play an important role in providing the quasars' fantastic energy.

Quarks are hypothetical building blocks of nuclear matter. Protons and neutrons are believed to consist of three quarks, which carry fractional electrical charges. Antiprotons and antineutrons consist of corresponding triplets of antiquarks. Quarks have not yet been produced in the laboratory, but there is some evidence of their existence.

The Princeton physicists discussed fusion reactions for quarks in a manner similar to Hans Bethe and Carl von Weizsacker 35 years ago when they discussed fusion reactions in the sun, long before these reactions could be observed in the laboratory.

The results were quite surprising. Quark fusion can convert 50 per cent of their rest mass into photons. The quark-burning should take place at a temperature of several million degrees centigrade. Ordinary fusion of hydrogen converts only 0.7 per cent of the rest mass into radiation. The authors suggested several tests to find quarks, but none have been found. [E. L. SCHÜCKING]

Biochemistry

In the spring of 1970, scientists proved that an enzyme called RNA-directed DNA polymerase used ribonucleic acid (RNA) to make deoxyribonucleic acid (DNA). This is a reversal of what previously had been thought to be a one-way reaction–DNA to RNA. The reversal has been commonly called Teminism, after biochemist Howard M. Temin of the University of Wisconsin, who first discovered it. The process has great implications and has triggered research throughout the world. See Close-Up.

Satellite DNA. In 1971, the base sequences of the chains of a satellite DNA were determined for the first time. As much as 30 per cent of the DNA in the cells of many plants and animals has satellite properties, such as density differences, that indicate it is different from the remainder of the organisms' DNA. This strange DNA, called satellite DNA, is found in species as different from one another as mice and men. In 1970, a team of biochemists at the University of Texas, Houston, reported that in a survey of 90 different mammals, they had found 48 of them whose cells contained satellite DNA.

Despite its widespread presence, however, satellite DNA has resisted all efforts to determine its function. All DNA is composed of two chains that are held together side by side. The links of the chains contain units called bases. Biochemist E. M. Southern at the University of Edinburgh in Scotland analyzed the so-called alpha-satellite DNA of the guinea pig, one of three known forms of satellite DNA found in the animal. Southern discovered the sequence of bases to be repetitive–cytosine, cytosine, cytosine, thymine, adenine, adenine–repeated over and over in one chain, and guanine, guanine, guanine, adenine, thymine, thymine repeated over and over in the other.

Most of the DNA in a cell is genetic material. Its bases are arranged in a code that specifies the production of enzymes and other proteins vital to cell chemistry. Through a complex decoding and synthesizing process, the bases provide instructions that determine the order in which amino acids are linked

Biochemistry
Continued

to form proteins. But it seems certain that the bases of alpha-satellite DNA do not direct the formation of any protein. It is improbable that the cell should need a protein that is simply a series of repeating amino acid sequences.

If satellite DNA does not specify the production of proteins, perhaps it plays some structural role in the chromosomes, of which DNA is a part. Evidence for such a role came in 1970 from experiments by groups headed by biochemists K. W. Jones at the University of Edinburgh and Joseph G. Gall at Yale University. They found that satellite DNA is concentrated in *centromeres* (regions of chromosomes that are about to divide in cells reproducing by mitosis, or cell division). Each centromere holds two chromosome halves together until the two new daughter cells are just about to be formed.

Perhaps all chromosomes contain satellite DNA in the centromere region. If so, this may explain both the chemistry and the function of satellite DNA. Repetitive base sequences are particularly effective in holding DNA chains together, and would be ideal for this purpose just before chromosome halves are pulled apart in a dividing cell.

Other questions would still remain, however. Not the least of them: Why do the cells of some species have five or six times as much satellite DNA as other species, even though they do not have that many more chromosomes?

Isolating acetylcholine receptor. In many nerves, stimuli are carried beyond the nerve ending as a substance called acetylcholine. This substance is released and diffuses into the receptor cell, such as a muscle fiber cell or another nerve cell, which the first nerve cell serves. Acetylcholine causes the receptor cell to transmit the stimuli further.

In 1971, biochemists R. Miledi, P. Molinoff, and L. T. Potter at University College, London, reported that a snake toxin, or poison, blocked this transmission in the electric muscle tissue of the Torpedo eel. The toxin, alpha-bungaro, is also known to block the acetylcholine reaction at the junction of nerve cells and muscle cells in other vertebrates. The British scientists used the toxin to isolate the protein that normally receives the acetylcholine at the cell membrane and carries it into the cell.

The snake toxin acts by combining with this receptor protein and blocking its receptor sites. Therefore, the scientists purified the toxin, attached radioactive iodine to some of it, and then added it to some of the receptor tissue. The toxin-receptor complex that formed was then easily located by testing chemical fractions of the tissue for indications of the radioactive toxin.

The receptor protein is composed of units each of which appears to have one binding site for a molecule of the snake toxin. The units form groups called polymers. The most common polymer contains four units and has four binding sites.

Myelin membrane proteins. Myelin membrane surrounds the central section of nerve cells. Chemically, it is relatively simple. For example, about 80 per cent of the myelin of the nerves in the central nervous system is composed of only two proteins.

One of the proteins is called the Al protein. When Al is injected into some animals, including monkeys, rabbits, guinea pigs, and rats, it causes experimental allergic encephalomyelitis (EAE), a paralytic disease. In this disease, the animal seems to become allergic to its own myelin. Scientists have studied EAE for about 20 years, because of its similarity to human demyelinating diseases such as multiple sclerosis.

Complete amino acid sequences for Al protein in cattle and human beings were reported in 1970 by biochemist Edwin H. Eylar of The Salk Institute, San Diego, Calif. Eylar's structures show the cattle protein to have 170 amino acids and the human protein to have 172. The two proteins differ at 12 points, including the one at which the human product has its two extra amino acids. Biochemist P. R. Carnegie of the University of Melbourne, Australia, has worked out a partial structure for human Al that differs from Eylar's structure at several points.

Biochemists Fred C. Westall of The Salk Institute and Arthur B. Robinson of the University of California, San Diego; and Eylar, Juanita Caccam, and Jesse Jackson of The Salk Institute reported in 1970 that the entire Al protein is not needed to induce EAE. They found that injections of either of two portions of the protein, one 14 and the other only 9 amino acids long, caused the disease.

Reverse
Reaction

DNA ⇌ RNA

In 1970, biochemists and scientists in related fields for the first time accepted the idea that ribonucleic acid (RNA) can be a template, or pattern, for the synthesis of deoxyribonucleic acid (DNA). Previously, most scientists had interpreted the so-called central dogma of molecular biology to mean that the characteristic DNA-to-RNA pattern of information transfer was never reversed. This pattern accounts for most genetic activity. Genes, which are made of DNA, direct the assembly of RNA molecules. Then, these molecules play a crucial role in producing enzymes and other proteins required by living cells.

However, some viruses, called RNA viruses, contain no DNA. Their genetic information is directly in RNA.

In 1964, I proposed that an RNA cancer-causing virus, called Rous sarcoma virus, reproduced its RNA through a DNA intermediate. (Most other RNA viruses reproduce their RNA through an RNA intermediate.) To me, this seemed the most logical conclusion to draw from some experiments I had done with the virus. Although I reported these experiments in support of my proposal, the evidence that proved my hypothesis to most scientists was the discovery of an enzyme in several RNA tumor viruses that was using RNA as a template for making DNA. The enzyme was first discovered in the spring of 1970 by Satoshi Mizutani in my laboratory at the University of Wisconsin and, at the same time, by molecular biologist David Baltimore at the Massachusetts Institute of Technology. The enzyme is called "RNA-directed DNA polymerase" or "reverse transcriptase."

All the implications of the discovery of DNA synthesis from RNA templates are still not clear. But, certainly among them are new avenues of cancer research. Thus, since its discovery, RNA-directed DNA polymerase has become the focal point of an enormous amount of experimental activity. The enzyme has been found in all RNA tumor viruses and in several RNA viruses that previously were not suspected of being tumor viruses.

In the fall of 1970, for example, molecular biologist Fu Hai Lin and virologist Halldor Thormar of the New York State Institute for Research in Mental Retardation, Staten Island, N.Y., showed that the enzyme was in Visna viruses, which cause a slowly progressing neurological disease in sheep. Other scientists added still more new viruses to the list of those containing the enzyme. It is still not clear whether the presence of the enzyme in these viruses indicates that they are cancer viruses, or that noncancer viruses also use the enzyme.

Experiments in my laboratory have uncovered evidence that the RNA viruses containing the RNA-directed DNA polymerase also contain several other enzymes that act on DNA. One of them is a DNA ligase, an enzyme that can rejoin the ends of broken DNA strands. The ligase might be used to join the DNA made from the viral RNA to the DNA of the host cell. This would not be surprising, because it has already been proved that a DNA cancer virus called SV-40 can insert its DNA into that of its host cell. It is this ability that has led scientists to hope that modified viruses may someday be used to cure some genetically determined diseases (see CELL FUSION AND THE NEW GENETICS).

Perhaps the most exciting report concerning RNA-directed DNA polymerase was made in December, 1970, by scientists at the National Cancer Institute (NCI) in Bethesda, Md. They detected RNA-directed DNA polymerase activity in human leukemia cells. To many, this seemed fairly good circumstantial evidence that an RNA virus was somehow involved in human leukemia. However, only a few months later, research groups from the NCI and the Institute for Cancer Research in Philadelphia, detected the enzyme activity in healthy human cells.

If this activity in uninfected cells is caused by RNA-directed DNA polymerase, rather than an abnormal reaction of another DNA polymerase, these results fit a theory I had proposed in the fall of 1970. I said that the manufacture of DNA from RNA probably plays a role in normal development, and that cancer not caused by viruses might be caused by derangements in this normal RNA-to-DNA process. At this early stage in our investigations, however, we can do little more than theorize about the relationship between RNA-directed DNA polymerase and cancer, and hope what we find will ultimately help prevent or cure the disease. [HOWARD M. TEMIN]

Biochemistry

Continued

In an effort to determine what portion of the Al protein causes EAE, the investigators determined the amino acid sequence of the two portions, and synthesized and tested 10 similar chains containing 11 amino acids each. They found that for a sequence to cause EAE, it had to contain the amino acid tryptophan followed by any four amino acids and, then, glutamine and lysine or arginine.

However, it appears that the tryptophan-containing sequence may not always be the only lethal site in the Al protein. Guinea pigs injected with Al in which the sequence has been modified do not develop EAE. But, rabbits injected with amino acids 43 to 91 (which do not include the tryptophan sequence) of the Al protein do develop the disease.

Also in 1970, biochemists Li-Pen Chao and Elizabeth Einstein at the University of California Medical Center, San Francisco, confirmed these results. They found that any chemical modification of the Al protein's tryptophan makes the protein safe for guinea pigs, but still active against rabbits. They also reported that guinea pigs injected with the modi-fied material become resistant to the effects of the total Al protein.

Parathyroid hormone. Two hormones, parathyroid hormone and calcitonin, control calcium metabolism. The parathyroid hormone increases the levels of calcium in the blood, and the calcitonin decreases it. In December, 1970, biochemists Bryan Brewer and Rosemary Ronan of the National Institutes of Health (NIH) in Bethesda, Md., reported that they had isolated cattle parathyroid hormone and determined its structure. The hormone is a single chain, 84 amino acids long.

In January, 1971, two groups of investigators, one from the NIH, the other from the Massachusetts General Hospital, Boston, reported that they had synthesized an end segment of the parathyroid hormone. The segment was 34 amino acids long. In tests of its biological potency, the synthetic segment equaled or exceeded a fragment of comparable size isolated from the natural hormone.

The potency of the synthetic segment is greater in the test tube than in test animals. This suggests that the portion

"I hope you're working on an antidote for my spell."

The Uncertain Structure of HGH

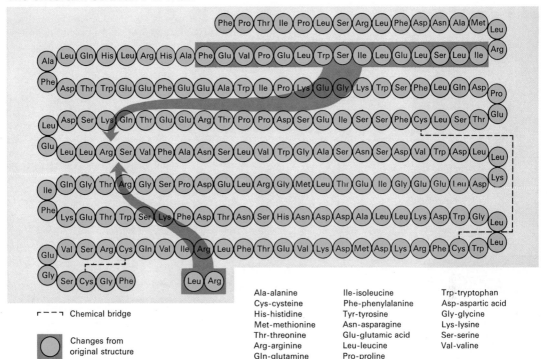

⌐ ⌐ ¬ Chemical bridge

○ Changes from
original structure

Ala-alanine	Ile-isoleucine	Trp-tryptophan
Cys-cysteine	Phe-phenylalanine	Asp-aspartic acid
His-histidine	Tyr-tyrosine	Gly-glycine
Met-methionine	Asn-asparagine	Lys-lysine
Thr-threonine	Glu-glutamic acid	Ser-serine
Arg-arginine	Leu-leucine	Val-valine
Gln-glutamine	Pro-proline	

Biochemistry

Continued

In 1970, Choh Hao Li reported synthesizing human growth hormone (HGH) to the specifications of a structure he had worked out in 1966. But Hugh D. Niall claimed the structure was wrong, and suggested two changes. In May, 1971, Li announced a revised structure similar to Niall's version.

Biology

See Biochemistry, Botany, Ecology, Genetics, Microbiology, Zoology

of the hormone that was missing in the synthetic segment may protect the whole hormone against rapid metabolic degradation in the body.

Human growth hormone (HGH). Biochemist Choh Hao Li and Donald Yamashiro of the University of California Medical Center, San Francisco, reported in 1970 that they had synthesized HGH. The structure they had followed was one reported in 1966 by Li, who was also the first person to isolate and purify the hormone. However, the scientists reported that their synthetic product had only 5 to 10 per cent of the physiological activity of the naturally occurring hormone.

Li's structure for HGH showed areas of great similarity to points in the structures of ovine prolactin (which promotes mammary growth and lactation in sheep) and human placental lactogen hormone (HPL). That is, the three have several areas in which identical or nearly identical amino acid sequences occur.

In March, 1971, biochemist Hugh D. Niall of Massachusetts General Hospital reported that comparisons of the three

structures suggested that the Li structure for HGH was incorrect. Niall proposed transposing the so-called tryptophan segment of 15 amino acids from positions 17 to 31 (where Li had them) to positions 77 to 91. He also suggested adding a leucine and an arginine (not included in Li's structure) between positions 76 and 77.

These changes result in a structure of 190 amino acids, instead of Li's 188. But, more important, the altered structure is closer than the original to those of ovine prolactin and HPL. In fact, Niall and co-workers have pointed out that the altered structure is about 80 per cent identical in amino acid sequence to HPL. Such close similarities in the internal structures of the molecules of these hormones strongly suggests that all three may have evolved from a common smaller chemical ancestor.

Li announced a revision of his structure for HGH on May 5, 1971, at the International Symposium on Growth Hormones. His revised structure proved to be close to Niall's alteration of the original. [EARL A. EVANS]

Books of Science

Here are 50 new science books of interest to general readers. The books, published in 1970 and 1971, were chosen by the director of libraries of the American Association for the Advancement of Science with the assistance of his professional reviewers.

Anthropology. *Never in Anger: Portrait of an Eskimo Family* by Jean L. Briggs. The author, an anthropologist, tells about her experiences living with an Eskimo family in northern Canada. (Harvard, 1970. 400 pp. illus. $15.95)

The Raft Fishermen: Tradition and Change in the Brazilian Peasant Economy by Shepard Forman. An interesting study of technological change among sailboat fishermen. This is especially good reading for social scientists. (Indiana, 1970. 173 pp. illus. $8.50)

The Slavs by Marija Gimbutas. A concise account of the beginnings and early history of Slavic peoples. (Praeger, 1971. 240 pp. illus. $10)

Thrice Shy: Cultural Accommodation to Blindness and Other Disasters in a Mexican Community by John L. Gwaltney. A blind anthropologist investigates life in a Mexican community, many of whose members are blind. (Columbia, 1970. 231 pp. $6.95)

Archaeology. *Oldest Man in America: An Adventure in Archaeology* by Ruth Kirk. Tells of the discovery of the earliest evidence of human life in America. The book goes back to the close of the last Ice Age, when men dwelt in caves. (Harcourt, 1970. 96 pp. illus. $4.95)

Astronomy. *Astronomy* by Donald H. Menzel. The author provides a thorough review of astronomy and its instrumentation and techniques. The book is a wide-ranging introduction for a beginning student or nonspecialist. (Random House, 1970. 320 pp. illus. $17.50)

Geology of the Moon—A Stratigraphic View by Thomas A. Much. Covers the shape and motion of the moon, explains remote sensing techniques, stratigraphy (the study of strata), and the relative and absolute ages of lunar materials. (Princeton, 1970. 324 pp. illus. $17.50)

This Island Earth edited by Oran W. Nicks. The earth seen from afar is vividly described and illustrated in color, thanks to perspective that was gained by the Apollo lunar missions. (NASA, SP-250, 1970; U.S. Government Printing Office. 182 pp. illus. $6)

Biology. *Freezing Point: Cold as a Matter of Life and Death* by Lucy Kavaler. Discusses how cold affects animals and plants, ways of adapting to cold, and experimental use of cold in medicine and for other purposes. (John Day, 1970. 416 pp. illus. $8.95)

The Friendly Beast: Latest Discoveries in Animal Behavior by Vitus B. Droscher. A skillful blending of classic and recent studies by anthropologists, zoologists, neurophysiologists, and comparative and experimental psychologists. Social behavior is emphasized. (Dutton, 1970. 248 pp. illus. $8.95)

The Seasons: Life and Its Rhythms by Anthony Smith. An explanation of the seasons in polar, temperate, and tropical areas, and discussions of migration, hibernation, internal clocks, and adaptive coloration. (Harcourt, 1970. 318 pp. illus. $12.50)

Botany. *Ingenious Kingdom: The Remarkable World of Plants* by Henry and Rebecca Northen. Tells of the diversity of the plant kingdom, focusing mainly on flowering plants, although other groups are discussed. (Prentice-Hall, 1970. 286 pp. illus. $7.95)

Chemistry. *Chemistry: A Cultural Approach* by William F. Kieffer. "Science and technology, specifically chemistry and its use by modern man, are part of our culture," says the author. His book describes chemistry in relation to all human thought and achievement. (Harper, 1971. 473 pp. illus. $10.95)

Communications. *The Information Machines: Their Impact on Men and the Media* by Ben H. Bagdikian. The book examines the impact of modern communication techniques and technology on society, including social innovations resulting from use of computers. (Harper, 1971. 366 pp. illus. $8.95)

The 26 Letters by Oscar Ogg. A complete and classic study that traces the fascinating history of the alphabet. The book includes new methods of printing and the new techniques involved in composition by photography, electronic typewriter, and computer. (Crowell, 1971. 294 pp. illus. $6.95)

Conservation. *The Politics of Pollution* by Charles J. Davis. An instructive account of pollution control from the viewpoint of a political scientist. The author presents helpful information on pollution-control standards, laws, funds for

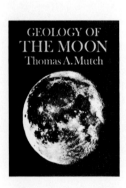

THE OLDEST MAN IN AMERICA

RUTH KIRK

GEOLOGY OF THE MOON
Thomas A. Mutch

Books of Science

Continued

achieving control, and law enforcement. (Pegasus, 1970. 243 pp. $6)

Drugs. *The Drug Epidemic: What It Means and How to Combat It* by Wesley C. Westman. A clinical psychologist who treats drug users gives a straight-talk account of the causes of drug dependence and narcotic addiction. He urges human renewal before urban renewal. (Dial, 1970. 173 pp. $4.95)

Earth Sciences. *A Century of Weather Service: A History of the Birth and Growth of the National Weather Service* by Patrick Hughes. An account of the growth and scope of weather services in the United States over the past 100 years, including many interesting anecdotes. (Gordon, 1970. 224 pp. illus. $10)

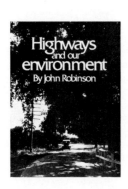

Invitation to Geology: The Earth Through Time and Space by William H. Matthews III. This intriguing work that reads like a novel takes the reader through the basic concepts of historical and physical geology. (Natural History, 1971. 158 pp. illus. $5.95)

The Negev: The Challenge of a Desert by Michael Evenari, Leslie Shanan, and Naphtali Tadmore. This book explains how botanists, geologists, archaeologists, agronomists, engineers, photogrammetrists, and historians collaborated in two decades of work to reconstruct an ancient agricultural system. (Harvard, 1971. 335 pp. illus. $15)

Ecology. *The Nature of Life: Earth, Plants, Animals, Man and Their Effect Upon Each Other* by Lorus and Margery Milne. A beautiful book about the evolving, ever changing earth and its plants and animals. (Crown, 1971. 320 pp. illus. $17.50)

The People Problem: What You Should Know About Growing Populations and Vanishing Resources by Dean Fraser. An interesting look at a problem that seems to be leading to inescapable disaster. (Indiana, 1971. 248 pp. $6.95)

Engineering. *Highways and Our Environment* by John Robinson. A thoughtful study of what is happening to the countryside because of the ever increasing complexity of highways and superhighways to accommodate more and more automobiles. (McGraw-Hill, 1971. 340 pp. illus. $24.50)

Man the Builder by Gosta E. Sandstrom. A history of building offering examples of various historical periods that express in their construction the social,

mental, economic, and technical capabilities of their age. (McGraw-Hill, 1970. 280 pp. illus. $16)

Forestry. *The Secret Life of the Forest* by Richard M. Ketchum. This book discusses physiology, diversity of trees, forestry management principles, and use of forest products. The colored illustrations are outstanding. (American Heritage, 1970. 108 pp. illus. $7.95)

Horticulture. *Gardening Indoors Under Lights* by Frederick H. and Jacqueline L. Kranz (revised edition). An up-to-date edition of a pioneering work useful to hobbyists, environmentalists, florists, and amateur gardeners. (Viking, 1971. 254 pp. illus. $7.95)

Medical Sciences. *The Mind of Man: An Investigation into Current Research on the Brain and Human Nature* by Nigel Calder. The author visited laboratories in eight countries to obtain a fascinating illustrated report. (Viking, 1971. 288 pp. illus. $8.95)

The Scalpel and the Heart by Robert G. Richardson. A chronological account of treating heart diseases by surgery. (Scribner's, 1970. 330 pp. illus. $8.95)

The Strange Story of False Teeth by John Woodforde. An interesting, comprehensive history of an important aspect of dental technology. (Universe, 1970. 137 pp. illus. $4.95)

Oceanography. *Deep Oceans* edited by Peter J. Herring and Malcolm R. Clarke. American and British scientists of various disciplines describe how they have studied the oceans, and what is needed for future research. (Praeger, 1971. 320 pp. illus. $18.50)

The Greatest Depths: Probing the Seas to 20,000 Feet and Below by Gardner Soule. An exciting, adventure-filled account of deep-sea exploration. (Macrae Smith, 1970. 194 pp. illus. $5.95)

Paleontology. *Mammoths, Mastodons, and Man* by Robert Silverberg. An exploration of the methods of paleontologists in various parts of the world. It describes how these scientists collect evidence and reconstruct prehistoric animals in order to determine their way of life. (McGraw-Hill, 1970. 223 pp. illus. $5.50)

Physics. *Sound: From Communication to Noise Pollution* by Graham Chedd. A clear, broad understanding of the current knowledge and technology of sound. (Doubleday, 1970. 187 pp. illus. $5.95)

The Splendor of Iridescence: Structural Colors in the Animal World by Hilda Simon. The artistry, physics, physiology, psychology, and philosophy of color are portrayed by a good artist in a magnificent book. (Dodd, Mead, 1971. 263 pp. illus. $25)

Psychology. *Body Language* by Julius Fast. An interesting, thoughtful exposition of human nonverbal communication such as gestures, facial expressions, and posture. (M. Evans, 1970, distributed by Lippincott. 192 pp. illus. $4.95)

On the Side of the Apes by Emily Hahn. A fascinating, detailed account of monkeys and apes used in scientific research, and of the scientists who study them. The book describes work at primate research centers in Texas, New Mexico, Puerto Rico, and Oklahoma. It shows how man has discovered many striking similarities between apes and man in the past 50 years. (Crowell, 1971. 239 pp. illus. $7.95)

Science and Religion. *This Little Planet* edited by Michael P. Hamilton. Three theologians and three scientists discuss how Judeo-Christian theology applies to the conservation of natural resources, to the ethical aspects of technology, and to pertinent aspects of ecology. (Scribner's, 1970. 241 pp. illus. $6.95)

Science in General. *The Invisible Pyramid* by Loren C. Eiseley. A remarkable excursion into the philosophy of science. The author speculates that modern technology may exhaust earth's resources and cause the end of man. He also discusses man's settlement on another planet. (Scribner's, 1970. 164 pp. illus. $6.95)

My Several Lives: Memoirs of a Social Inventor by James B. Conant. The title and contents of this autobiography suggest that people should deliberately and radically alter their careers at intervals to retain their youthful creativity. (Harper, 1970. 717 pp. $12.50)

Pieces of the Action by Vannevar Bush. These reminiscences provide a candid and humorous look at the accomplishments of American science during World War II. Bush also shows how an engineer's approach can be used to tackle many problems. (Morrow, 1970. 345 pp. $8.95)

Zoology. *Animals in Migration* by Robert T. Orr. The story of the migrations of all kinds of animals along with a treatment of environmental factors plus a rich bibliography. (Macmillan, 1970. 318 pp. illus. $12.50)

An Artist's Safari by Ralph Thompson. A superb book of drawings and natural history notes made during travels through Tanzania and Kenya. (Dutton, 1970. 148 pp. illus. $19.50)

Eagles (World of Animals series) by Leslie Brown. The book treats the major groups of eagles, describing their habits including nesting and breeding, as well as hatching and growth. Well illustrated. (Arco, 1971. 96 pp. illus. $3.95)

Gorillas (World of Animals series) by Colin P. Groves. With excellent photographs, Groves interprets the life story of the gorilla. There are also comparative studies of gorillas, other primates, and man. Included are stories of monsters in mythology that probably were gorillas. (Arco, 1971. 96 pp. illus. $3.95)

The Last of the Loners by Stanley P. Young. Nine stories about wolves told by a man who studied and hunted them. (Macmillan, 1970. 316 pp. illus. $9.95)

Owls: Their Natural and Unnatural History by John Soarks and Tony Soper. Two owl enthusiasts present their conservationist-ecological information. Mythological lore, the "unnatural history," occupies about 12 pages. (Taplinger, 1970. 206 pp. illus. $5.95)

The Shark, Splendid Savage of the Sea by Jacques-Yves and Philippe Cousteau. Actual encounters with various species of sharks are described. There are notes on the form and structure of sharks as well as experimental information. The photographs are unique. (Doubleday, 1970. 277 pp. illus. $7.95)

Turtles and Their Care by John Hoke. This book tells how to provide a balanced environment for turtles, and thereby learn about ecology. All of the necessary information is here. (Franklin Watts, 1970. 89 pp. illus. $3.75)

The Wasps by Howard E. Evans and Mary Jane West Eberhard. Here is information on social and solitary wasps gained through years of careful research. The book should interest many readers. (Michigan, 1970. 271 pp. illus. $3.45)

The Year of the Seal by Victor B. Scheffer. This life history of the Alaskan fur seal told from the seal's point of view gives insights into ways to protect this valuable animal. (Scribner's, 1970. 218 pp. illus. $7.95)　　　[HILARY J. DEASON]

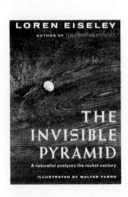

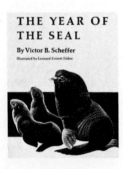

THE YEAR OF THE SEAL

By Victor B. Scheffer

Illustrated by Leonard Everett Fisher

Botany

The first indisputable remnants of angiosperms from before the Cretaceous Period, which began 130 million years ago, were found in 1970. Angiosperms are plants whose seeds are enclosed in ovaries. Paleobotanists William D. Tidwell, Samuel R. Rushforth, and A. Daniel Simper of Brigham Young University; James L. Reveal of the University of Maryland; and Homer Behunin of Utah found fossilized logs, roots, and stems in the Arapien shale formation near Redmond, Utah. The formation is known to have originated during the Jurassic Period, which began 180 million years ago.

The fossils came from two palm plants. The scientists described the plants as a new genus, which they named *Palmoxylon*. The plants were classified as two different species., *Palmoxylon pristina* and *Palmoxylon simperi*.

Many botanists have expected such a find. Although angiosperms made up only 5 per cent of early Cretaceous plants, they were extremely diverse. Such diversity could only have evolved over many millions of years.

Chromosomal drift. Botanist Walter H. Lewis of Washington University discovered that the number of chromosomes varies in the spring beauty, *Claytonia virginica*. The spring beauty is an herb that grows only 2 or 3 inches high and bears tiny pink flowers.

Chromosomes are the structures that contain the genes of a living organism. They are located in the cells, and their number per cell is constant in most creatures. For example, each human body cell contains 46 chromosomes. Lewis found that some spring beauty plants had as few as 12 chromosomes per cell, while others had as many as 191 chromosomes per cell.

He also discovered that the chromosome number of some spring beauties varied from one part of the plant to another. For example, he examined root cells and pollen-producing cells in 13 plants and found that eight plants differed in the number of chromosomes in these parts. Two plants had one extra chromosome in their pollen-producing cells, two had two extras, and three had three extras. One plant had pollen

Touching the giant hogweed, shown in Kew Gardens in London, *below*, and in Summit, N.J., *below right*, may cause rash, blisters, and even lasting scars.

Botany

Continued

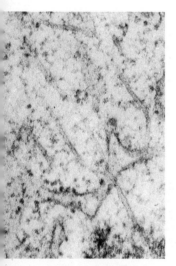

Minute rigid tubes called microtubules have been observed in many kinds of cells. Their function may be to move material around within the cell.

mother cells with two different chromosome counts, 31 and 33. Its root cells had 28 chromosomes.

In addition, Lewis found that the average chromosome number of entire populations of spring beauties could change considerably from year to year. He studied a small population of the plants near Carthage, Tex., during the 1961, 1967, 1968, and 1969 annual flowering seasons. In all the years but 1967, the average chromosome number of the population stayed fairly constant – about 29.5 in 1961 and 1968, and 30.5 in 1969. In 1967, however, the average chromosome number was only about 26.

Lewis called this variation chromosomal drift and speculated as to its cause. "There is no ready explanation for this temporary change in 1967," he said, "but coincidentally one climatic factor varied during that season only. For the critical six-month growing period, roughly from November through April, rainfall was only 39.25 centimeters (cm) –about 15.5 inches– in the Carthage area, whereas, for the other three years, it varied from 76.25 to 97 cm – about 30 to 38 inches – which was about normal for a 20-year period. Is a precipitation of less than one-half the expected amount during a time of maximum growth and development responsible in some way for this marked chromosomal loss?"

Giant hogweed. An article in the British medical journal *Lancet* in 1970 revealed for the first time that contact with *Heracleum mantegazzianum*, one of several species of plants known as giant hogweed, causes a rash and blisters that may leave scars on human skin. The plant grows wild extensively in England, often reaching a height of 15 feet in 3 months. Children have been its main victims because they often play in weedy areas. They are attracted to the plant because its thick stems are hollow, making them ideal pea shooters and make-believe telescopes.

The plant has also been known to grow on the East Coast of the United States, although it is not widespread there. There is evidence that the plant does its damage only when there is moisture on the skin touching it or on the plant itself.

Ozone intake. Botanists Saul Rich and Harley Tomlinson of the Connecticut Agricultural Experiment Station in New Haven developed a method to measure the intake of ozone by plants. Ozone is a form of oxygen that contains three, instead of the normal two, atoms per molecule. Because ozone is one of the principal air pollutants that damages plants, the fact that plants can absorb appreciable quantities of it and thus take it out of circulation is interesting and important information. The ozone formed when sunlight reacts with automobile exhaust fumes is one of the main sources of the gas.

Laboratory studies by Rich and Tomlinson and subsequent computer analysis showed that when air polluted with a typical concentration of ozone passes over a forest, approximately 50 per cent of the ozone is removed in one hour, and four-fifths of it is removed in eight hours. Tree leaves absorb the ozone through their stomata, tiny valvelike openings that control the leaves' air intake. Thus, ozone intake by plants increases during the day, when the stomata are open. Fortunately, this is also when most of the ozone is formed.

Parthenocarpy, the production of fruit without pollination, was chemically induced in cucumbers in 1971 by R. W. Robinson, J. D. Cantliffe, and S. Shannon of the New York State Agricultural Station, Geneva. Parthenocarpy leads to production of seedless fruit. This is important, because plants normally slow or stop their production of any further flowers and fruits when they produce seed. Thus, every good gardener knows that he must remove faded flowers before they "go to seed" if he wishes his plants to continue blossoming. But, the inhibitory effect of seed production is especially troublesome in farm crops in which continuous production of fruit is highly desirable. To solve this problem in cucumbers, strains have been bred that have a gene for the production of parthenocarpy.

The new method, however, offers a means of inducing parthenocarpy in strains of cucumbers that do not have the parthenocarpy gene. The chemical used is morphactin, a fluorene derivative. It need merely be applied to the leaves of young cucumber plants that are beginning to flower. The chemical causes the flowers' ovaries to develop into fruit without first being pollinated, as is normal.　　　[WILLIAM C. STEERE]

Chemical Technology

Innovators in chemical technology were particularly busy during 1971 with problems of the environment. But they did not neglect new engineering materials and other more technical developments.

Disappearing plastics. The growth of plastics as a packaging medium is probably matched only by the quantity of them that lie around as solid wastes. From an environmental standpoint, plastics that disintegrate after use would be ideal.

Researchers at the University of Toronto in Canada and the University of Aston, near Birmingham, England, mix additives that are sensitive to ultraviolet rays in resins to be polymerized into plastic packaging material. When discarded, such packages turn into powder after being exposed to one or two months of direct sunlight. Fortunately, ordinary window glass screens out most of the ultraviolet rays, so packages stored behind them will not decompose prematurely.

In another approach, Austin Science Associates of Austin, Tex., uses a polymer that breaks down on contact with water – at a rate set by the choice of chemical additive. The ultimate self-destructing package, which the Austin group is working on, would have a water-degradable plastic core between layers of plastic that are degradable by sunlight. Water-degradable medicine capsules and pellets of agricultural fertilizer were also being developed by the Austin group in 1971.

Disposing of today's plastics poses another problem. When they are burned in an ordinary commercial incinerator, they give off undesirable gases, particularly hydrochloric acid vapor, which can corrode the incinerator. In Baton Rouge, La., Ethyl Corporation is making an easy-to-incinerate plastic bottle from a material developed by Standard Oil of Ohio. The bottles are said to burn without creating objectionable gases and produce an ash that can be easily disposed of.

The new bottles are said to be break-resistant, 80 per cent lighter than conventional bottles, and quick to chill. Producers expect the soft drink industry

 is credited vertically: Booth © 1970 The New Yorker Magazine, Inc.

"Would it be technically feasible, Hussan, to make pots that would disintegrate after a few years?"

to provide a large market for the new containers.

Cleaning air and water. Important technological progress against pollution is being made in other areas. Techniques have been developed to scour virtually all mercury from waste water, and there are several new ways to remove sulfur dioxide from power plant stacks before it becomes an air pollutant. One company, Rollins-Purle, Incorporated, Wilmington, Del., is building plants throughout the United States to treat industrial wastes.

George and Company, a firm in Liège, Belgium, is using cryogenics to reclaim metals from scrapped autos. A car is compressed into a cube and doused with liquid nitrogen, which makes the iron-bearing metals especially brittle. A shredder then shatters the ferrous components into small chips that can easily be removed from the other metals in the rest of the cube.

Activated carbon is a special form of coal with a highly porous internal structure that is used to remove impurities from liquids. Granules of the material act like tiny sponges, trapping dissolved wastes in their tiny pores when liquid is pumped through a bed of activated carbon. Standard Oil of Ohio has developed an improved version that is made from the carbon in petroleum rather than from coal. The new product is harder and, because each sand-sized carbon granule is round, is easier to use.

Researchers estimate that the vast number of pores gives a pint of the petroleum carbon granules a total surface area of about 112 acres. When the pores eventually become clogged with waste matter, the carbon is removed from the filter bed and exposed to intense heat. The heat burns out the waste material and leaves the carbon unharmed, ready for another service cycle. The new activated carbon is suitable for treating difficult-to-handle wastes.

Protein through chemistry. Efforts continue to find cheaper, more convenient ways to supply protein to human beings and livestock. British Petroleum Company began operating a Scottish plant in late 1970 that can generate 4,000 tons of protein from petroleum per year. By the end of 1971 it expected to have a plant four times that size in operation near Marseilles, France.

Northern Illinois Gas Company announced plans to build the first U.S. plant to make protein-rich commercial poultry feed from natural gas. The plant, to be built near Chicago, is to have a capacity of 10,000 tons of feed per year that will be sold at from 6 to 10 cents per pound. The British Petroleum and Northern Illinois Gas processes are based on microbes that eat the hydrocarbons and convert them into protein.

Because protein-forming microbes can feed on almost any form of cellulose, a variety of feedstocks has been developed. Soybean protein is already well established. But Plains Coop Oil Mill Company, of Lubbock, Tex., began, in mid-1971, to build the first commercial plant for making food-grade cottonseed-protein flour. Because its tiny, red-orange gossypol cells are mildly toxic to nonruminant (single stomach) animals, humans have been unable to use cottonseed protein. The Texas firm will use a "liquid cyclone" technique developed by the U.S. Department of Agriculture.

Protein from dried bagasse, the cellulose-rich pulp left after the juice is extracted from sugar cane, is being fed to chicks and mice in a test by Louisiana State University. Efforts to make paper from bagasse have found only limited success, and much bagasse has been burned as sugar-refinery fuel. In the new test, bagasse is treated with a culture of microorganisms to prepare an animal food of about 75 per cent protein. A pilot plant is being operated at the Mississippi Test Facility of the National Aeronautics and Space Administration near Bay St. Louis, Miss. Researchers have also conducted some experiments using waste newspaper and computer printout paper.

A different approach yields protein from algae for Kohlenstoff Biologische Forschungsstation E.V., a West German research institute. Algae is grown in an agitating tank and nourished with fertilizer and carbon dioxide in the exhaust of the agitator motor. The algae is harvested in a centrifuge and dried to produce a digestible green powder containing about 50 per cent protein.

Fuel from garbage. University of Virginia researchers in January, 1971, began to study ways to convert solid municipal waste, about half of which is cellulose, into protein, possibly by a combination of nuclear and biological

Phosphates Leave the Laundry

Are detergents that contain phosphates dooming the lakes and rivers of the industrial nations? That question is the basis of a scientific controversy with political and emotional overtones that came to a head in 1970 and extended into 1971. During that period, many communities in the United States and Canada outlawed the sale of phosphate-containing detergents after mid-1972. Understandably, the detergent industry began a massive search for phosphate-free replacements.

The phosphates in modern synthetic detergents are called "builders." They improve the performance of the active detergent components by removing heavy metals such as calcium and magnesium from hard water. The active ingredients in detergents are synthetic organic compounds called surfactants, which form films around the dirt and grease particles that are rinsed away. They replace the laundry soaps made from animal fats in predetergent days. The surfactants make up as much as 50 per cent of a detergent, and in 1970 they were being used at the rate of about 2.5-billion pounds a year.

Conservationists argue that phosphates should be removed from detergents because they contribute heavily to the accelerating eutrophication – the result of the accumulations of nutrients in the nation's largest bodies of water. Eutrophication is normally a slowly occurring process in which natural phosphates, nitrates, carbon dioxide, and some organic compounds fertilize algae and aquatic weeds to the extent that they ultimately exhaust the oxygen supply, smother fish, promote plant decay, and eventually convert the lake or stream into a slimy bog.

Scientists agree, however, that detergent phosphates are not the only cause of the accelerated eutrophication. Also implicated are phosphates from human wastes and from agricultural fertilizer run-off, as well as nitrates and organic compounds that end up in the water. All may be involved to different extents in different bodies of water.

For instance, Professor Walles T. Edmondson of the University of Washington has shown that phosphates did the damage to the severely eutrophied Lake Washington, near Seattle. In 1968, communities on the shores of Lake Washington responded to his findings by directing their sewage lines away from the lake into the depths of Puget Sound. By 1971, Lake Washington had returned to a purity it enjoyed in 1933.

Phosphates also have been linked to the eutrophication of the Great Lakes. But nitrates may be the major cause in bays and other coastal waters or in certain lakes. Organic compounds are generally believed to play a less serious role in eutrophication.

Whatever the cause, midwestern cities such as Chicago, Detroit, and Akron are gradually prohibiting the sale of most commercial detergents because they are the easiest source of eutrophication to bring under control.

It is generally agreed that a return to the soap-flake days will not solve the problem, however, because they leave a very undesirable discoloring residue on modern fabrics washed in hard water. In addition, detergent manufacturers insist that there is not enough animal fat available to meet the cleaning needs of the world's fast-growing population.

Thus, the detergent industry is looking at substitute builders of various chemical compositions, and attempting to develop new surfactants that work well in washing machines without the help of phosphates. Among the most promising phosphate replacements was nitrilo triacetic acid (NTA). But tests by scientists of the National Institute of Environmental Health Sciences (NIEHS) in late 1970, indicated that large amounts of NTA – in conjunction with heavy metals, such as cadmium and mercury – damage the embryos of rats and mice. Because of the danger to pregnant women, the detergent industry stopped putting NTA into detergents.

New detergents containing alkalies came on the market in 1970, and perform well, according to most housewives. However, animal tests proved that the corrosive alkalies cause eye damage and are extremely poisonous. As a result, they are marketed with appropriate labels warning of the potential danger to human beings.

The relationship between phosphate detergents and eutrophication is still far from clear. However, it seems likely that phosphates will play a significantly smaller role in the laundry products of the future. [JOHN F. HENAHAN]

means. They speculate that ionizing radiation such as gamma rays from nuclear power plant wastes might help break down the cellulose into glucose. Simple fermentation methods could complete the transformation into protein.

Another class of municipal waste, garbage, can become a new source of pipeline gas, according to an official of the U.S. Bureau of Mines. Herman Feldman of the bureau's Pittsburgh Energy Research Center said in May, 1971, that the process is technically and economically feasible. With the addition of hydrogen, as much, or more, methane can be produced from garbage as can be generated from oil shale or lignite (brown coal), Feldman says. Source materials for the garbage-based process are plentiful at population centers, where methane is needed.

Reclaiming water from raw sewage is moving from dream to reality through a new reverse osmosis process at a pilot plant in San Diego. The plant, with a capacity of 20,000 gallons per day, began operating in October, 1970. Effluent diffuses when passed through a series of tubular membranes, leaving undesirable components behind. In past attempts, feed pumps wore out rapidly and solid materials in the sewage damaged the membranes. The San Diego plant screens the sludge and adds a special protective coating to the membranes. A second stage would be necessary to produce drinkable water, but the plant now produces water suitable for most industrial and agricultural uses.

Progress in paper. Is paper still paper if it is not made from rag or wood pulp? Whatever it is called, a new printing medium based on plastics is threatening conventional printing paper markets in the same way that plastics have seized much of the packaging market from paper.

Plastic paper is made in two ways—all-plastic films and synthetic pulp that is mixed with natural pulp. Japan has made big strides in developing plastic films by coating normally transparent polystyrene or polyvinyl chloride films with pigments. The principal advantages of plastic papers, which can help offset their normally higher cost, are strength and moisture resistance.

Established papermakers find the second approach, supplementing normal wood pulp with a synthetic pulp, more appealing than plastic film. Crown Zellerbach Corporation is now operating a pilot plant in Camas, Wash., to make synthetic pulp from ethylene. While this pulp can be used to make paper directly, it is more likely to be blended with natural wood pulp to create specialty papers. Making pulp from chemicals instead of wood avoids some significant pollution problems.

Glass technology also progressed during the year, with glass fibers being used as concrete-reinforcing material. The alkali of concrete normally disintegrates glass, but a British group has developed a fiber that resists this action. One member of the group, Fiberglass, Limited, believes fibers up to 4 inches long mixed into concrete eventually will provide total support and replace steel rods as reinforcement.

Glass ceramics that are unusually easy to machine are being produced by Corning Glass Works of Corning, N.Y. Excellent insulators and very strong, the ceramics may be shaped with conventional metalworking tools, and can serve continuously at temperatures as high as 1500° F. These qualities suit them for such uses as crucibles, extrusion dies, and printed circuit boards.

Fibers also figure in a new, electrically conductive composite. The material, developed by firms in Austria, Belgium, Liechtenstein, and West Germany, is made by impregnating polytetrafluoroethylene with particles of graphite, then bonding the mixture to glass fiber. Grids of the material have already been used to heat ramps at a parking garage in Vienna, Austria, and for space heating in a Belgian museum.

A Swedish firm, Bulten-Kanthol AB, is making a paint that keeps things warm. It contains small granules of metallic powder. The coating resists an electric current passed through it and generates heat depending on the amount of powder in the mixture and the amount of current applied. The mixture is seen as a de-icing product.

Liquid thread may be the answer to how to create seams in fabrics at high speed. Eastman Chemical Products, Inc., has developed a hot-melt adhesive of linear polyester that permits rapid joining of pillowcase seams and sheet and towel hems.   [FREDERICK C. PRICE]

Chemistry

Chemical Dynamics.

The period 1970-1971 may be remembered as the year that the laser became firmly established as a tool of the chemist. The laser has led the chemist to an astonishing variety of information about the submicroscopic world of molecular collisions and molecular structure.

Using a laser beam to excite molecules in a gas permits the chemist to study the excited molecules, all of which have a completely specified energy. This amounts to the realization of the chemist's dream of a hypothetical submicroscopic imp—called a "Maxwell's Demon"—who could sit inside a gas sample and separate the molecules according to their energies, making the gas hot on one side and cold on the other.

The laser beam, because of its single, sharply defined wave length, causes a molecule to jump from a low level of vibrational and rotational energy to only one specific excited level. And, by choosing the polarization of the laser beam (the orientation of the oscillating electric field associated with the light wave), the experimenter also determines the orientation of the excited molecules.

Once the molecule absorbs laser radiation, it can lose its energy in only three ways: (1) collisions where the molecule reacts to form new compounds; (2) collisions in which the energy is transferred to excite another molecule; and (3) spontaneous radiative decay, whereby the molecule acts as a small radio antenna and radiates away the energy.

In spontaneous radiative decay, the radiation emitted by the molecule has several precise wave lengths. These correspond to the various steplike vibrational-rotational energy levels to which it can make transitions. These specific wave lengths, or "fluorescence" lines, identify the energy levels involved.

The chemist uses the lines to find what processes have affected the excited molecule before it radiates. In this regard, the radiative lifetime of the excited molecule, typically less than a millionth of a second, serves as an "internal clock." It permits only a certain average number of collisions to occur while the molecule is excited. By measuring the intensity of the collision-induced fluorescence

This argon-ion laser at the National Bureau of Standards in Boulder, Colo., joins a host of other lasers in use by chemists. Monochromatic laser light will excite molecules to a specific energy level for study.

lines as a function of gas pressure, which directly affects the collision rate, it is possible to determine how rapidly the molecule loses its energy through collisions. For these reasons, laser-induced molecular fluorescence is an exceptionally sensitive means of investigating both the internal structure and the collision processes of excited molecules.

Laser-induced fluorescence was first observed from the dipotassium molecule (K_2) in 1967. The list of molecules whose laser-induced fluorescence has been studied had grown by 1971 to include: iodine (I_2) and various organic dyes by Jeffrey I. Steinfeld and co-workers at Massachusetts Institute of Technology (M.I.T.); bromine (Br_2), barium oxide (BaO), nitrogen dioxide (NO_2), and chlorine dioxide (ClO_2) by Herbert P. Broida and co-workers at the University of California, Santa Barbara; and the diatomic molecules of lithium (Li_2), sodium (Na_2), cesium (Cs_2), and sodium potassium (NaK) at Columbia University.

Many of these compounds occur only in very minute concentrations in the gas. Molecular fluorescence is, therefore, an extremely sensitive method for detecting and monitoring the presence of trace compounds. There are many important practical applications. One may be the development of a "laser radar" capable of measuring the impurity levels of gases that contaminate the atmosphere. Another may be laser-sensor systems capable of performing faster and more accurate medical laboratory analyses.

Similar developments took place in the infrared portion of the spectrum. Pioneering work in this field was carried out in the laboratories of Bradley C. Moore at the University of California, Berkeley; Jeffrey I. Steinfeld at M.I.T.; and George W. Flynn at Columbia University. It is possible to obtain high-precision measurements of molecular transitions by placing the gas sample inside the cavity of an infrared laser.

Picosecond pulses. Not only are lasers gaining wide acceptance as an ideal means of populating specific molecular energy levels, they also are allowing chemists to investigate nature's processes on a time scale never before possible. A pulsed laser, such as a flash-lamp-pumped ruby rod, normally emits a single burst of light that lasts only 2 to 20 *nanoseconds* (billionths of a second).

However, research scientists at the United Aircraft Company Laboratories in Hartford, Conn., found in 1966 that they could cause the laser to emit a long train of much shorter pulses by placing a dye cell inside the laser cavity. The cell causes the many slightly different frequencies produced by the laser to be in phase, and this "mode locking" produces a series of "beats" of intensity, or pulses. Using conventional light detectors, the researchers at first could only determine the duration of these pulses to be less than 1 nanosecond.

But in 1967, a research team at Bell Telephone Laboratories found an extraordinarily simple means of measurement. The pulse train is sent through a dye solution that is transparent to the laser frequency but fluoresces strongly at twice the frequency of the laser. The pulse train is then reflected back on itself by placing a mirror at the far end of the dye cell. Occasionally, two photons will be simultaneously absorbed and re-emitted by the dye. This appears as twice the energy, and thus twice the frequency of the laser signal. When two pulses overlap, however, this two-photon process is greatly enhanced. Thus, a photograph of the dye cell reveals a weak streak of fluorescence through the dye and regularly spaced bright spots where the pulses overlap. The 1-millimeter length of each bright spot corresponds to a pulse lasting about 3 *picoseconds* (trillionths of a second).

One vibration of a typical molecule takes a few picoseconds; one rotation takes 10 to 100 times longer. In 1971, Peter M. Rentzepis at Bell Labs reported using picosecond pulses to follow the rearrangement of energy between vibrations and rotations in a large molecule after it absorbed a photon. Kenneth B. Eisenthal at International Business Machines Corporation, San Jose, Calif., used a train of picosecond pulses to measure directly such rearrangements at different molecular orientations as well as ultrafast intermolecular energy transfer.

Because picosecond pulses are on the same time scale as molecular motions in solids or liquids, this technique also shows promise as a means of probing chemical processes that are of biological importance.   [RICHARD N. ZARE]

Chemistry
Continued

Structural Chemistry. Chemists in 1971 announced the structural secret of how actinomycin D acts. The drug is one of a series of antibiotics that are particularly effective in controlling various bacteria. The antibacterial action of actinomycin D is due to its ability to form a chemical complex with the deoxyribonucleic acid (DNA) of the bacteria. The resulting structure prevents the bacteria from producing ribonucleic acid (RNA) and thus from reproducing itself.

Previous investigations suggested that the combining of actinomycin D with DNA involves chemical units called deoxyguanosine groups. Guanosine (G) is one of the four fundamental units of the genetic code that make up a DNA chain. There is no indication that actinomycin D forms strong links with the other three units, adenosine (A), cytidine (C), or thymidine (T).

Two models had previously been proposed to explain how actinomycin binds to DNA. The first model presumes that a hydrogen bond forms between the actinomycin molecule and the guanosine group. The second model supposes that a portion of the actinomycin molecule is physically locked between two layers of the DNA helix. During 1970 and 1971, chemists Henry Sobell of the University of Rochester and Christer Nordman of the University of Michigan, working together, provided impressive evidence that the actual bonding of actinomycin to DNA has features of both these models.

Rather than working on the extremely complex molecule of DNA itself, Sobell and Nordman focused their attention on the deoxyguanosine (dG) group. Sobell succeeded in preparing single crystals containing molecules of deoxyguanosine and actinomycin D in a 2:1 ratio. Using X-ray diffraction methods, Sobell and Nordman derived the complete structure of this 2:1 complex, locating the positions of the atoms in all three of the molecules.

The actinomycin D molecule is composed of two *polypeptide* (protein) chains joined by a "chromophore"—in this case, three six-membered rings joined together. Hydrogen bonds between the

Actinomycin's Action

The chromophore group makes actinomycin D effective against bacteria. The group wedges against the DNA helix, forcing it to unwind slightly and preventing formation of messenger RNA.

two polypeptide chains force the chromophore ring group to protrude from the rest of the molecule. In the crystal, this ring group was sandwiched between the rings of the two dG molecules. The stacked ring systems are stabilized by hydrogen bonds from the two polypeptide chains to the guanosine molecules and by the electrical forces of attraction between the stacked rings. It is interesting to note that the entire complex structure has an almost exact horizontal axis of symmetry.

The structure of these crystals suggests a model for the binding of actinomycin to DNA. The chromophore of actinomycin could be trapped between strands of the DNA helix in places where two dG groups are adjacent. This would force the helix to unwind slightly at that spot, and thus prevent the formation of RNA.

Of the four units of DNA, apparently dG alone has the proper size and shape to fit against the actinomycin molecule and form the hydrogen bonds. The activity of actinomycin is thus restricted to those portions of a DNA double helix that contain two adjacent dG units.

Protein structure. Max Perutz of Cambridge University, England, completed his pioneering studies on the crystal structure of the hemoglobin molecule in 1971. His detailed mapping of the positions of the atoms has led him to understand many of the properties of this vital molecule.

Richard E. Dickerson at the California Institute of Technology prepared a map of the structure of the enzyme trypsin. Trypsin is a "proteolytic" enzyme, which breaks proteins down into smaller pieces as the first step of digestion. Knowledge of the detailed geometry of the trypsin molecule should be of great help in understanding how it performs this important job.

Many groups of scientists are hard at work trying to unravel the detailed structure of transfer ribonucleic acid (tRNA). This is a key substance in the synthesis of proteins from amino acids in living cells. X-ray diffraction photographs were prepared for several of the different crystalline forms of tRNA during 1971.

Unfortunately, these photographs were not detailed enough to permit structural chemists to completely deter-

mine the structure of tRNA. However, preliminary findings by Alexander Rich of the Massachusetts Institute of Technology and by Muttaiya Sundaralingam of the University of Wisconsin suggest that tRNA contains segments of a helix that may be similar to that found in DNA. It seems likely that there will be a major breakthrough in our structural knowledge of these key cell molecules within the next few years.

The mysterious octahedron. A particularly vexing question for structural chemists has been that of the geometry of xenon hexafluoride (XeF_6). Xenon, a noble gas, has eight electrons in its outer shell. In XeF_6, six of these electrons are used in forming bonds with the six fluorine atoms grouped around the xenon atom, leaving two unshared electrons. Thus, seven atomic orbitals, regions of electron concentration, are needed to accommodate the electrons—one orbital for each of the covalent bonds, and one for the unshared electron pair.

Chemists have theorized as to how these seven orbitals are arranged about the xenon atom. In particular, they question whether the repulsions between the six crowded fluorine atoms would be great enough to force them into forming a perfect geometrical figure—an octahedron, with eight faces and six corners—about the xenon atom despite the presence of the seventh orbital. Earlier electron-diffraction investigations of XeF_6 gas suggested that the molecule does not have octahedral symmetry, but the exact details of how the fluorines are arranged about the xenon atom could not be deduced.

In 1971, Robinson Burbank of Bell Telephone Laboratories, Murray Hill, N.J., reported success in growing crystals of two modifications of XeF_6. He used X-ray diffraction methods to determine their structures. Interestingly, Burbank found neither crystal to contain whole XeF_6 molecules. Instead, the structures are built up from XeF_5^+ groups linked to one another by F^- ions. The XeF_5^+ grouping requires only six orbitals about the xenon atom—one for each fluorine atom and one for the unshared pair—and thus can have the octahedral geometry. A similar structure occurs in PCl_5, which crystallizes as a mixture of tetrahedral PCl_4^+ and octahedral PCl_6^- ions. [RICHARD E. MARSH]

Chemistry

Continued

Chemical Synthesis. Researchers took a major step toward developing a new insecticide in 1971. Hopefully, it will eliminate the environmental hazards posed by other insecticides. Since man began farming, crops have frequently been attacked and often tragically destroyed by insects. To protect crops, farmers have applied two types of insecticides directly to crops and both have had unfortunate side effects. The complete and simplified laboratory synthesis of a key insect hormone in 1970 points the way toward commercial development of a third type of insecticide based upon synthetic insect hormones.

The first type, which dates back perhaps 100 years, was based on elements such as arsenic, which were known to be toxic to human beings and animals. Farmers applied these first crude insecticides to plants by spraying or dusting. Later, the insecticides were either washed off by rain or were biologically degraded. These materials were not effective insecticides, however, and large quantities were used with often unpredictable results.

The synthesis chemist began to furnish more effective insecticides after World War II. This second type of insecticide was most often chlorinated hydrocarbon derivatives, such as DDT. These compounds are exceedingly toxic to a wide range of insects. However, they also killed insects that are helpful to man and farm animals.

Furthermore, this second type of insecticide was not biodegradable, and this allowed the toxic chemicals to accumulate in large quantities in streams, lakes, and eventually the oceans. The accumulation has now affected the growth and very survival of many species of birds and wild animals throughout the world.

Although the continued use of chlorinated hydrocarbon insecticides is most undesirable, completely discontinuing their use would seriously decrease farm productivity. And this would occur at a period of maximum human population growth. How can we selectively control destructive insect populations without damaging useful insects, birds, animals, and man himself? The answer to this

A Clue to an Enzyme

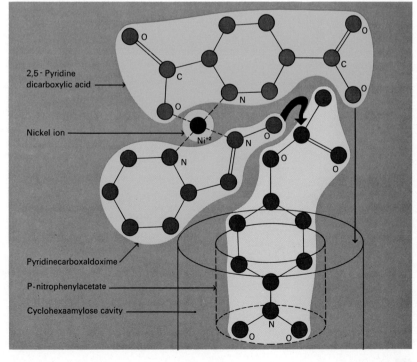

A man-made molecule that breaks down *p*-nitrophenylacetate advanced the chemists' search for a synthetic enzyme. The cavity in the cyclohexaamylose holds one end of the molecule, allowing an attack (arrow) on the other end by the nickel ion complex.

2,5-Pyridine dicarboxylic acid ⟶

Nickel ion ⟶

Pyridinecarboxaldoxime

P-nitrophenylacetate

Cyclohexaamylose cavity

dilemma may be at hand due to recent work by synthesis chemists.

In 1956, a golden oil extracted from male Cecropia moths was found to contain a hormone known as juvenile hormone. All insects secrete this hormone at certain stages in their lives. Juvenile hormone directs the growth and metamorphosis of insects from larva to pupa to adult. It is an absolute requirement for an insect to progress through the usual larval stages of development.

However, in order for a mature larva to change into a sexually mature adult, the juvenile hormone must cease to flow. If juvenile hormone is applied to insects toward the end of their larval development, most resulting adult insects will die or be deformed. In either case, they will be unable to damage crops.

In 1965, Herbert Roller and co-workers at the University of Wisconsin isolated pure juvenile hormone from the golden oil of the Cecropia moth. The physiological activity of the pure hormone was many times that observed in the crude extract. This high activity indicated, in fact, that only 1 gram of pure juvenile hormone applied to a crop could conceivably result in the death of about 1 billion maturing insects, such as meal worms.

The determination of the structure of juvenile hormone followed by its chemical synthesis could thus provide a third type of insecticide. This insecticide would not harm plants and animals. And it would not harm helpful insects except at specific, predictable stages in their development.

In 1967, Barry M. Trost, also at the University of Wisconsin, successfully determined the structure of juvenile hormone. His work was followed by several crude syntheses. These efforts finally culminated in a simplified stereospecific synthesis in 1970 by Elias J. Corey and co-workers at Harvard University. The novelty of Corey's synthesis lies in the simple preparation of a single intermediate chemical that may be simply converted to either of the two useful forms of juvenile hormone, designated as C_{17} and C_{18}JH.

Synthetic enzymes. Chemists have long used catalysts to speed up reactions that would otherwise occur slowly or not at all. The most effective catalysts are the enzymes that trigger the chemical reactions in living things. Enzymes are extremely complex protein molecules, and their action is not clearly understood. They have never been synthesized in the laboratory. Thus, the secret of their catalytic action has not been applied to man-made catalysts.

In 1971, Ronald C. D. Breslow of Columbia University advanced the search for a synthetic enzyme. He synthesized a molecule that displays many of the properties of natural metal-containing enzymes. While the enzyme model is admittedly crude, it represents one of the first successful attempts to duplicate the function of naturally occurring enzymes in the laboratory.

The synthesis of this model enzyme, Breslow finds, rests upon a molecule of cyclohexaamylose, a unique polysaccharide, or high-molecular-weight sugar, which can bind certain simpler molecules in a cavity in its structure. Such a cavity is associated with the specific binding properties of natural enzyme systems.

While cyclohexaamylose itself does not constitute an enzyme model, coupling its binding site with a transition metal ion complex does. The complex is derived from a nickel ion (Ni^{+2}) and pyridinecarboxaldoxime. By itself, this complex is a unique catalyst. It will break down, or hydrolyze, simple esters to their component carboxylic acids and alcohols. This simple nickel complex speeds up hydrolysis of the ester p-nitrophenylacetate by 250 times compared with the uncatalyzed rate.

Chemical attachment of the nickel ion-pyridinecarboxaldoxime complex to cyclohexaamylose further increases the hydrolysis rate of p-nitrophenylacetate by a factor of 4. Breslow takes this as evidence that the nonreacting portion of p-nitrophenylacetate is held within the cavity of cyclohexaamylose in such a way that the ester group is oriented toward the nickel ion catalyst.

Breslow obtained additional proof of this mechanism of catalyzed ester hydrolysis when he observed that the enzyme model is inhibited by other molecules that can bind within the cyclohexaamylose cavity. The foreign molecules apparently prevent the p-nitrophenylacetate molecule from becoming properly oriented, so that rapid hydrolysis cannot occur.   [M. FREDERICK HAWTHORNE]

Communications

The rate of advance in communications technology continued or even accelerated during 1970 and 1971, compared to recent years. Noteworthy changes occurred, not only in communications systems and devices, but also in the types of services offered.

Communications satellites. On Jan. 25, 1971, the Communications Satellite Corporation (Comsat) launched the first of its new generation satellites, Intelsat 4. It was placed in synchronous orbit 22,300 miles above the Atlantic Ocean to serve stations in North and South America, Africa, and Europe. This 3,000-pound satellite carries 6 earth-directed antennas and 12 transponders. Each of these devices receives, shifts the frequency, amplifies, and retransmits to earth stations a single broadband radio signal carrying hundreds of voice channels. Four of the antennas, two for transmitting and two for receiving, can cover any earth station from which the satellite is visible. The other two provide steerable "spot" beams for concentrating the radiated signals in a smaller area, such as a section of the United States or Europe.

With this array of equipment, Intelsat 4 can transmit from 3,000 to 9,000 voice channels, or up to 12 television channels, depending on how its radiated power is divided and directed. In July, 1971, Comsat reduced its prices for overseas transmissions, reflecting the economies of the Intelsat 4s.

In addition to the Intelsat network, many proposals for local systems were being made. In 1971, Comsat proposed a general-purpose system within the United States. Comsat and the American Telephone and Telegraph Company (A.T.&T.) jointly proposed a system for use in conjunction with existing land-based telephone and video networks. These two systems would use stationary satellites, each having as many as 24 transponders. Comsat's system would provide 14,400 voice channels, the joint proposal 10,800 channels.

In addition, six other system proposals were submitted to the FCC by mid-1971. The most ambitious of these was by Fairchild Industries, Incorporated. Its 120-transponder satellite would have capacity for all prospective customers plus extensive free, public-interest services such as television to remote schools,

public broadcasting, and telephone service to remote parts of Alaska.

For technical and economic reasons, it is generally agreed that only a few of the proposed systems could coexist. The FCC asked for a staff recommendation by the end of 1971 on the best approach to a domestic system.

Undersea cables were used more in 1971, supplementing the growth of satellite communications. Total overseas calling increased by over 30 per cent during 1970 alone. With the first five transatlantic cables operating nearly at capacity, new cable systems with capacities as high as 3,600 voice circuits were under development in both Europe and the United States.

In domestic communications, trials of a new coaxial cable system, the L-5, began in late 1970 between Cedar Brook and Netcong, N.J. With amplifiers spaced every mile along the route, the 22 coaxial tube cable will be able to carry 90,000 two-way voice messages, $2\frac{1}{2}$ times the capacity of the L-4 system presently in use. The first commercial service on the L-5 system is scheduled between Pittsburgh and St. Louis in late 1973.

Picturephone and high-speed data-transmission services, becoming increasingly popular, will require the enormous capacity of the L-5 system and others even larger. Picturephone is now commercially available in Pittsburgh, Chicago, and Washington. With present techniques, each intercity Picturephone connection requires the equivalent of from 100 to 400 voice circuits. To permit the transmission of more simultaneous Picturephone signals, along with many thousands of telephone channels, Bell Laboratories' researchers are working to reduce the frequency band required.

The principal approach is based on the fact that the visual message is highly redundant. That is, very little signal information changes from one video frame to the next. If all the information in one frame could be stored and compared to the next one, the picture could be updated by transmitting only the differences. In laboratory tests, this approach has already reduced the required bandwidth to one-third its former size. And researchers expect further improvements to reduce it to one-fourth or less.

A number of techniques are under study as possible "memories" for the

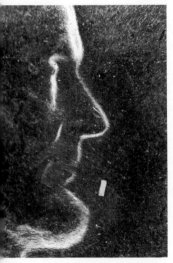

A semiconductor laser that was designed at Bell Laboratories covers only a small part of a Lincoln penny. It can operate at room temperature with a power source no larger than a flashlight battery.

Communications

Continued

video picture storage. About a million bits of information must be stored, so that cost per memory cell must be held to a small fraction of 1 cent if the memory is to be economically feasible. Furthermore, the entire memory content must be reviewed and renewed for each video frame, or about 33 times per second. Thus the system must be very fast.

Optical systems. Even with such band reduction schemes, pictures and high-speed data transmission will require much more capacity and much broader bands than are now available. Optical transmission offers one promising avenue to broader bands. The electromagnetic waves of light could make available bands many times wider than those now used. The band of the visible light spectrum, for example, is 370,000 times wider than the band assigned to the commercial microwave systems. Two noteworthy developments have significantly advanced the commercial prospects of such frequencies.

Until recently, plans for optical transmission envisioned laser beams in pipes with lenses to keep the beams focused and lenses or mirrors to change directions. But improvements in tiny glass fibers, announced in August, 1970, by Corning Glass Works, Corning, N.Y., have brought optical transmission closer to use for short distances in large cities. In these new flexible light guides, a core glass with a given refractive index is coated with a layer of glass of slightly lower index. The new fibers, extremely pure and transparent, promise to carry signals a mile or more before they must be amplified. Separate channels can be transmitted on the individual fibers of a bundle just as separate electrical signals are carried in a multiple-wire cable in present transmission methods.

Both Bell Telephone Laboratories and Britain's Standard Telecommunications Laboratories announced late in 1970 that they had developed solid state lasers capable of continuous operation at room temperature and requiring only the power of flashlight batteries. The lasers appear to open the way to suitable and economic optical-signal generators for a system using glass fiber light guides.   [EUGENE F. O'NEILL]

Computers

The far-reaching impact of computers was displayed by several new applications in only one month in 1971.

One of these was at a health-testing center, which began operations in March under the direction of the Health Insurance Plan of greater New York City. There, a computer conducts an automated physical examination that begins by taking the patient's medical history. It asks up to 1,000 carefully programmed questions to gather detailed information about the patient's health. Nurses and technicians then perform a series of tests with specially designed, high-precision equipment. The tests include, for example, an electrocardiogram, measures of blood pressure and of breathing capacity, and blood and urine chemical analyses.

The test results are stored in the computer system, and the physician can consult the computer printout of the test results before he sees the patient. The physician then completes the physical examination, advises the patient on his health, and prescribes treatment, if any is needed. Such an automated testing system can detect any one of a variety of disorders at its earliest stage, with a minimum use of the physician's time.

Identification system. A computerized New York State Identification and Intelligence System was also installed in March in New York City. The record-keeping system holds over 7 million fingerprints and corresponding identification and case-history data. A communication network, linked to the computer system, serves more than 3,000 law-enforcement agencies in the state.

The fingerprint of a suspect can be transmitted electronically from any agency to the computer center in minutes. After the print is classified by a fingerprint expert, the computer sorts through millions of stored fingerprints in seconds to ascertain if the individual has a record on file. If there is such a record, it is immediately sent by facsimile to the agency.

Waste disposal. Also in March, environmental scientists and computer experts at the State University of New York, Stony Brook, began using a set of computer programs that are expected

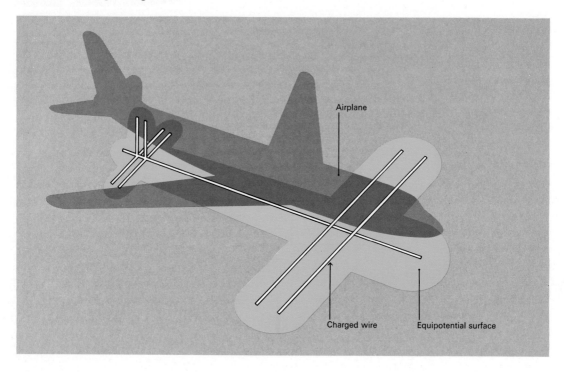

Computers

Continued

A computer at the Environmental Science Services Administration draws imaginary surfaces of equal voltage potential for aircraft. The computer then uses the model to predict whether or not the aircraft will attract a lightning bolt if it flies in a thundercloud.

to greatly improve the collecting and disposing of garbage in New York City. The programs schedule over 10,000 sanitation workers to match their work shift to collection demands, assign sanitation truck routes, set a pickup timetable, and schedule the 50 garbage barges that haul refuse from eight marine transfer stations to landfill areas. The computer programs promise to eliminate gaps in service, which will save the city millions of dollars.

Finally, the Columbia Broadcasting System television network and the Memorex Corporation introduced a computerized system for editing films and tapes for television and motion pictures. The system collects and files all "takes" of a motion picture or a television program, and makes them available to an editor on request. Sitting at a console with several viewing screens, the editor can compose and edit the materials to his artistic satisfaction by choosing scenes from the vast amount of photographic material stored on magnetic disks. The editor can select a specific scene in a fraction of a second. He

presses a light-pen on a word displayed on the viewing screen to direct the system to record, play back, fade-in, or fade-out a scene. The finished print is recorded on tape.

In December, 1970, the Bunker-Ramo Corporation of Stamford, Conn., established a nationwide computer system for quotations of over-the-counter stocks and commodities. Market quotations for 2,000 over-the-counter stocks are now available to more than 1,000 brokers. The system enables the brokers to enter price quotations instantly.

New computers. Almost every major computer manufacturer announced a new series of computers in 1970 and 1971. Most of the new models provide larger and faster main memories, more compact integrated logic circuits, large-capacity and high-speed disks, and more input-output channels.

Typical of these new machines was the System 370 introduced by the International Business Machines (IBM) Corporation in the fall of 1970. The Model 155 and Model 165 of the 370 series are medium- to large-scale, general-purpose

Computers

machines, employing uniform semiconductor integrated circuits for logic and arithmetic operations at *nanosecond* (billionths of a second) speed. They have magnetic-core main memories that can store up to 3 million *bytes* (units of storage equivalent to two decimal digits). Each model features a buffer memory that holds large blocks of data and instructions ready for transmission to the central processor at high speed. The machines also can handle up to 15 different programmed tasks simultaneously.

The smaller IBM Model 145, announced late in 1970, is the first medium-size, general-purpose computer to use a semiconductor main memory in place of the conventional ferrite core memory. The semiconductor memory is composed of silicon integrated-circuit chips, each containing 128 *bits* (binary digits) of storage. The Model 135, announced in March, 1971, was designed for teleprocessing applications. The machine has eight communication lines linked directly to a central processor, and two high-capacity data channels capable of transferring 2.4 million bytes per second. The semiconductor main memory, similar to the one used in the Model 145, contains up to a quarter million bytes.

The popularity of low-cost, small computers continued. The Data General Corporation in November, 1970, introduced three all-semiconductor, general-purpose mini-computers. The largest, named Supernova SC, features a memory system that contains 4 thousand 16-bit words operating at a cycle time of 300 nanoseconds. The computer's semiconductor memory retains its information after each readout, which allows for an overlap of instructions to speed the computation. In most other computers, the material must be restored to the core memory before it is readied for a second readout.

A subminiature computer the size of a desk telephone was reported by the Bunker-Ramo Corporation in January, 1971. The computer uses only 35 watts of power, yet it takes less than 5 *microseconds* (millionth of a second), to complete simple instructions. Multiplication and division takes 30 to 40 microseconds.

The IBM System 7, introduced in October, 1970, is a control computer using integrated circuits for both memory and logic functions. It is designed to monitor and control industrial and laboratory processes. The machine can take a quarter of a million readings per second and analyze them almost instantly. IBM also announced a portable audio terminal, which can be carried in an attaché case, that enables a businessman to dial his computer from any telephone. Using a keyboard, the businessman can ask questions of the computer, which it answers in audible English composed of words from a prerecorded vocabulary.

Technology. Although computer hardware has been highly proficient in recent years, there were some important technological developments in 1970 and 1971, particularly in memory hardware. The mass production of semiconductor main memories for the IBM System 370 machines breaks the tradition of using ferrite cores for large-capacity, random-access memories. The semiconductor memory is composed of tenth-inch square silicon chips, each containing 128 storage cells.

Metal-oxide-semiconductor (MOS) technology still proved to be the best for medium speed, large-capacity computers. An MOS memory system can operate either as a reusable random-access memory or as a group of fast-access shift-registers. And because of the structural simplicity of MOS technology, such memory systems can be fabricated to sell at less than one cent per bit.

The low-cost electronic memory systems were developed to replace rotating electromechanical devices, such as disks and drums, by using magnetic-domain memory devices and semiconductor charge-transfer devices. They were developed by Bell Laboratories.

Recent advances in the engineering of magnetic-domain materials has made it possible to achieve a storage density of over a million bits per square inch. The fabrication of such devices, including the deposition of a thin oxide layer and a permalloy pattern over the oxide, achieves high yield at low cost. Semiconductor charge-transfer devices, based on charge-storage and charge-transfer at the surface of silicon, also accomplish the utmost in structural and fabrication simplicity at low-cost production rates. The signal level of the charge-transfer memory system is directly compatible with current logic circuits. [ARTHUR W. LO]

297

Drugs

The effectiveness of American drug products came under attack in 1970 and 1971. In January, 1971, Henry E. Simmons, director of the Bureau of Drugs of the Food and Drug Administration (FDA) said: ". . . of the 16,000 therapeutic claims evaluated by joint panels of the National Academy of Sciences (NAS) and the National Research Council (NRC), approximately 10,000, or 60 per cent, were found to lack medical evidence of efficacy as defined by law."

Combinations came under particular censure, especially those containing antibiotics. Combination drugs comprise an estimated half of the 3,500 individual drug products available on the American market, and about 40 per cent of the 200 best-selling drugs.

In December, 1970, the FDA made public a list of 356 drug products that were rated ineffective. By January, 1971, 173 of them had been removed from the market place.

Because so many drug combinations were among those removed, controversy has developed over appropriate methods for evaluating the effectiveness of such preparations. Strict guidelines requiring proof that each active ingredient in a combination drug contributes to the drug's effectiveness were developed at the request of FDA Commissioner Charles C. Edwards. The new guidelines, proposed February 17, will also require the pharmaceutical industry to make appropriate studies before introducing new fixed-dose combinations.

AMA drug evaluations. The Council on Drugs of the American Medical Association (AMA) published *AMA Drug Evaluations* in 1971. The book, which is distributed to AMA members, gives brief evaluations of older drugs and mixtures as well as descriptions of new single-entity drugs. The council repeated its position that relatively few drugs in fixed-dose combination have been studied sufficiently to establish their safety and efficacy. For this reason, it does not recommend most drugs in fixed combinations for therapeutic use.

Statements by the council are based on the best information available from scientific literature, unpublished data, and consultants. Although, in general, the council has arrived at judgments concerning safety and efficacy that closely parallel those of the FDA, it has not felt constrained to limit its recommendations entirely to those approved by the FDA.

Diabetes pill controversy. A major debate during the past year centered around a study called the "University Group Diabetes Program" (UGDP). This study was undertaken in 12 widely dispersed clinics to determine the long-range benefits of certain oral drugs for the treatment of diabetes. The 10-year study showed that the death rate was higher among those patients who took tolbutamide, an antidiabetes pill, than among those who took insulin or even those who took no drug at all.

Tolbutamide was the first of the oral drugs for diabetes. It came on the market in the late 1950s. The study reported that 89 deaths occurred in a group of 823 patients, and that about one-third of the number were in patients who took tolbutamide. Of these deaths, 87 per cent were diagnosed as due to cardiovascular causes. Although the UGDP report clearly asserted that tolbutamide increased cardiovascular risk, this has been challenged by a group of more than 30 diabetologists. These doctors took the unusual step of denouncing the study in a letter to the editors of *The New York Times*. Regardless of the final judgment, it appears that patients in the group studied received no substantial benefit from tolbutamide.

New drugs introduced during 1971 included:

■ Matulane (procarbazine-hydrochloride), developed by the Roche Laboratories for the treatment of Hodgkin's disease, a form of cancer of the lymphatic system that is a leading killer among young adults in the United States. Hodgkin's disease strikes about 8,000 Americans each year and kills about 5,000. Matulane produced a significant response in 57 per cent of 480 patients studied. Most of these patients had developed resistance to more traditional methods of therapy.

■ L-asparaginase (L-asparaginase aminohydrolase), an enzyme used for cancer therapy. Certain cancer cells require an external source of L-asparagine for the synthesis of protein. Thus, destroying L-asparagine with the new drug reduces the amount of L-asparaginase available to tumor cells. A wide variety of

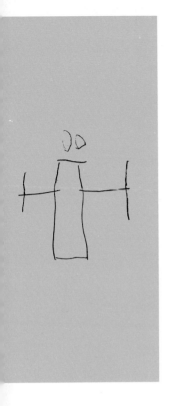

Drugs

Continued

Drawings by a young child with "minimal brain dysfunction" before, during, and after treatment with amphetamines show a remarkable improvement. For reasons not yet fully understood, these stimulants calm overly active children so they can concentrate for longer periods of time.

Earth Sciences

See Geochemistry, Geology, Geophysics, Meteorology, Oceanography

dangerous side effects prevent all but the most limited use of the drug to treat leukemia. However, it may find some use when employed in conjunction with other cancer drugs.

■ Rifampin (rifampicin), introduced as an antituberculosis remedy. It also demonstrated exciting possibilities as an antiviral agent. Drugs that combat viruses with no dangerous side effects have been difficult to find. Most drugs that block viral growth also damage cells. Rifampin apparently does not.

Cancer and viruses. With growing evidence that viruses cause certain forms of cancer, (see INTERNAL MEDICINE) the discovery of rifampicin may aid investigations in cancer chemotherapy. Indeed, a rifampicin derivative, N-demethyl rifampicin, has been identified by the National Cancer Institute Biometrics Research Laboratories, Bethesda, Md., as an inhibitor of a newly discovered enzyme found in viruses known to cause cancer in animals. This discovery may represent an important breakthrough in the development of a new cancer treatment for human beings.

Amphetamine control. Controversy continued over amphetamines as acceptable therapeutic agents. The general view emerging is that amphetamines have a limited therapeutic usefulness – particularly in areas of psychiatry for the treatment of hyperkinetic children (those afflicted with excessive muscular movement), and in the treatment of narcolepsy (abnormal sleepiness). But their use as an appetite-suppressant is probably not justified.

In May, 1971, the U.S. Department of Justice, concerned over widespread use of amphetamines by young people, proposed that amphetamines be placed in the same drug category as morphine and similar agents. This listing would place the same legal constraints on the use and manufacture of amphetamines as is placed on morphine.

Unfortunately, strong legal restrictions on importing, manufacturing, and distributing narcotics has not controlled heroin addiction in the United States. It may well be that increasing the legal restrictions relating to amphetamines will be equally futile.   [ALLAN D. BASS]

Ecology

A scorpion wired with heat-sensing devices helped to reveal how some desert creatures escape extreme heat.

Ecological research increased in intensity and diversity in 1970 and 1971, under the impetus of increasing environmental concerns. The broad-scale integrated ecological work of the International Biological Program had hundreds of ecologists studying large areas such as forests and grasslands. See PROBING THE PAWNEE GRASSLAND.

The Ecological Society of America—the professional society for ecologists—joined with 42 universities and research institutions to establish an Inter-American Institute of Ecology. The institution will study large-scale ecological problems that are national and global in scope. These kinds of studies are beyond the present capabilities of colleges and universities.

Energy relations. Ecologists John M. Teal and John W. Kanwisher of the Woods Hole Oceanographic Institution at Woods Hole, Mass., reported a study of energy exchange in the grasses of a salt marsh. Their main objective was to test the accuracy of theoretical calculations of energy loss.

The plant life of salt marshes along the East Coast of the United States consists of almost pure stands of two species of grass of the genus *Spartina*. The scientists found that these grasses radiated the sun's energy (heat) back and forth between their leaves, which absorbed some of it. The rest of it returned to the air through evaporation, convection, and radiation. But, more important, the scientists' measurements of heat loss agreed closely with the theoretical calculations they were testing.

Ecologist Neil F. Hadley of Arizona State University studied two creatures—the scorpion and the black tenebrionid beetle—living under the extreme conditions of the desert. Hadley was particularly interested in finding out how the two might use micrometeorological conditions—the "climate" in their burrows—to survive. Wires that monitored temperature were attached to the surface of the animals' bodies, and some were inserted in their bodies.

Unlike warm-blooded animals, which use sophisticated temperature-regulating mechanisms in the body to survive, the scorpion and beetle must develop behavioral methods of dealing with even relatively small temperature changes in their environment. Hadley found that they moved up and down in their burrows to escape extremely high surface temperatures and humidities low enough to desiccate them. In a typical experiment, the surface temperature at midday was 65° C. (150° F.). A scorpion, however, stayed down 8 inches in its 16-inch-deep subterranean retreat to enjoy a relatively cool 42° C. (108° F.).

Microcosms. Ecologists are increasingly using experimental microcosms to unravel interactions between factors in whole ecosystems. An ecosystem is a distinct area defined by its characteristic living creatures and the environmental factors they require. An ecological microcosm is a miniature ecosystem or portion of it. Ecologists Martin Witkamp and Marilyn L. Frank of the Oak Ridge National Laboratory, Oak Ridge, Tenn., studied microcosms to analyze some of the factors involved in the decomposition of materials that are on the forest floor.

The microcosms used consisted of clear-plastic containers in which there were sand, millipedes, and leaves of the tulip poplar tree. The leaves contained the radioisotope cesium 137, which could easily be traced through its radioactivity.

Each round container was 10 centimeters (cm) (about 4 inches) in diameter and 7 cm (about 2¾ inches) high. Three population sizes of millipedes—three, six, and nine—were used for the tests. There were also three temperatures and three levels of simulated rainfall used. For 14 weeks, the millipedes ate some of the leaves, and naturally occurring bacteria decomposed leaf material both from the leaves and from the millipede excrement. Then the microcosms were checked.

In all of the microcosms, most of the cesium 137 freed from the leaves by decomposition had been absorbed by the sand. Most of the freed potassium had been drained off with the "rain," which was allowed to run out through a spout in the bottom of the containers. Most of the freed magnesium stayed in the decomposed leaves, and most of the carbon dioxide created by all the living organisms had escaped into the atmosphere of the microcosms.

The experiment also showed that, although the millipedes inadvertently consumed many bacteria along with the

"I'll tell you what *I'm* doing to protest environmental pollution. I'm taking in carbon dioxide, but I'm *not* giving off oxygen."

"We're lucky. This stream could be next to a paper mill instead of a brewery."

"And now a few words in favor of the . . ."

"... SST."

"I think the public is becoming better educated; I felt positively guilty about dumping that last lot."

"Hello! We can't be far from civilization."

Are Metals A Menace?

In early February, 1971, scientists at the Food and Drug Administration (FDA) testing swordfish found that 87 per cent of them contained more mercury than federal guidelines allow. FDA officials asked the swordfish industry to hold all shipments for analysis, confiscated thousands of pounds of the fish, and recommended that the public stop eating them. This and earlier reports of contaminated tuna, salmon, and other seafood introduced the public to yet another form of pollution—metal pollution.

When does a metal become a pollutant? Virtually every living and nonliving thing contains many metals in widely differing concentrations. Our bodies, for example, contain nearly every metal element, and we need many of them to maintain our health. Metals are pollutants only when they are in the wrong place, in the wrong concentration, at the wrong time. Under these conditions, they often adversely affect our environment and our health. Yet, it is often man who is the cause of metal pollution.

Each year, man discharges as much mercury into all bodies of water as is carried to the oceans by all the world's rivers. The rivers get the mercury in the form of mercury compounds washed from the soil and also eroded by wind and rain from natural ores. Man puts a much greater quantity into the atmosphere in the smoke from burning fuels.

Surprisingly, however, these amounts are only a small fraction of the mercury found in nature. For example, the oceans contain so much natural mercury that even if we could eliminate the quantity added by man, mercury concentrations in fish from the sea might still be greater than the FDA considers safe. But man's mercury additions are usually more concentrated than nature's, and are, therefore, considerably more dangerous.

Mercury accumulates in the body by combining with enzymes, interfering with their activity in many vital life processes. The danger of mercury metal vapor absorbed through the lungs or skin has been known for centuries. Many workers exposed to the vapor in mines or in industries where mercury metal was used, showed the common signs of mercury poisoning—lack of appetite, muscular weakness, and tremors and other nervous system disorders.

But until more than 100 persons were poisoned near Japan's Minimata Bay between 1953 and 1960, no one realized that mercury could be a problem in the diet. A plastics factory was discharging mercuric chloride and methyl mercury into the bay, and all fish and shellfish, the principal food in the Minimata area, became contaminated.

A normal diet contains 10 micrograms (about a millionth of an ounce) of mercury per day. The Japanese afflicted by mercury poisoning consumed more than 200 micrograms per day. In Sweden, many lakes have been contaminated by the discharge of mercury compounds from papermaking factories and farm chemicals. Residents near such lakes for whom fish is a staple food have a daily mercury intake of 44 micrograms.

It is the extreme insolubility of mercury and most of its compounds that led us to think we could safely dispose of them in water. But, now we know that the sediments of lakes and streams contain bacteria that convert these insoluble substances into methyl mercuric chloride, which is readily absorbed by organisms living near the bottom. These organisms are eaten by others that are eaten by fish. Finally, the fish are eaten by human beings.

Human beings have also inadvertently eaten lead, another metal that has become known as a major pollutant. The sweet taste of the lead compounds in some paints has caused many a child to poison himself with flakes of paint that peel from the walls. But, such paints have largely given way to nontoxic paints, and today we face potentially more dangerous lead pollution.

Almost all gasolines have tetraethyl lead added for better performance. As a result, engine exhausts spew tiny particles of lead compounds into the air. There have been no recorded instances of fatal poisoning from this lead so far. However, a small fraction of the lead that enters the body is not readily excreted. Therefore, continuous exposure will cause the concentration in the body to increase slowly, and a toxic level may ultimately be reached.

On the other hand, studies show that the particles from engine exhausts are deposited in a comparatively small area along streets and highways. Also, airborne lead particles contribute no more than one-tenth of our daily lead intake. The remainder comes mainly from food.

The average concentration of lead in the blood of such persons as tollbooth and parking-lot attendants who are continually exposed to automotive exhausts is .26 parts per million (ppm). This is little different than that for city dwellers −.25 ppm. Both these values are far higher than that for persons who live in the suburbs−.13 ppm. And persons from rural areas have only .11 ppm.

However, even the higher concentrations are far lower than the level at which lead may interfere with metabolic processes. Even if unfavorable weather conditions should increase air pollution for days, the problems due to lead would be minor compared to those caused by other pollutants such as sulfur dioxide.

Clearly, determining whether and when a metal pollutant is dangerous to health is a complex business. And although we know little about mercury and lead pollution, we know even less about the effects of other metals in the environment. At high enough concentrations, such common metals as copper, zinc, aluminum, and chromium are poisonous to fish.

Cadmium and arsenic, which are extremely poisonous to man, are widely distributed in the environment and are being redistributed by industry. For example, large concentrations of arsenic are added to the phosphates in detergents, and much of this is discharged into streams with sewage effluents. Because arsenic is not easily removed from these wastes, it may contaminate the water supply of cities downstream that reuse the water.

What little is known at this time concerning metal pollution points to the need for extensive research. We must determine how metals affect man, estimate what part of present levels is the consequence of man's activities, and set standards for the use and discharge of the metals.   [DAVID N. EDGINGTON]

Ecology

Continued

leaves, the decomposition activity of the bacteria was not reduced. They merely grew and multiplied faster. This is similar to what happens when cattle graze a pasture. The grass grows thicker and faster than it did before.

Macrocosms. Some ecologists develop and utilize what might be termed macrocosms. These are usually complete, but relatively small, ecosystems rather than just a segment of an ecosystem. Ecologists Donald J. Hall, William E. Cooper, and Earl Turner of Michigan State University described some of the first of their findings of a three-year macrocosm study of the interactions that control population sizes within an ecosystem. The macrocosms were a unique set of 20 similar ponds, each about 350 feet long and 350 feet wide.

The study focused on the *zooplankton* (microscopic animals), bottom insects, and fish, as well as the plants and the water chemistry of the ponds. The initial objective was to see what happens when the concentrations of chemical nutrients such as nitrogen and phosphorus are manipulated along with the density of invertebrate predators that prey upon zooplankton and bottom insects. These predators include relatively large ones, such as diving beetles and dragonfly larvae, and small ones, such as *copepods* (tiny crustaceans), and hydra. The scientists were also interested in finding out how fish eating the zooplankton and bottom insects affected their populations.

They found that adding nutrients to a pond generally increased its production of zooplankton, but had little effect on the number of the other pond residents. Increasing the number of invertebrate predators did not affect the total number or mass of zooplankton or bottom organisms. It did, however, reduce the numbers of some individual species.

When fish fed on the zooplankton, the tiny animals' diversity and size distribution were greatly altered. But the total number of zooplankton was reduced by the fish only when nutrient levels in the pond were low. The total weight of the fish population was clearly directly related to the amount of nutrients that were present in the ponds.

IF ALLOWED TO SURVIVE
THIS GRASS WILL PRODUCE
ENOUGH OXYGEN FOR TWO
STUDENTS TO BREATHE
FOR ONE SEMESTER

Ecology

Continued

An ecological "keep off the grass" sign appeared on newly sodded area at the University of Iowa.

Forest regeneration. In 1970, ecologist Jay C. Gashwiler of the U.S. Bureau of Sport Fisheries and Wildlife reported on a study of the effects of small animals, especially rodents and birds, on regeneration of Douglas fir and western hemlock. He measured the loss and survival of seeds of these two important timber species from 1960 to 1967.

To isolate the effects of individual species of animals, Gashwiler used large, elaborate enclosures that excluded all but the animals he wanted to enter. Each enclosure was 35 feet square and 5 feet high with screening on the sides. When birds were to be excluded, he placed chicken wire across the top of an enclosure.

Gashwiler found that under normal conditions only 10 per cent of the Douglas fir seed survived from the time that seed normally starts to fall until germination occurs the following year. He estimated that mice and shrews destroyed 41 per cent of the seed. Birds and chipmunks took 24 per cent of the seed, and disease, insects, and other factors accounted for 25 per cent.

About 22 per cent of the western hemlock seed survived. Mice and shrews destroyed an estimated 22 per cent of the seed. Birds and chipmunks took 3 per cent, and disease, insects, and other factors accounted for 53 per cent. The high mortality rates of these seeds shows clearly that large amounts of seed are needed to restock forests that have been cut or burned.

Water-release by trees. In 1970, ecologists J. R. Kline, J. R. Martin, C. F. Jordan, and J. J. Koranda of the Argonne National Laboratory and the Lawrence Radiation Laboratory measured transpiration, or water release, from trees. How to measure transpiration by trees under field conditions has long been a challenging problem. Smaller plants can be totally enclosed and water in samples of air from the enclosure can be measured to determine transpiration rates.

The scientists worked with three trees in the rain forest of eastern Puerto Rico. They bored small holes around the trees' trunks and fed in water containing radioactive tritium, a form of hydrogen.

Ecology

Continued

Transpiration occurs almost exclusively through the leaves. So, by periodically collecting leaves from the tops of the trees and measuring their tritium content with Geiger counters, the scientists were able to calculate transpiration rates.

Average transpiration rates ranged from 1.75 to 372 liters (about 1.85 to 394 quarts) per day per tree. The great variation was caused by varying weather conditions. The lowest transpiration rates were on rainy days, the highest on hot, dry days. The new technique has considerable potential. It is easy to perform, does not damage the tree, and should enable field ecologists to determine much more accurately than any other known techniques how much water forests need.

Population and aggression. Crowding, with associated increases in aggressive behavior has been considered by population ecologists as one of the major factors in regulating population sizes, especially those of small mammals. Charles J. Krebs of Indiana University reported in 1970 on a two-year study in which he measured aggression and exploratory activity, and related them to population size. He studied 1,215 voles—mouselike rodents—of two species. The animals lived in seven populations in southern Indiana. A population is a group of animals in which individuals live close enough to one another to meet and interact.

Krebs measured exploratory activity by placing animals in a testing box with squares marked off on the floor. The more squares an animal entered in a fixed time, the greater his exploratory activity. He measured aggressiveness by allowing pairs of voles to fight in a neutral fighting area, away from territory either animal might consider its own.

Exploratory activity did not seem to be related to population size. Aggressiveness, however, changed significantly with population size in both species. The animals from the largest populations were the most aggressive. This supports the hypothesis that aggressive behavior may act to regulate population size. Aggressive animals tend to reduce population size by killing or driving out other members. [STANLEY I. AUERBACH]

Education

Leading scientists and science teachers expressed concern in 1971 about the way science is taught. Their comments came at the 19th annual convention of the National Science Teachers Association (NSTA), in March.

Willard J. Jacobson, chairman of the Department of Science Education at Teachers College, Columbia University, pointed out that science and scientists are blamed for many of society's problems, and that many young people are rejecting science. Since this ". . . bodes ill for the future . . . it is important that we make a critical examination of the past, take a hard look at the present, and then project an imaginative view of the future."

Convention participants suggested the following actions to improve science education:

First, present science in a broader philosophical perspective. Because science is one of the ways in which man tries to understand himself and the universe, it is not the exclusive domain of the brilliant researcher. It belongs to, and can be experienced by, everyone.

Second, emphasize the relationships of science and society to give a more comprehensive approach to problem solving. Many of society's ills will not be cured until man deals with their educational as well as scientific dimensions. For instance, the world now has the knowledge and technology to solve many of our environmental difficulties, including the population problem. Obviously, the way to make progress is by teaching man how to take advantage of the technology.

Finally, train teachers to be curious, investigative, and active so they can stimulate student curiosity. The particular style of the teacher is not as important as his attitude. The authoritarian teacher, who dictates a series of established truths that are to be memorized, ignores the most important dimensions of science—creativity and discovery.

The Tilton program. Excellent science curriculums do not develop overnight. For example, in 1967, an interdisciplinary, problem-solving approach to the problem of water pollution was used as an activity in the science curric-

Education

ulum at University School in Cleveland. The success of this approach subsequently led to its presentation at the 1969 annual conference of the National Association of Independent Schools. Observers from the Ford Foundation were so impressed that, during the summer of 1969, the foundation funded the first student-teacher training project for further development of the water-pollution curriculum at Tilton School, a private preparatory school in Tilton, N.H. The success of the Tilton project, in turn, led to a second water-pollution program at Tilton in the summer of 1970.

After further revisions, the Tilton program was formally published in May, by the Environmental Agency. Entitled *A Water Pollution Activities Guide*, it is not yet available for distribution. The Tilton program will probably expand to other areas of study in environmental science and use the same partnership in learning where teachers and students share equally in the training, learning, and writing of curriculums.

The draft form of the Tilton program, *A Curriculum Activities Guide to Water Pol-*

lution and Environmental Studies, is available free of charge from the Training Grants Branch, Federal Water Quality Administration, U.S. Department of the Interior, Washington, D.C. 20242. It is 560 pages long and includes field activities arranged in four sections: hydrologic cycle, human activities, ecological perspectives, and social and political factors. Questions are provided which enable teachers to lead students into initiating and continuing their activities in guided discovery. There are behavioral objectives to aid the teacher in evaluating student performance. The guide also includes methods of data analysis, details of water-analysis procedures, glossary, annotated bibliographies, suggestions for student projects, and program implementation.

The guide is written so that teachers without training in environmental science may use the curriculum effectively. William Schlesinger, a hydrologist and ecologist who helped design the Tilton program, says, "All in all, the guide gives the teacher an initial package to institute water-pollution environmental

Projected Imbalances in Scientific Professions

Profession	Estimated 1968 jobs	Projected 1980 needs	Estimated 1980 supply
Chemists	130,000	200,000	
Dentists	100,000	130,000	Significantly below requirements
Dietitians	30,000	42,100	
Physicians	295,000	450,000	
Physicists	45,000	75,000	
Engineers	1,100,000	1,500,000	Slightly below
Geologists and geophysicists	30,000	36,000	
Pharmacists	121,000	130,000	Slightly above
Life scientists	168,000	238,000	
Mathematicians	70,000	110,000	Significantly above
Teachers, elementary and secondary	2,170,000	2,340,000	

Source: Bureau of Labor Statistics

A 13-year-old works with the cardboard punch-outs from the pages of *Our Man-Made Environment: Book Seven*. This new book on urban architecture and planning is designed to teach pupils how to view their surroundings.

Education

Continued

studies in his school—a package embodying the open-ended approach and partnership in learning, in that it leaves many questions open for discovery and enlargement by the teacher as well as the student." Unfortunately, to many teachers this is a major drawback because they will not let themselves be learners, or because they feel they must have expertise in ecology.

The "open" classroom is an informal, individualized, student-centered approach to education that has become popular in British primary schools. In 1970, the system was being tried in many areas in the United States—for instance, California, New York, and the District of Columbia—at both elementary and secondary levels. Charles E. Silberman, author of *Crisis in the Classroom* (1970), has recommended the approach as a possible solution to "grim" and "joyless" classes.

In science, the open classroom relies primarily on a laboratory or activity-stimulated approach. It offers students a maximum choice of subject material and often stresses an open-ended ap-

proach to problem solving, in which they invent ways to investigate a problem that may have several possible answers. The teacher interacts in a non-authoritarian way with individuals or small groups of students.

Classwork stresses the broad dimensions of science and scientific processes, rather than the facts. What is evidence? How do we get it? What do we do with it? Essential teacher characteristics are spontaneity and enthusiasm. Success depends on skillful teachers who are adept at guiding the students to discovery by posing the right questions at the right times.

Critics attack both the theory and the practice of such informal education in science because it does away with orderly and uniform classrooms. They claim that the open classroom fails to equip children with basic skills and facts, and that it will result in placing students at a disadvantage in selective examinations for jobs and college entrance. There is no objective evidence available yet either to support or negate these assertions.

Education

Some curriculums (such as the Tilton guide, the Science Curriculum Improvement Study, and the Biological Sciences Curriculum Study) emphasize many elements of the open classroom. Yet sales of the traditional, content-oriented texts remain high, probably due to the specific nature of college entrance examinations.

Pregraduate publication. The master's or doctoral thesis formerly represented a student's first (and, unfortunately, often the last) experience in research and publication. But, in 1970, predoctoral research and publications prepared by professor-student teams were more common. This trend indicates that professors are using the manpower that eager, energetic, and intelligent students can provide. The professors get more research done and students play a more active role in the scientific community.

Public environmental education. Earth Week, 1971, was a far different event from Earth Day, 1970. The fervor of the 1970 rallies and rhetoric was replaced by what Michael McClosky, executive director of the Sierra Club, described as a "higher level of consciousness." Students provided the primary energy behind public concern for the environment, and in 1970 more adults became active in educating the public on what must be done against pollution. Earth Week was marked by grass-roots action in which millions of students and citizens became involved in locally oriented environmental drives. These included bicycle days, cleanup drives, and walk-to-work days.

As further evidence that public education on the environment was taking hold, support of national environmental organizations rose spectacularly during 1970. The Wilderness Society's membership increased from 50,000 to 72,000. Friends of the Earth were signing up at a rate of 2,000 a month, and comparable increases were reported by other major environmental groups. New nature centers were being built, both privately and by groups such as the National Audubon Society, and existing centers were expanding many of their educational facilities. [LINDA GAIL LOCKWOOD]

Electronics

Deep cutbacks in government spending hit the electronics industry hard in 1970 and continued into 1971. Large weapons programs were reduced and major space exploration programs were curtailed as the Apollo program neared its end. Although business seemed to improve in mid-1971, there was widespread unemployment among engineers, particularly in the northeastern states and southern California. Many small companies, especially semiconductor and computer manufacturers, were forced to close, or reduce operations.

U.S. electronics firms began to seek new, nondefense markets to take up the slack. Integrated circuits were available at attractively low prices, due in part to overproduction in anticipation of orders that never materialized. New uses for them were being frantically explored.

Electronic calculators. It became possible to produce, at relatively low cost, versatile calculators that add, subtract, multiply, and divide silently and almost instantaneously. Large-scale integrated circuits (LSI's), particularly metal oxide semiconductor (MOS) types, can contain most of the required arithmetic circuits on a few tiny silicon chips. The calculators range from pocket-sized, battery-operated units to desk-top units that can be programmed for scientific and engineering use.

Ironically, by supplying most of the LSI-MOS chips, American technology permitted Japan to dominate the calculator field. Almost 70 per cent of the calculators sold in the United States by late 1970 were made in Japan. Chances of American industry recapturing this market are slight, but the growing proliferation of calculators helped the beleaguered U.S. semiconductor firms. It triggered a tremendous upsurge in MOS technology.

In February, 1971, Mostek Corporation of Carrolton, Tex., announced that it was producing a complete calculator logic circuit on a single silicon chip less than 1/5 inch square. The chip is intended for a four-function, 12-digit calculator being produced by Japan's Busicom Corporation. The chip contains more than 2,100 transistors. An American firm, SCM Corporation, redesigned

Electronics
Continued

its Marchant 1 calculator using only two chips, supplied by American Micro-Systems, Incorporated.

Electronic timepieces. Another new and beguiling application for integrated circuits is the wristwatch. Electronic timepieces have potentially greater accuracy and longer life than their mechanical counterparts. Electronic watches are now priced between $650 and $2,000. However, if demand increases, they could conceivably be available for between $50 and $75 within two years.

Eventually, electronic watches will have no moving parts, but most of those produced so far are motor-driven. The battery-powered integrated circuits precisely control the speed of the motor. The electronic circuitry and motor must be designed to consume very little power, so that the battery will last at least a year. Consequently, a family of integrated circuits that consume only millionths of a watt – called micropower circuits – has been developed.

The ultimate goal of electronic watch-makers is to replace not only the motor but also the traditional watch face and hands with an all-electronically, illuminated digital display. One U.S. firm, Hamilton Watch Company, introduced such a watch in 1970 called Pulsar and priced at $1,500. Its face has many light-emitting diodes (LED's) arranged to form numbers. To conserve power, the diodes are normally turned off. But when the wearer presses a little button, the time is instantaneously displayed on the watch face.

The numbers game. Many other consumer applications must provide a numerical display. LED's, such as those used in the Hamilton Pulsar, are beginning to appear in instrument panels and calculators. Although still more costly than glowing gas readout devices, LED readouts are more readable and last longer. Because they are all solid state, they are compatible in size and performance with integrated circuits and operate at a low voltage.

LED readouts are made in two forms, hybrid and monolithic. In the hybrid, diodes of gallium arsenide or gallium arsenide phosphide are arranged in an array (usually 5 x 7) on a ceramic sub-

A fleeting pulse of laser light is captured by a new Bell Labs camera that has a one hundred-billionth second shutter speed. From tip to tip, the pulse measures about 1/5 inch.

Electronics

Continued

strate and wired individually. In the monolithic, the diodes are formed together on a semiconductor chip. When voltages are applied to the appropriate diodes of either type, they emit light (usually red), providing an instantaneous and changeable numerical display. The monolithic display is not yet as advanced as the hybrid.

Although the LED has a head start, a new display technique, the liquid-crystal cell, is developing rapidly and may soon be used extensively in calculators, watches, and electronic instruments. A liquid-crystal cell consists of two parallel glass plates with a drop of liquid crystal material sandwiched between them. Electrodes are attached to the inside surfaces of the glass plates.

When a voltage is applied to the liquid crystal, it becomes opaque. When the voltage is removed, the crystal immediately becomes transparent.

Liquid-crystal devices are passive— that is, they do not generate any light of their own. They can be designed, however, either to transmit or reflect light. For example, if both glass plates are transparent, the device can operate as an electronically controlled window. The opacity depends on the voltage applied, so that a photocell could be used to regulate the amount of light transmitted.

Reflective cells are used in number displays. Etched segments on one composite glass sandwich, each an independent cell, will display the numerals 0 to 9 when charged in the appropriate pattern. The new devices potentially offer extremely low cost, simplicity, ruggedness, and no limitation on size.

Variations on the liquid-crystal concept promise a host of useful applications. Photochromic dyes—materials that change color in response to different light wave lengths—can be used for electronic or optical color control when added to a liquid-crystal cell. Another type of liquid crystal, called cholesteric, has a "memory." It will retain its opacity for many hours after the applied voltage has been removed.

Researchers at RCA Laboratories in Princeton, N.J., have built an assortment of experimental devices using liquid crystals. These include electronic

Electronics
Continued

clocks, window shades, and aircraft cockpit panel displays. Seiko of Japan has developed a prototype of an all-electric clock with a liquid-crystal display, and is working on using the principle in watches.

Optical communications. The communications industry has spent much effort adapting the laser to communications systems. The reason for the interest in optical transmission is obvious. Theoretically, a single beam of light could carry 100 million simultaneous two-way telephone calls and still have ample capacity for all Picturephone and television broadcast signals. To utilize the laser, however, the electronics industry must produce the optical equivalents of the components and functions required for today's electric and microwave systems, such as modulators, filters, couplers, switches, detectors, and multiplexing devices.

Scientists at Bell Telephone Laboratories are exploring the possibility of using techniques developed for integrated circuits to build optical wave guide components that will perform the functions.

Basically, the components are designed on principles similar to microwave systems, but they are much smaller for the extremely short optical wave lengths.

Instead of hollow metal wave guides, for instance, optical integrated circuits are made of thin, transparent glass films only a millionth of an inch wide and a fraction of that thick, which have been deposited on a polished glass substrate. The proper relationship of the film's index of refraction to that of the substrate causes total reflection within the film (see COMMUNICATIONS). With photolithographic techniques, it is feasible to produce an entire single-channel amplifier for a voice, data, or video signal in less than a square inch.

Optical communications also came closer to reality through improvements in lasers. In August, 1970, Bell Labs scientists announced that they had achieved continuous operation of a solid state laser at room temperature. Previously, the excessive heat generated by the high electric current required for laser action had permitted only pulsed operation. [SAMUEL WEBER]

Energy

No spectacular advances were reported in the broad field of energy in 1970 and 1971, but there were some notable accomplishments. Continuing the trend of more than two decades, most of these were in the United States and Russia.

Russia announced the successful demonstration of a nuclear-thermionic system that could be used for long manned missions in space. Accounts, which appeared in the Soviet press in April, 1971, gave few details, simply claiming that "several kilowatts" had been obtained from each of a number of nuclear-reactor fuel pins that were capped with thermionic converters.

A thermionic converter is essentially a metal cathode and anode separated by a rarefied gas and enclosed in a capsule. When great heat, such as that provided by a nuclear fuel pin, is applied to the cathode, electrons boil off its surface. They pass through the gas to the anode, creating a voltage difference. Theoretically, this conversion of heat into electricity can be 40 per cent or more efficient at high enough temperatures. This, and the fact that no moving equip-

ment is required, is an advantage in such space operations as orbiting laboratories or launching stations.

The Russians' small nuclear reactor probably was fueled by from 50 to 60 slender pins. Each pin probably contained the common uranium 238 highly enriched with the uranium 235 isotope. U.S. scientists presumed that the objective was about 10 kilowatts (kw) of electricity from each pin. Output evidently fell below this mark.

If the inferences of U.S. observers are correct, the experiment also provided an insight into Russia's energy research and development "philosophy." It reflected a firm commitment to get new technology out of the laboratory and onto the drawing board as quickly as possible in the hope of achieving practical applications sooner.

The opposite point of view is increasingly prevalent in the United States, where large developmental investments are seldom thought justified until all relevant technology has been so thoroughly explored that a construction fiasco becomes highly unlikely.

SNAP 27, a radioisotopic
thermoelectric generator,
was left on the moon by
Apollo 14 astronauts
to provide power for
an experiment package.

Energy

Continued

The major U.S. nuclear-thermionic effort is carried on by Gulf General Atomic, Incorporated, for the Atomic Energy Commission (AEC). There is little chance that it will lead to construction of a nuclear-thermionic reactor until research and development is far advanced and the extent of the large power source needs of the civilian and military space programs are established.

The only other country known to be vigorously working on thermionic technology is West Germany. The outcome of a proposal to build a large, expensive experimental system there probably hinges on the availability of U.S. funds.

Nuclear power in space. The year also saw the first acknowledged Russian use of nuclear power in a space mission. It was used in the unmanned lunar rover Lunokhod, which landed on the moon in November, 1970. The application was a simple one. A radioisotope-fueled gas heater system maintained the temperature of the vehicle's instruments between 14° F. and 85° F.

The Russians did not identify the isotope used. But, it was probably pluto-nium 238, which has been used exclusively for such purposes in the U.S. space program. Lunokhod's successors will probably have similar heaters, and radioisotopic thermoelectric generators (RTG's) will probably be substituted for the solar cells that are Lunokhod's sole electric power source.

The RTG's powering the U.S. Apollo lunar-surface experiment packages that were left on the moon by the Apollo 12 and Apollo 14 astronauts continued to function perfectly, generating more than 70 watts, and showing almost no sign of weakening. They are fueled by plutonium 238. Although the two missions were 15 months apart – November, 1969, and February, 1971 – the output of the generators was nearly identical.

The unmanned vehicles to be used in the Pioneer F and G Jupiter fly-by missions in 1972 and 1973 will make much more extensive use of isotopic power sources. Each will have four RTG's designed to produce a total of 120 watts after three years in space. Also, 11 small isotopic heaters, producing 1 thermal watt each, will protect the instruments.

Energy

Nuclear explosives. Russia pushed efforts to develop nonmilitary uses of nuclear explosives. Detailed press accounts hailed past successes and revealed ambitious plans for a wide variety of experiments and applications.

By late 1970, Russia had reported 11 completed projects involving 15 explosions. Nuclear explosives had been used successfully to create water reservoirs, suppress gas-well blowouts, or fires, and increase both the production and the production rate of oil wells.

Perhaps the most notable Soviet achievement described was the extinguishing of a gas-well fire that reportedly had burned out of control for several years. The fire had caused the loss of more than 1 million cubic meters of gas per day and had created large quantities of hydrogen sulfide gas, posing a serious health hazard.

The well was sealed and the fire suppressed by detonating a nuclear explosion equivalent to 30 kilotons of TNT. The device was placed close to the drill hole and more than 7,900 feet under the ground.

Other uses of nuclear explosives reported to be under development in Russia included: increasing production from natural gas reservoirs, dam and canal construction, removal of earth and rock lying over ore in open pit mining, the fracturing of ore for underground mining, and the creation of underground storage reservoirs.

Interestingly, Russia seems to have the same broad objective in using nuclear explosives as in nuclear thermionic experiments. That is, the scientists try to bring the technology into the service of the national economy as fast as possible, even if all the techniques involved have not been thoroughly tested.

The United States, the only other country carrying on nuclear explosives development for civilian purposes, concentrated on prospects of adding to the dwindling reserves of natural gas that it is economically feasible to recover. Nuclear detonations can greatly increase flows from submarginal gas fields, making exploitation practical.

While no new experiments were conducted, evaluation of the most recent one, the September, 1969, Rulison test in Colorado, went far to confirm the usefulness of the technique. The quantity of gas released from the well during 110 days in 1970 and 1971 equaled an estimated 36 years of production at the flow rate previous to the explosion, and the potential for continued high production was promising. Equally important, radiation levels in the gas, principally from tritium produced by the explosion, fell off markedly with time.

However, it was unlikely that gas from the well could be marketed soon. This was primarily because of public concern about any increased exposure to radioactivity, no matter how slight.

MHD generators. In Russia, long the acknowledged leader in this field, completion of the first semi-industrial scale magnetohydrodynamic (MHD) generator ever built was announced in April, 1971. In an MHD generator, power is produced by passing a conducting fluid—a liquid, gas, or molten metal—through a magnetic field. Theoretically, MHD generators can convert heat into electricity at as high as 60 per cent efficiency as compared to the approximate 40 per cent efficiency of modern conventional generators.

Designed to produce 25 megawatts (Mw) of electricity, the Russian MHD generator is operated in conjunction with a 50-Mw conventional steam generating plant. The steam plant uses the hot gases (upward of 3000°F.) which have passed through the MHD section. The fuel of the plant is natural gas.

Soviet specialists also revealed plans to design a 1,500-Mw station of the same type. They estimated that with an overall conversion efficiency of 50 per cent—well below the theoretical maximum—the generating costs of the plant would be lower than those of any conventional unit in Russia.

In Germany, a relatively small MHD plant (1 Mw) was built to test the usefulness of such plants in generating power for utility systems during peak consumption periods. It went into operation in April, 1971. In Japan, an MHD pilot plant was scheduled for completion sometime in 1972.

After a long hiatus, there were stirrings of renewed interest in MHD generators in the United States. However, it was not clear whether government or industry was prepared to finance an effort comparable to those already underway elsewhere.   [JOHN H. STUMPF]

Genetics

Human chromosomes stained with quinacrine hydrochloride, become patterned in such a way that each can be identified accurately. These are chromosomes of a person with Down's syndrome, a type of Mongolism that is caused by one too many of chromosome 21.

In 1971, a new staining technique made it possible to positively identify and distinguish human chromosomes from one another. Except for a few special cases, detailed observation of chromosomes in any species has been extremely difficult. In humans, even the correct number of chromosomes was not known until as late as 1957.

The standard technique for examining human chromosomes has been to use shadowgrams, which provide only silhouettes of the chromosomes. Shadowgrams reveal only overall size and shape. This does not allow many of the chromosomes to be distinguished from each other, because their differences in size and shape are indistinguishable.

The new technique was developed in the laboratories of biophysicist Tobjörn O. Caspersson of the Karolinska Institut in Stockholm, Sweden. It involves staining the chromosomes with special fluorescent dyes that are derivatives of quinacrine. The dyes are complex organic substances, one of which is quinacrine hydrochloride, the common antimalarial drug also known as atebrin. Quinacrines

have a high affinity for certain portions of chromosomes. Caspersson and his co-workers observed that the positions of these portions and, therefore, the patterns of fluorescence they create when stained are characteristic for each chromosome. Other scientists have extended this method and it is now possible to identify each of the 23 pairs of human chromosomes by the pattern of fluorescence it exhibits.

This will help answer some questions that have long baffled geneticists. The great potential of the technique was illustrated by several reports of research in which it figured prominently. For example, in March, 1971, a group of scientists from the MRC Clinical and Population Cytogenetics Unit of Western General Hospital, Edinburgh, Scotland, used the technique in experiments that identified the chromosome involved in chronic myeloid leukemia. A form of Mongolism called Down's syndrome is known to be caused by having three, instead of the normal two, of chromosome number 21. Chronic myeloid leukemia, on the other hand, is caused by loss of a

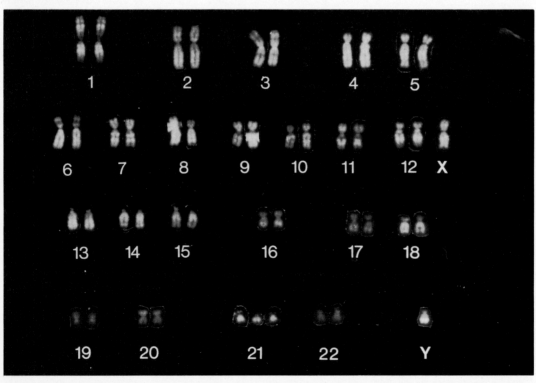

Genetics

Continued

portion of a chromosome. This chromosome was also thought to be number 21.

By using quinacrine hydrochloride to stain chromosomes from the cells of people with these maladies, the scientists could clearly see that the same chromosome was not involved in both conditions. Chronic myeloid leukemia is caused by a partial loss of chromosome number 22.

Prenatal diagnosis. Liberalized abortion laws in several states improved prospects for preventing inherited disorders in 1970 and 1971. In many cases, women who might deliver a deformed or incurably sick baby could legally choose to terminate their pregnancy. One goal of medical genetics is to identify fetuses that will produce such children.

By 1971, it was possible to diagnose in the fetus over two dozen genetic diseases and a large number of disorders that result from chromosome defects. In the method used for such detection, a sample of the amniotic fluid, which surrounds the fetus, is removed through a needle inserted through the prospective mother's abdomen. The fluid consists of liquid (reflecting in part the metabolic processes of the fetus) and a few cells shed by the fetus.

The medical geneticist allows the cells to grow and multiply in a nutrient solution for a few days. Then he examines the chromosomes. One irregularity he might find is three copies of chromosome 21. This would indicate a fetus that would develop Down's syndrome, which cannot be treated. Many of the detectable inherited diseases, however, are caused by faulty genes and show no visible abnormality of chromosome structure. These can be detected either by chemical analysis of the liquid part of the amniotic fluid or by chemical or enzyme studies of the fetal cells.

Often, however, it may be found that the fetus has normal chromosomes and genes in spite of a high risk of abnormality indicated by the family histories of the prospective parents or their own chromosomal or genetic makeup. In such cases, the parents are relieved of the great tension that would certainly otherwise last throughout the woman's pregnancy.   [H. ELDON SUTTON]

Geochemistry

Tests completed in 1971 show what appears to be indisputable evidence of extraterrestrial amino acids in a stone meteorite. This meteorite fell on Sept. 28, 1969, near Murchison, Australia. The research was completed by a group of scientists from the National Aeronautics and Space Administration's Ames Research Center in California; the University of California, Los Angeles; and Arizona State University.

Amino acids are organic compounds found in all living cells, and are essential building blocks for peptides and proteins. Scientists have known for more than 10 years that amino acids can also be produced in the laboratory by irradiating mixtures of methane, ammonia, and water with ultraviolet light. Yet they did not know if amino acids occur naturally elsewhere in the universe. The moon probably contains no organic compounds, so meteorites are the only extraterrestrial objects available for study on earth.

Five amino acids were found in the Murchison meteorite, but they differ in abundances from those in living matter on earth. In addition to the five common amino acids, the researchers found substantial amounts of two other amino acids that are never found naturally on earth, but which have been observed in some of the laboratory irradiation experiments.

They also found that, unlike amino acids found in terrestrial living matter, the meteorite amino acids lacked optical activity. Optical activity is related to the symmetry of molecules. It shows when polarized light is passed through the substance. If the amino acid has optical activity, the plane of the polarized light will rotate either to the right or to the left. Amino acids in living organisms are always optically active; those synthesized by irradiation in the laboratory are optically inactive.

The scientists also measured the amount of carbon 12 and carbon 14 isotopes in carbon compounds in the meteorite and found that their ratio of abundance differed from that found in common organic deposits, such as ocean sediments and petroleum. Finally, the hydrocarbons in the meteorite were

Amino Acids Found in Meteorite

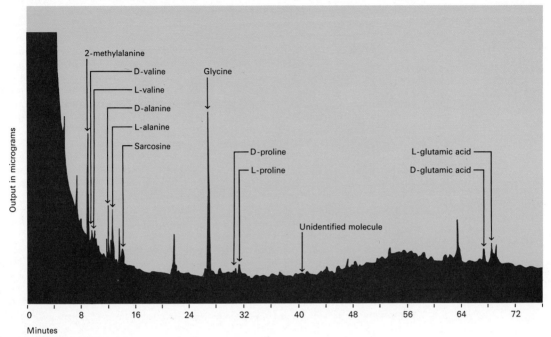

Source: NASA's Ames Research Center

Geochemistry

Continued

Gas chromatogram of hydrolyzed water taken from the Murchison meteorite shows roughly equal quantities of the left-handed (L) and right-handed (D) forms of four amino acids. This equality is a characteristic of amino acids that are formed in outer space rather than on earth.

about as abundant as the hydrocarbons obtained in the laboratory when electrical discharges are passed through pure methane gas.

The evidence thus indicates that the Murchison meteorite amino acids were of extraterrestrial origin and not due to contamination after the meteorite landed on earth. It also indicates that the amino acids were not produced by living things. Most meteorites are believed to originate from the asteroidal belt that lies between the orbits of Mars and Jupiter. They assumed their present chemical compositions about 4.6 billion years ago. At that time, conditions in the asteroidal region must have favored the formation of complex organic molecules, including amino acids.

Lunar geochemistry. Much information gathered from the lunar rocks recovered by the Apollo 12 astronauts was published in 1970 and 1971. (See THE MOON'S UNFOLDING HISTORY). Chemically and mineralogically, the Apollo 12 crystalline rocks and soil were similar to the Apollo 11 rocks. However, the Apollo 12 rocks contained less ti-

tanium than did the Apollo 11 rocks.

The Apollo 12 rocks, though very old, were found to be younger than the Apollo 11 rocks. Dimitri A. Papanastassiou and Gerald Wasserburg at the California Institute of Technology (Caltech) found rubidium-strontium ages of 3.3-billion years for Apollo 12 rocks, compared to ages of 3.65 billion years for Apollo 11 rocks. Uranium-lead ages showed the same trend, although not the same ages as had been given by the rubidium-strontium method.

One unusual rock from the Apollo 12 collection, Rock 12013, was a breccia consisting of light- and dark-colored fragments of igneous rocks welded together. It contained about 10 times as much barium, potassium, rubidium, thorium, uranium, zirconium, and the rare earths as the crystalline rocks at the Apollo 11 and 12 sites.

The abundance of rubidium and uranium and their decay products, strontium and lead, enabled scientists to measure the age of Rock 12013 with great accuracy. It is the oldest of the lunar crystalline rocks yet measured,

Geochemistry

Continued

and may help to explain the age results on the various lunar soil samples. Using the uranium-lead method, Mitsunobu Tatsumoto of the U.S. Geological Survey found the rock to be about 4.4 billion years old.

Despite their different chemical compositions, the light and dark fragments in Rock 12013 were almost the same age. Scientists at Caltech and at the Goddard Space Flight Center at Beltsville, Md., found rubidium-strontium ages of 4.0 to 4.1 billion years old. The Caltech group found additional evidence that the 4-billion-year age was for the time of metamorphism, when the rock fragments that originally crystallized 4.5 to 4.6 billion years ago were combined into a single rock.

A number of investigators noted that the Apollo 12 soils had a chemical composition that was quantitatively similar to what a mixture of material from both the dark portions of Rock 12013 and the neighboring crystalline rocks would be. In addition, Norman Hubbard, Charles Meyer, Jr., and Paul Gast of the National Aeronautics and Space Administration (NASA) Manned Spacecraft Center and Henry Wiesmann of the Lockheed Electronics Corporation discovered glass fragments in Apollo 12 soil samples that had chemical compositions like the dark material in Rock 12013. The fragments came from coarse fractions of the soil and had diameters of about 1 millimeter (0.03937 inch). The investigators showed that mixtures of 25 to 65 per cent of the glass fragments with the local crystalline rocks could account quantitatively for the chemical composition of the soil samples. Chemical analyses of individual grains of fine fractions of the soil are yet to be made.

Scientists theorized that Rock 12013 and other high-potassium rocks might be material from the lunar highlands that had rolled across the lunar surface, or they may be older rocks that once laid beneath the younger crystalline rocks and came to the surface when meteorites struck the moon and formed the craters. It may be significant that the Apollo 12 site was along a ray of material ejected from the crater Copernicus.　　[GEORGE R. TILTON]

This is a slice of Rock 12013 brought from the moon by the Apollo 12 astronauts. Dated at about 4.6 billion years of age, it is the oldest moon rock yet studied.

Geology

Two latitudinal belts of volcanoes have accounted for most of the volcanic gas and dust hurled into the stratosphere during the past 200 years, according to John F. Cronin of the U.S. Air Force Cambridge Research Laboratories, Bedford, Mass. He reported in November, 1970, on a worldwide survey of volcanic eruptions, from which he concluded that volcanoes in these belts have thrown three to four times as much material into the stratosphere as all other volcanic areas. The stratosphere begins at an altitude of about 6 miles in the polar regions and more than 10 miles near the equator.

One belt is just below the Arctic Circle. It includes the volcanoes of Kamchatka Peninsula in Siberia, the Aleutian Islands, Alaska, and Iceland. The greatest eruption within this northern belt during the past century was that of Bezymianny on Kamchatka in 1956. This volcano sent a cloud of dust and gases to an altitude of at least 25 miles.

The 1883 eruption from the volcano Krakatoa in Indonesia sent material even higher. Its clouds contained an estimated 5 cubic miles of volcanic rock. Krakatoa, in Sunda Strait between Java and Sumatra, is in the second major belt. This broad belt, centered along the equator, includes a series of volcanoes in Indonesia, the Philippines, Central America, the Galapagos Islands, and the Caribbean Sea.

According to Cronin, volcanic eruptions powerful enough to lift material into the stratosphere result from massive explosions of steam, as often occurs when island volcanoes become active and when volcanoes erupt in areas where there is an abundance of water. Cronin also noted that major eruptions tend to occur in groups. Seven occurred from 1950 to 1956, culminating in the Bezymianny explosion, but no eruption cloud reached the stratosphere during the following seven years. Discharges into the stratosphere began again in 1963, with violent eruptions of the volcanoes Agung on the Indonesian island of Bali, Trident in Alaska, and Surtsey, a new island formed south of Iceland. They have since occurred at Taal in the Philippines in 1965, Redoubt in Alaska in

Intensity Zones of the Quake in Peru–1970

The Peruvian earthquake of May 31, 1970, was centered 22 miles off the coast of Chimbote. Its full effect, nearly 8 on the Richter scale, was felt at many inland towns. The aftershock pattern indicated that the initial break occurred 25 miles deep. It also indicated that the rupture proceeded southward for 100 miles.

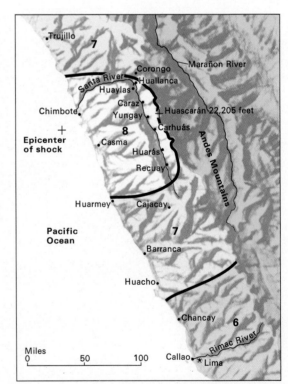

Geology

Continued

1966, Fernandina in the Galapagos Islands in 1968, and Hekla in Iceland in 1970.

Cronin says that three of these latest eruptions may have put large quantities of water into the stratosphere. This could have an effect on chemical composition and aerosol generation, but the nature of such stratospheric changes is not yet known. See Is MAN CHANGING HIS CLIMATE?

San Fernando earthquake. On Feb. 9, 1971, the densely populated San Fernando area of southern California was severely shaken by an earthquake of 6.6 Richter scale magnitude. The quake killed 64 persons, and tens of thousands of lives were endangered by partial collapse of a large earth-fill dam. Centered in the Santa Susana-Sierra Madre fault zone along the base of the San Gabriel Mountains, the earthquake caused displacements of the ground surface in an area 12 miles long and 1 to 3 miles wide. They ranged from a fraction of an inch to about 5 feet.

Geologists were puzzled immediately after the earthquake because most of the fault breaks did not correspond closely in position to faults shown on published maps. Richard J. Proctor, chief geologist of the Metropolitan Water District of Southern California, pointed out, however, that the breaks resulted from movements along faults that were known but not included in the published record. Nearly all of these faults had been fairly well defined during recent geologic investigations of routes proposed for major tunnels in a new water distribution system for southern California. Although inconspicuous, the faults were youthful geologic features that had displaced alluvial deposits during the past 3 million years. The evidence included distinctive variations in the distribution of ground water. See GEOPHYSICS.

Precambrian North America. The past 600 million years of earth history, with its complex story of volcanic activity, dislocation and drifting of continents, and evolving forms of life, is far from satisfactorily understood by geologists. Even less satisfactory, however, is the state of our knowledge concerning the events of Precambrian time, the ap-

The Peruvian city of Huarás was in the area of greatest tremor in the 1970 earthquake. With its buildings of adobe, Huarás was virtually destroyed.

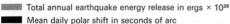

Correlation Between Polar Wobble and Earthquakes

▬ Total annual earthquake energy release in ergs × 10²⁶
▬ Mean daily polar shift in seconds of arc

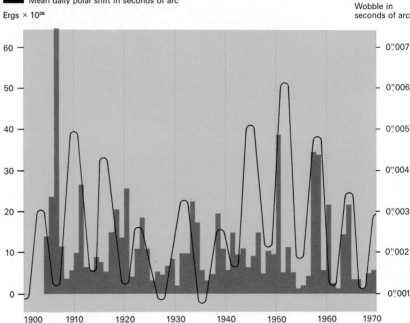

A relationship between large earthquakes (over 7 on the Richter scale) and the earth's polar wobble, especially during the past 20 years, was demonstrated by Charles Whitten, chief geodesist, NOAA. Whitten predicted major earthquakes in 1971, when the earth's wobble reaches the maximum in its 7-year cycle.

Geology

Continued

proximately 4 billion years of earth history that preceded this.

Precambrian rocks bear the imprint of many changes in their environment, and it is extremely difficult to establish the sequence of their development and absolute ages. Most of these rocks are concealed beneath younger rocks, and few of those that are exposed have yielded identifiable fossils. To obscure the record still further, rocks belonging to the first billion years of terrestrial history have not yet been found. Our only clue as to what they may have been like comes from rocks brought back from the moon's surface. See THE MOON'S UNFOLDING HISTORY.

Despite such difficulties, geologists have had increasing success in interpreting ancient rocks and in dating them by the process of measuring their radioactivity. Reports of progress came during two Penrose Conferences of the Geological Society of America, one convened in September, 1970, by Preston Cloud of the University of California, Santa Barbara, and the other in June, 1971, by Fred Barker of the U.S. Geo-

logical Survey. In both sessions, nearly 100 geologists exchanged ideas, data, and interpretations of Precambrian times in North America.

The oldest dated rocks on this continent are gneisses found in Minnesota that are approximately 3.6 billion years old. These gneisses, which are metamorphic rocks, evolved from earlier materials whose nature and history is not yet fully understood. Surveys of older Precambrian rocks, including many that have been penetrated by deep drilling, show that North America was at least half its present size 2.5 billion years ago.

Major periods of igneous activity, some of it bringing molten rock to the surface of the earth, cluster about ages of 2.6, 1.7 to 1.9, 1.2, and 1 billion years ago. Rocks including a partly oxidized banded iron formation, which indicate ancient biological oxygen production, range in age from 1.8 to 2.2 billion years. The subsequent widespread appearance of red beds of rock, colored by iron oxide, may have been caused by significant accumulations of oxygen in

Geology

the Precambrian atmosphere. Evidence of worldwide glaciation more than 620-million years ago was also cited at one of the conferences as a potential means for correlating the ages of some rocks throughout the world.

Exploration for oil and gas. The demand for petroleum continues to expand, though present methods of searching for oil and gas are somewhat limited. According to Kenneth H. Crandall, former vice-president of Standard Oil Company of California and now a geologist at Stanford University, exploration is still somewhat restricted in the United States by federal regulatory agencies. However, he noted in an article published early in 1971 that the federal agencies have begun to show a less restrictive attitude, which is welcomed by the petroleum industry. He also noted that geologists now have a better understanding of global techtonics, which should increase chances for new oil discoveries.

Environmental geology. The U.S. Geological Survey and the Department of Housing and Urban Development have combined efforts on a $5-million, 3-year research and demonstration project in the San Francisco Bay region. It is directed toward better understanding of the effects of the environment on man. According to A. Gordon Everett of the Geological Survey, the research is aimed at understanding problems of slope stability, the danger of earthquakes and ground motion, physical and chemical behavior of various solid and liquid materials, resource use and protection, natural hazards, and disposal of solid and liquid wastes. In March, 1971, the Geological Survey also announced plans for environmental and resource studies in six other major urban areas – Denver, Hartford, Pittsburgh, Seattle, Tucson-Phoenix, and Washington-Baltimore.

Projects in environmental geoscience, ranging from educational programs to urban geology and resource studies, are also underway at 19 state geological surveys and several universities. They are intended to provide the kinds of information necessary for better environmental management.   [RICHARD H. JAHNS]

Geophysics

The San Fernando, Calif., earthquake of Feb. 19, 1971, although not of major magnitude, was significant because it occurred in a densely populated area near Los Angeles. It registered 6.6 on the Richter scale, and caused the collapse of a portion of the Van Norman Dam. As a result, 80,000 persons had to be evacuated from their homes while water was drained from the reservoir behind the dam.

It was the most scientifically monitored earthquake in U.S. history. Data gathered during and following the earthquake by government agencies and university groups helped geophysicists locate its epicenter, determine the type of faulting that occurred underground, and aided further scientific studies related to the understanding of earth movements.

Measurements of the surface faulting indicated that the San Gabriel Mountains had been uplifted and thrust slightly toward the northern margin of the San Fernando Valley. This major motion was accompanied by extremely strong tremors.

The earthquake provided the first comprehensive test of building practices in and near the epicenter of an earthquake. Modern buildings generally withstood the quake well, but in regions of extremely large ground movements, some buildings were severely damaged and a few collapsed. Especially distressing was the fact that four hospitals were damaged so severely they could no longer continue to function.

Experts from the National Academy of Sciences and the National Academy of Engineering surveyed the damage and recommended revisions in the building codes to provide adequate damage-control features. The experts also recommended that old dams be strengthened, and that future dams be designed with the experience of the San Fernando earthquake in mind. Several bridges and freeway overpasses collapsed, blocking roads. These must be built to withstand greater shock. It was also clear that land-use policies in the future should take into account the probability of natural disasters and community developers should resist

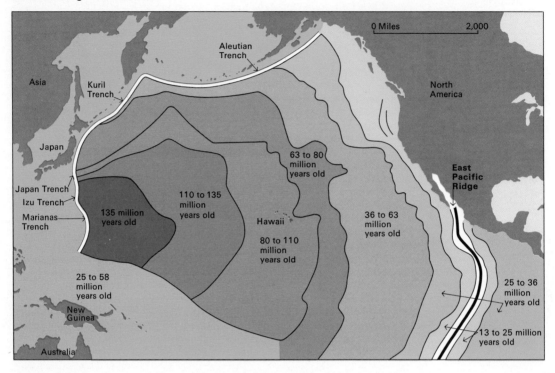

0 Miles   2,000

Aleutian
Trench

Asia

Kuril
Trench

North
America

Japan

Japan Trench

Izu Trench

Marianas
Trench

135 million
years old

110 to 135
million
years old

63 to 80
million
years old

Hawaii

80 to 110
million
years old

East
Pacific
Ridge

36 to 63
million
years old

25 to 58
million
years old

New
Guinea

Australia

25 to 36
million
years old

13 to 25 million
years old

Geophysics

Continued

Sections of the bottom
of the Pacific Ocean
have been dated by
geophysicists who have
examined fossils in
sediment cores from
holes drilled in the
sea floor. Data shows
that the floor spread
both westward and
eastward from the
East Pacific Ridge.

pressures to develop hazardous locations. See GEOLOGY.

Plate tectonics. The theory that sections of the earth's crust move laterally to one another to produce continental drift was further developed during 1970. Mawia Barazangi, Bryan Isacks, and Lynn Sykes of the Lamont-Doherty Geological Observatory at Columbia University analyzed shear waves from deep earthquakes. They found that the lithospheric plate in the western Pacific Ocean that thrusts under the island arc in the Tonga seismic zone is an extension of the undisturbed lithosphere that lies under the Pacific Ocean to the east.

Isacks and Peter Molnar, also of Lamont-Doherty observatory, analyzed more than 200 deep and intermediate-depth earthquakes and found that they were caused by the stresses created as the lithospheric plates descend into the mantle in the various seismic zones around the Pacific Ocean. Their study, published in March, 1971, suggests that the lithosphere sinks into the underlying asthenosphere under its own weight, but

encounters resistance which produces some quakes below about 188 miles. The results also indicate that contortions and fractures occur in the descending lithospheric plates.

Sykes studied groups of earthquakes on the mid-Atlantic Ridge and other parts of the mid-ocean ridge system where two lithospheric plates are believed to butt. He found that, while most of the larger earthquakes occur along fracture zones to the side of the ridge, nearly all of the minor earthquakes occur along the ridge itself. These earthquakes are accompanied by faulting and may prove to be a means of detecting underseas volcanic activity.

Direct measurements of sea floor spreading were attempted in Iceland by Robert Decker of Dartmouth College. He had established an extremely accurate optical triangulation network across the island in 1967. After the volcano Hekla erupted in May and June, 1970, he examined many of his stations and found that measurable horizontal movements had taken place. There also appeared to be downward movements

Geophysics

Continued

in the fault areas. In contrast, studies in Hawaii suggest the vertical movements there are upward, caused by the pressure of hot magma beneath the earth's crust.

William Dickinson of Stanford University studied the sedimentary, metamorphic, and igneous rock sequences that form the island arcs and deep-sea trenches. He found that the sequence of rock types and their structural locations could be explained by the concept of underthrusting lithospheric plates, and did not conflict with that theory.

The magnetism in the rocks of the North Atlantic Ocean bottom, which is evidence of sea floor spreading, was studied in 1971 by Lamont-Doherty's Walter Pitman, Manik Talwani, and James Heirtzler. They projected the spreading process back in time to find that the European and North American continental masses were joined 190 million years ago; there was no ocean between them. Charles Drake, in a dissenting view, suggested that a small ocean, perhaps a third the size of the present North Atlantic, existed there 190 million years ago. He said this was more compatible with the fossil evidence than the theory of no ocean at all.

Also in 1971, petroleum geologist Arthur Meyerhoff of the American Association of Petroleum Geology in Tulsa, Okla., pointed out a number of irregularities in the current theory of sea floor spreading. These included the distribution of animals and various climatic indicators. For example, 95 per cent of all *evaporites* (sedimentary deposits formed by the evaporation of seawater) are in areas under today's dry-wind belts. This shows that the lower atmosphere has circulated in much the same pattern for 1 billion years. This is physically impossible unless the rotational axes of the earth and the positions of the continents and ocean basins have remained stable over that period of time. The data from the Deep Sea Drilling Project, however, continued to verify the theory of sea floor spreading. See Drilling into the Ocean Floor.

Man-made earthquakes continued to occupy the attention of seismologists. These quakes were first noticed as aftershocks from nuclear tests in Nevada, and

"The Van Allen belt is falling! The Van Allen belt is falling!"

Moving Plates on the Pacific Floor

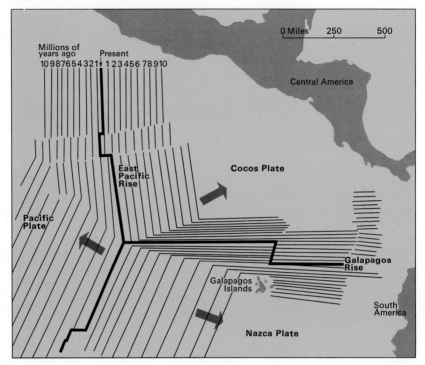

Data used to create this chart of Pacific Ocean floor is believed to confirm theory that three plates of the earth's crust are being pushed apart in area southwest of Central America. The heavy lines indicate upwelling of lava along the ocean rises, while lighter lines show the increasing ages of floor on either side of rises.

Geophysics

Continued

after pumping liquid wastes into a deep hole near Denver. It is possible that the natural strain that is constantly accumulating in such earthquake-prone areas as California might be intentionally relieved by a large number of relatively small man-made shocks before it builds to a single, large, destructive one.

Most of the earthquake activity associated with human endeavors appears to occur along known fault zones. Barry Raleigh, Jack Healy, and John Bohn reported an increase in the number of shocks in the Rangely, Colo., oil field, which occurred after an oil reservoir was flooded with water. The principal seismic activity occurred at a depth of from 1 to 2½ miles along a vertical fault.

Moonquake studies continued, and several types were identified. Some appeared to be associated with such events as the impact of the empty lunar modules from the Apollo 11 and 12 missions, while others were probably natural moonquakes. A correlation was found between moonquakes and the period of minimum earth-to-moon distance. This suggests that many of the moonquakes

are triggered by the gravitational pull of the earth, a phenomenon similar to tidal action on earth that is caused by the gravitational pull of the moon.

Geodynamics project. The National Academy of Sciences has established a U.S. committee to design a program for this country's participation in the long-term Geodynamics Project, sponsored by the International Union of Geophysics and the International Union of Geological Sciences. A major part of the U.S. program will be devoted to the plate tectonics. The U.S. effort, which began on Jan. 1, 1971, will fall into three categories: (1) determining the state of the earth prior to the last period of extensive sea floor spreading, (2) studying the state of the lithosphere and the nature of its present activity, and (3) theoretical and experimental investigations to produce momentum and thermodynamic equations relating to the earth.

Coupled with these studies will be investigations of intraplate movements and other motions that have no apparent relationship to the plate tectonics hypothesis. [CHARLES DRAKE]

Medicine

Dentistry. At a meeting of the International Association for Dental Research in Chicago on March 21, 1971, researchers reported the results of preliminary tests of a new plastic sealant developed to prevent tooth decay.

Throughout history, man has been plagued with dental decay. In the past, this decay has been treated on a restorative basis; that is, attempting to repair the damage once it has occurred. Recently, however, dentists have come to realize that a preventive approach would be far more practical. This has given rise to a new division within the dental health sciences, that of preventive dentistry. The reports on the new plastic sealant indicate a major breakthrough in this area.

During a year of testing, the material proved to be dramatically effective in reducing the chances of decay starting. It adhered to the teeth, was durable, and was simple, and painless to apply. Compared to the cost of fillings, the treatment is relatively inexpensive. The clear plastic material seals pits and fissures not easily reached by the toothbrush.

The tooth to be treated is thoroughly cleansed with brushes and pumice, and a phosphoric acid solution is applied to prepare the surface. The dentist then "paints" the plastic material onto the tooth. An ultraviolet-light "gun" is directed at the tooth. This hardens the new plastic material into a dense, clear, and highly protective shield.

Further research may make this one of the most efficient measures to prevent tooth decay in children, and for mass public health programs, handicapped persons, and those not able to receive regular dental care.

New dental glue. A group of scientists at the University of Utah reported in January, 1971, on their attempts to synthesize one of the most amazing glues in nature—that of the tiny barnacle. Barnacle glue solidifies instantly on contact with water, and thus it is particularly useful for dentistry. If the glue can be synthesized, it could eliminate the drilling that is necessary to anchor gold and silver fillings. The research is funded by the National Institute of Dental Research.   [ROBERT J. HILTON]

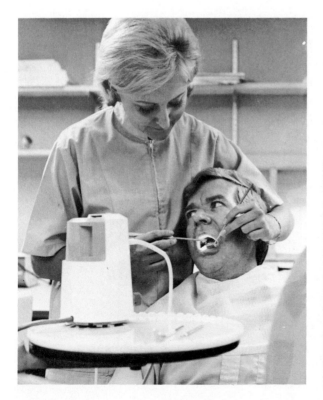

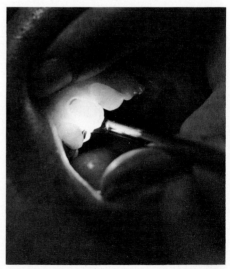

Transillumination—a technique for examining teeth with light—has been improved with the use of a flexible optic fiber light source, *left*. Dentist can detect decay, fillings, and abscesses as each transmits a different amount of light. Transillumination of upper-right cuspid, *below*, shows silicate filling (dark area near gum).

Medicine

Continued

Internal Medicine. The search for the causes of cancer and better methods of diagnosing and treating it was advanced in 1971. The first isolation of a human cancer virus was announced in July by a research team led by virologists Leon Dmochowski and Elizabeth S. Priori of the M. D. Anderson Hospital and Tumor Institute in Houston. The team isolated the spherical-shaped virus known as Type-C after a decade of research on 174 different human cancers. They finally succeeded on the 175th try. The researchers said at least five more years of testing would be required to determine if the virus can indeed cause cancer in a healthy person.

Blood test. In January, Chloe Tal, senior lecturer at the Hebrew University-Hadassah Medical Center in Jerusalem, reported developing a new blood test for early diagnosis of cancer. The new test determines the presence in the blood serum of a distinct protein, called the T-globulin, which has been found only in cancer patients and in pregnant women. A positive result from the test indicates either pregnancy or cancer.

Serum samples of Hadassah University Hospital patients, some of them confirmed cancer cases, were selected and tested. Of 520 patients, 356 had positive results. When these results were checked against hospital records, the data confirmed 350 verified cases of cancer, 3 suspected but unverified cases, and 3 pregnancies. The tests had detected 27 different types of cancer—including leukemia, Hodgkin's disease, carcinomas of the breast, thyroid, stomach, lung, kidney, prostate, female genital tract, and nasopharynx, and lymphosarcoma of the lung.

Breast cancer virus. In March, an international team of scientists led by Dan H. Moore and Jesse Charney of the Institute of Medical Research in Camden, N.J., and scientists in Detroit, Mich., and Bombay, India, reported detecting particles in samples of human milk that have identical characteristics of a type of virus that can cause breast cancer in mice.

It was clear that a number of experiments must be performed to determine if particles in human milk are indeed

The sound of blood flowing through the vessels is picked up by a sensitive microphone, recorded on magnetic tape, and analyzed by phonoangiography—a new method of locating and estimating size of arterial obstructions.

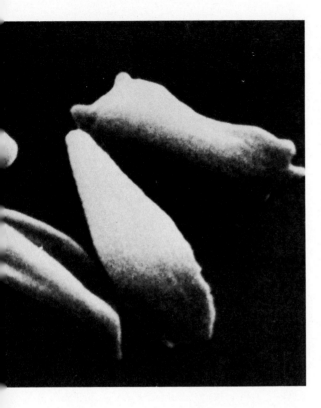
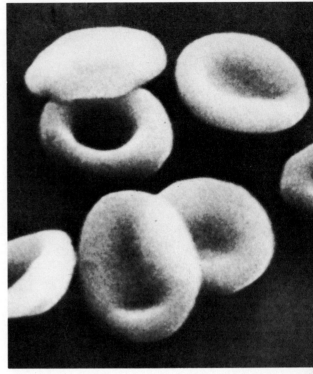

Medicine

Continued

Researchers found a drug that turns the crescent-shaped red blood cells of sickle cell anemia, *above*, into normal red blood cells, *above right*. The elongated sickle cells block narrow capillaries and deprive tissues of oxygen. The disease, which affects Negroes, can be fatal.

breast cancer viruses. An editorial in the British science journal *Nature* pointed out that "the sort of evidence that is needed will include proof that human breast cancer cells contain virus-specific antigens or viral nucleic acids, that the cancer cells produce the virus, that the virus can transform cells in culture or induce breast tumors in animals, and, most convincing of all, that it is possible to vaccinate against the virus and lower the incidence of the cancer as a result. . . ." There is also the possibility that a screening procedure for the human virus in milk might provide an important additional factor for the early diagnosis of breast cancer.

Molecular detection. Another possible major advance in the diagnosis of cancer in human patients was reported in March, 1971, by biophysicist Raymond Damadian of the State University of New York Downstate Medical Center. Damadian has successfully applied a method called nuclear magnetic resonance (NMR) to distinguish cancerous tissues from healthy cells. In NMR, radio waves are directed at individual

atoms within living cells in strong magnetic fields. At the right wave length, an atom will absorb the energy briefly and then emit energy to return to a resting state. The wave length varies for hydrogen, oxygen, and other atoms. The NMR characteristics of each type of atom change according to the kind of molecule the atom is in, or even to subtle changes within the molecule.

Damadian concentrated in his experiment on protons of hydrogen atoms in water molecules within cells. He measured the "relaxation time" of these protons—that is, the time required for a proton to give up the added electromagnetic energy. He discovered that there is enough difference in relaxation time in the protons of cancer cells to distinguish them from normal cells.

Cancer immunization. Ludwik Gross, chief of cancer research at the Kingsbridge Veterans Administration Hospital in the Bronx has succeeded in immunizing guinea pigs against leukemia, using a technique that may someday provide an approach to cancer prevention in human beings.

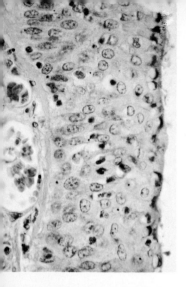

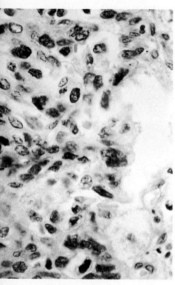

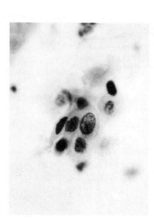

Sputum samples from smokers contain cells, *right*, that reveal condition of lung tissue, *left*, enabling doctors to alert patients before onset of cancer. Sequences show how cells match tissue in normal, in irregular, and in cancerous lungs, *top to bottom*.

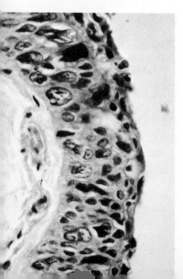

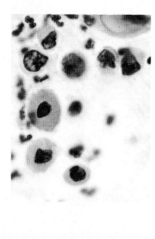

Dr. Gross discovered 20 years ago that injection of cancer-cell extracts under the skin of mice effectively produced immunity to cancer. His new work, announced by the American Cancer Society in November, 1970, uses the same principle. By injecting extracts of leukemia cells under the skin of guinea pigs, which are normally susceptible to the lethal disease, Dr. Gross made them highly resistant to later injections of leukemia cells. At present, however, the technique does not appear applicable for the prevention or treatment of leukemia in man.

Vaccine for leukemia? A preliminary report of a three-year study, published in the British medical journal *Lancet* in November, 1970, revealed that children under the age of 15 in Quebec Province who were vaccinated against tuberculosis appeared to have only half the death rate from leukemia of non-vaccinated children.

The study, at the Institute of Microbiology of the University of Montreal, was initiated when prior animal experiments showed that the tuberculosis vaccine, BCG, afforded some protection against leukemia. "This is the first time that an apparently protective effect of BCG against leukemia has been shown in man," the report stated. In the non-vaccinated population of about 340,000, there were 191 deaths from leukemia. This contrasted with 96 leukemia deaths among roughly 407,000 who had been vaccinated with BCG.

BCG is not thought to exert any specific effect on leukemia. Instead, it appears to enhance the activity of an individual's natural immunologic defenses, thus helping to destroy leukemic cells.

If the incidence of leukemia can be lowered in human populations vaccinated with BCG, the implications are profound. Apart from the obvious immediate benefit in providing some protection against leukemia, the Canadian work strongly reinforces the likelihood that immunotherapy will become of major clinical importance in cancer.

Sickle cell disease cure? Physicians conducting research into possible cures for sickle cell anemia expressed "guarded optimism" in 1970 after the announcement that a treatment for the incurable hereditary blood disease had been found. Sickle cell anemia, first identified

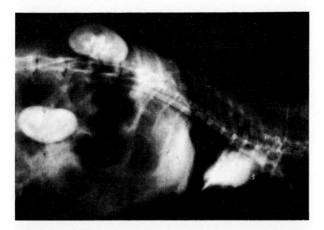

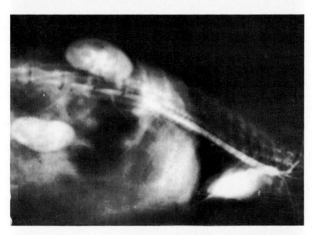

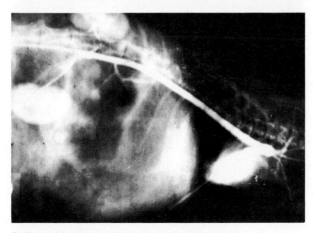

Radiographic plates, or X rays, showing progress of dye through the blood vessels and organs of a rabbit, were taken from three slightly different angles. Up to 24 such radiographs can be combined and then viewed on a special screen. This radiology technique, developed by Dr. R. Hoppenstein of New York City, permits three-dimensional viewing of X rays to see behind dense body structures.

in 1910, is a disease of the blood protein, hemoglobin. In the United States, it affects Negroes almost exclusively. The disease kills at least half its victims before the age of 20, and cripples most of the rest long before death. One in 400 American Negroes has the disease, and 1 in 10 carries the sickle cell trait.

The treatment, announced by pathologist Robert M. Nalbandian, at Blodgett Memorial Hospital in Grand Rapids, Mich., consists of intravenous injections of high concentrations of urea, a body waste chemical, in a sugar solution. The result, he said, is a conversion of the crescent-shaped red blood cells into a normal round shape and alleviation of the disease's crippling pain.

The treatment was based on work performed by Makio Murayama, a research biochemist at the National Institutes of Health in Bethesda, Md.

Paul R. McCurdy, whose work with 11 patients was included in Nalbandian's progress report on 25 patients in four hospitals, has pointed out that the "approach is chemically rational and very promising. But the road to possible treatment for sickle cell anemia has been paved with good, sound ideas, none of which has fulfilled its original promise."

Infusions of urea are not without danger. Patients have become dehydrated as a result of therapy. McCurdy, associate professor of medicine at Georgetown University School of Medicine and a medical officer in hematology at District of Columbia General Hospital, cautioned that a "Niagaralike diuresis," (excessive urination) amounting to as much as 20 or 30 liters in 24 hours, accompanies intravenous urea infusion. Unless fluid intake can be maintained, a patient will develop severe dehydration possibly leading to convulsions, and even death.

Cardiovascular system. A team of biomedical researchers led by Robert S. Lees and C. Forbes Dewey, Jr., of the Massachusetts Institute of Technology's General Clinical Research Center has developed a new, simple, noninvasive method of diagnosing diseased arteries. The new technique, called phonoangiography, was developed from engineering procedures and from data originally used to characterize noise produced by the flow of liquid through rigid pipes. It consists of listening to the sound of blood

Medicine

Continued

flowing in a patient's arteries. When the blood passes through an artery narrowed by atherosclerosis, the normal smooth flow breaks up and becomes turbulent, producing a characteristic sound.

The researchers announced in October, 1970, that the turbulent sound produced in the arteries can be picked up with a sensitive microphone, and that the sound is loudest near the narrowest part of the artery. By recording and analyzing the sound, the researchers can estimate the extent of narrowing.

Phonoangiography presently works only in arteries close to the surface of the skin. The sound from deeply buried arteries is too diffused by intervening layers of tissue to be useful. Improved sensitivity in pickup, recording, and analysis, however, should eventually make it possible to study deeper vessels, including the coronary arteries.

Red blood cell stimulant. In March, 1971, researchers at the University of Chicago and the Atomic Energy Commission's Argonne Cancer Research Hospital reported they had isolated erythropoietin, the hormone that stimulates red blood cell formation. The achievement, which has important implications for kidney disease patients, was the work of biochemist Eugene Goldwasser of the Pritzker School of Medicine at the university, and Charles Kung, senior scientist at Argonne.

Erythropoietin has been known to exist for at least 50 years, but in such minute traces that it has never been isolated in pure form until now. The hormone must be in a pure state for researchers to study its chemical nature and use it in clinical trials. Once its chemical structure is known, it may be possible to synthesize the hormone.

Erythropoietin was separated from the blood of anemic sheep, which contains an abundance of the material. Even so, blood from 150 sheep is required to produce 200 millionths of a gram. Anemic humans also produce the substance.

Kidney disease patients eventually may benefit greatly from the achievement, because they do not produce erythropoietin. If it were possible to treat them with pure erythropoietin, they might be able to lead a relatively normal life, much like a diabetic patient taking insulin. [THEODORE F. TREUTING]

New intrauterine devices (IUDs) made of copper have been found to be nearly 100 per cent effective contraceptives. They are also free of the troublesome side effects of earlier IUDs.

Surgery. Coronary arterial disease remained the greatest threat to health in the highly civilized world in 1971. In the United States alone, the disease causes more than 500,000 deaths each year, about half of which are sudden or unexpected. Countless other people suffer from complications of the disease, including angina pectoris and congestive heart failure.

The dangers associated with atherosclerotic coronary arterial disease are caused by a deficiency of blood in the heart muscle or by changes in the heart caused by such a deficiency. In recent years, cardiovascular surgeons have tried to prevent or correct the course of the disease by surgical procedures. They restored circulation by repairing or detouring around diseased coronary arteries. In 1971, surgeons stepped-up work on the surgical approach and pursued research into new methods.

Revascularization of the heart with the use of by-pass grafts using veins from other parts of the body, and gas endarterectomy (clearing a clogged artery with carbon dioxide gas injections), have gained wide medical acceptance. Both procedures have provided an excellent increase in blood supply to the heart. The heart performs better, the pain of angina is relieved, and subsequent heart attacks are fewer. See NEW LIFE LINE FOR THE HEART.

To perform the by-pass operation, the surgeon removes a length of saphenous vein from the patient's thigh. He sews one end of the vein to a clogged coronary artery beyond the site of obstruction or narrowing, and the other end to the aorta. Soon after the graft has been applied, the blood flows normally through the new vessel from the aorta to the heart muscle.

Gas endarterectomy is performed on a completely blocked right coronary artery. Carbon dioxide gas is injected through a fine-gauge needle to clear out obstructions and restore blood circulation. The same patient may require both a by-pass graft and carbon dioxide gas endarterectomy.

Complications associated with heart failure, in which pumping is impaired, also can be corrected surgically. Such complications include: (1) extensive damage to the heart muscle causing poor contraction or irregular, uncon-

Balloonlike device restarts heart after new procedure called cold brain surgery. Device is slipped under the breastbone through a small incision and then inflated and deflated rhythmically to massage heart back into action.

Medicine

Continued

trollable beating, (2) subsequent aneurysm, or ballooning out, of the scar that forms in the injured muscle, (3) insufficiency of the mitral valve of the heart because of rupture or dysfunction of muscles in the heart's ventricles, or (4) splitting of the ventricular septum, the wall that separates the two ventricles of the heart.

In the case of malfunctioning of the mitral valve, the valve can be replaced with a new artificial device. Also, defects in the ventricular septum can often be repaired with a Dacron patch.

Diagnosis of the cardiac disorder and localization of clogging in diseased coronary arteries are essential to surgical success. Cardiac catheterization – using a catheter to obtain blood samples and detect abnormalities – and coronary arteriography – filming the passage of a dye through the coronary arteries to determine sites of blockages – provide this information in patients with atherosclerotic coronary arterial disease.

A patient with acute heart failure and shock must have assisted circulation while diagnostic studies are being performed and preparations for surgery are completed. Assisted circulation can be provided by a femoral-vein-to-femoral-artery by-pass, with use of the heart-lung machine. Or, doctors may insert an intra-aortic balloon, which inflates and deflates, assisting the heart and providing better circulation.

Heart transplantations continued to decline. The virtual abandonment of the operation is a strong contrast to the fanatical enthusiasm and rash of such operations that followed the first human heart transplant in 1967. The high fatality rate undoubtedly is largely responsible for the decline. By July 1, 1971, 59 teams in 20 countries had transplanted 173 hearts to 170 human recipients. Only 25 of the recipients were still alive. Of these, 20 had lived longer than one year. The major critical problems that faced physicians at the time of the first human cardiac transplantation – control of the life-threatening rejection mechanism and limited availability of donor organs – are still unsolved.

For these reasons, many surgeons believe a workable artificial heart is ur-

Medicine

gently needed to support the cardiac recipient who rejects his transplanted heart. Obtaining hearts for transplantation is difficult because of the stringent criteria for suitable donors. The ideal donor is a young, healthy victim of an accident who died of brain damage but whose heart was not affected.

Despite the generally poor results of cardiac transplantation, at least 5 patients have survived longer than 30 months. In our experience at the Baylor College of Medicine and The Methodist Hospital in Houston with 12 patients, the 2 survivors – one a man 50 years old and the other, a 16-year-old boy – are leading relatively normal lives almost three years after undergoing cardiac transplantation.

Artificial kidney machines are prolonging the lives of patients with badly damaged kidneys. The machines were once available only in the larger hospitals and medical institutions. Now, simple, reliable, and inexpensive machines can be used in the home. Patients with acute kidney failure can be supported until satisfactory function is restored. Those with irreversible kidney disease can be kept alive indefinitely or until a kidney transplantation is possible for them.

Each dialysis, or treatment with the machine, lasts several hours, and the procedure is repeated two or three times a week, or more often. The patient's blood circulates through the artificial kidney, where waste products are removed and the cleansed blood is returned to the body.

At first, separate silicone plastic tubes were put into an artery and vein and then brought out through the skin. However, infection or occlusion usually developed within a year after insertion. The tubes had to be removed and new ones inserted in a different location. Additional dangers were bleeding and the possibility of the tubes accidentally becoming disconnected.

In 1970 and 1971, physicians devised a more satisfactory method. An arteriovenous fistula, or tubelike passage, is created by sewing an artery and vein together in the forearm or the leg so that blood flows directly from the artery

New absorbable synthetic suture at left contrasts with old-style black silk suture at right. The new suture is flexible, inert, and extremely strong. The suture is completely absorbed by the body within about 60 days.

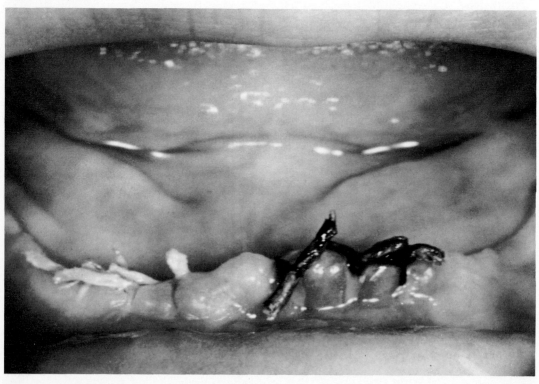

Medicine

Continued

A permanently implanted, self-powered infusion pump will administer injections of insulin or other drugs for a lifetime automatically. It can be refilled by a syringe through a membrane at its center.

into the vein. The vein becomes so distended from the increased pressure and flow of blood that it can easily be punctured with large needles. The blood can then be passed readily between the patient and the artificial kidney. The needles are removed after completion of the treatment.

Patients with arteriovenous fistulas can receive long-term, uncomplicated dialysis in the hospital or at home. They can use the arm or leg involved without danger of infection, clotting, or bleeding from accidental disconnection.

Intravenous hyperalimentation, a method of maintaining adequate nutrition in patients with prolonged intestinal disorders, was perfected in 1971. Patients with this disorder cannot digest food properly. The usual method of intravenous infusion of a solution of sugar and electrolytes to provide calories and vital salts is adequate for only a few days. This would be the time normally required to recover intestinal function after major abdominal operations or other brief gastrointestinal illnesses. These solutions do not, however, provide enough calories and protein to nourish the body adequately for extended periods of time.

In 1971, Stanley Dudrick, associate professor of surgery, University of Pennsylvania School of Medicine, Philadelphia, perfected a method of intravenous hyperalimentation that provides adequate electrolytes, protein, and calories for these patients. A catheter placed in the vena cava of the heart puts the solution into the blood. The large size of this vein and the high volume of blood flowing through it rapidly dilutes the solution. Adequate amounts of glucose provide a normal caloric intake. To avoid the allergic response of the body to foreign protein, protein is provided in the solution as short polypeptides, chemical chains of a few amino acid molecules, that are derived from more complex proteins. Iron can be provided by intramuscular injections or blood transfusions when required. This method maintains adequate nutrition in adults and actually has supplied sufficient nourishment to allow normal growth of infants.   [MICHAEL E. DEBAKEY]

Meteorology

A catastrophic storm that struck the Ganges-Brahmaputra River Delta in East Pakistan on Nov. 13, 1970, provided the biggest meteorology news of the year. U.S. meteorological satellites had observed the storm as it was developing, and U.S. scientists informed the government of East Pakistan. But, there was no time to evacuate the doomed area. Thus, at least half a million people –perhaps many more– were drowned.

The counterclockwise-rotating vortices of tropical cyclones can generate great waves in the open sea. In shallower coastal waters the cyclonic winds may pile up the water in a so-called storm surge. The narrow northern end of the Bay of Bengal sometimes allows the water to accumulate to a great height, especially when the tide and surface wind are in the same direction.

In the November storm, winds of up to 120 miles per hour drove a 25-foot-high wall of water toward the mouth of the Ganges River, inundating the densely populated low-lying islands. It was the worst recorded natural disaster of the 20th century.

Air pollution. The concern of meteorologists that the proposed U.S. supersonic transport plane (SST) might damage the atmosphere was widely publicized and helped end the project. In March, 1971, Congress cut off funds to develop the SST, and two months later, squelched an attempt to revive it.

As early as July, 1970, 50 scientists attending a meeting in Williamstown, Mass., sponsored by the Massachusetts Institute of Technology had publicly called attention to the risk of artificially changing the atmosphere. In a publication entitled "Man's Impact on the Global Environment," the scientists noted that the SST might change the balance of solar radiation and thus the world's climate by changing the particulate and water vapor composition of the stratosphere.

In a separate study, physicist James E. McDonald of the University of Arizona warned that SST's might reduce the ozone layer in the stratosphere. This layer protects the earth from the sun's ultraviolet radiation. A reduction might bring an increase in skin cancer.

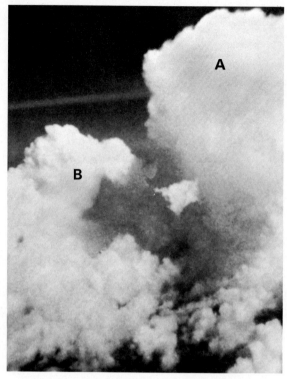

Meteorology

Continued

A Tale of Two Clouds:
The cloud at left begins
to puff up 26 minutes
after being seeded with
silver iodide crystals.
In the second picture,
the same cloud (A)
reaches an early
cumulo-nimbus stage
and tops its unseeded
neighbor (B). In the
third picture A and B
grow and move closer.
Finally, they reach full
cumulo-nimbus stature
and begin to merge. This
merger of a seeded
cloud with its unseeded
neighbor brought heavy
rain in southern Florida
on July 16, 1970.

The scientists stressed that although the potential effects of SST flights were great, more research was needed to determine to what extent these changes would occur. In 1971, the World Meteorological Organization (WMO) announced plans to establish a global network of stations to measure "background" air pollution—the pollution other than that coming from major sources of local concentrations. WMO is an agency of the United Nations with headquarters in Geneva, Switzerland.

Meteorologists in the nations that belong to WMO were to set up stations in areas free from local air pollution to monitor long-term changes in the chemical composition of the atmosphere around the world. WMO is particularly interested in carbon monoxide, carbon dioxide, and sulfur dioxide. Many independent stations which have been in service for several years—such as those the United States maintains in Antarctica and on Mauna Loa in Hawaii—became part of the new WMO system.

Weather modification research continued to focus on seeding clouds with silver iodide crystals to cause rain. In April, 1971, Joanne Simpson and William L. Woodley of the Experimental Meteorology Laboratory of the National Oceanic and Atmospheric Administration (NOAA) reported a particularly successful series of cloud-seeding experiments. They seeded cumulus clouds and found that seeded clouds rained more than three times as much as comparable unseeded clouds. The results were best on days on which few natural rain clouds were present.

The scientists carried out their experiments from 1968 to 1970. They used an evaluation technique in which the growth of a cumulus cloud, seeded or unseeded, is calculated on a computer from a theoretical model.

The severe drought that afflicted Florida, as well as the southwestern United States, led Florida officials in April, 1971, to plead for more NOAA cloud seeding.

Cloud seeding has also been used in efforts to alter precipitation patterns in storms, though such control is not always easy, as an analysis of the seeding

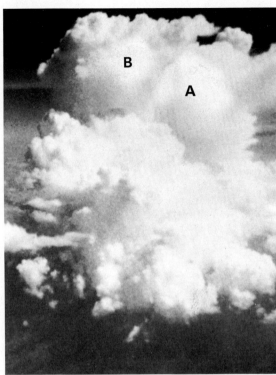

Meteorology

Continued

of Lake Erie snowstorms showed. Edmond W. Holroyd III and James E. Jiusto, meteorologists of the State University of New York, Albany, reported that heavy seeding of storms in 1968 and 1969 failed to redistribute snowfall on the shores of Lake Erie as expected. The goal had been to seed the storm with silver iodide in order to create smaller snowflakes that would be carried farther inland from the lakeshore by the wind. But the small flakes collected into larger flakes, which fell faster than desired.

Project Stormfury is an effort to find out if hurricanes can be controlled by cloud seeding. The project, operated by NOAA's National Hurricane Research Laboratory in Miami, was hampered in 1970 because there were no hurricanes available for experimentation in the project area. As a result, scientists were considering moving the project to an area of the Pacific Ocean in 1972 to take advantage of the many typhoons there.

Meanwhile, Stormfury scientists made much progress in programming computers with detailed data on the growth of a hurricane and the effects of seeding. This made it possible to predict, to a limited extent, the effect of seeding a hurricane or a typhoon. More experiments are needed to obtain sufficient data for more reliable predictions.

GARP. Planning for the Global Atmospheric Research Program (GARP) continued during 1970 and 1971. GARP is scheduled to begin about 1976. It will attempt to collect the most comprehensive sample possible of meteorological data throughout the world. It will use satellites, constant-level balloons, and other instruments and techniques. The data will be used for weather prediction programs that will determine the limits of atmospheric predictability.

Meteorological satellites. The National Aeronautics and Space Administration (NASA) launched its Nimbus 4 satellite in April, 1970. From June through November, 1970, Nimbus 4 tracked 25 giant balloons containing weather instruments that had been released into the stratosphere from Ascension Island in the South Atlantic Ocean.

NOAA's first operational satellite was

Meteorology

Continued

Steel balls, *above*, were rocketed into the air to trigger lightning in an October, 1970, test in New Mexico. They may be used before spacecraft launches in the future to prematurely discharge lightning that might otherwise damage the craft. *Right*, a U.S. weather satellite 900 miles high took this photograph of the disastrous cyclone that killed an estimated half-million persons in East Pakistan in 1970.

placed in orbit on Dec. 11, 1970. The new satellite, known as ITOS-A (Improved Tiros Operational Satellite) or NOAA-1, carried cameras for both day and night cloud photography. Radiometers to measure surface temperatures were also on board.

Applications Technology Satellite 3, (ATS-3) another NASA satellite, launched on Nov. 5, 1967, continued to provide time-lapse motion pictures of clouds. It is a geostationary satellite orbiting the earth at the same speed that the earth turns. It is positioned 22,000 miles above the mouth of the Amazon River where it can monitor a large part of the Western Hemisphere.

ATS-3 also enabled scientists to see, for the first time, the development of thunderstorms, tornadoes, and hurricanes. For instance, Tetsuya Fujita, meteorologist at the University of Chicago, used a technique known as image enhancement to film the growth and circulation of the most intense portion of the hurricane with the help of ATS-3. Fujita worked with the picture images taken from the original transmission.

Meteorologist Verner E. Suomi of the University of Wisconsin, inventor of the ATS-3 camera system, has also shown that radiometric soundings of the atmosphere are feasible from geostationary satellites. The temperature of the air at various heights is determined from the infrared radiation emitted by the air. These measurements are being made by orbiting Nimbus satellites.

Numerical weather prediction, or predicting with the help of high-speed electronic computers, was used increasingly during 1970 and 1971. Researchers began to concentrate on small-scale weather phenomena, such as showers, squalls, fronts, and heavy rain bands.

Several countries have begun to produce computerized forecasts for their local areas with the use of these so-called fine-mesh computer grids in the hope of predicting small-scale weather systems.

The National Meteorological Center of the U.S. National Weather Service in Suitland, Md., began use of a fine-mesh grid in the summer of 1971. Meteorologists believe that at least a year is needed to evaluate the system.   [JEROME SPAR]

Microbiology

An exciting insight into the mechanisms by which the enzyme RNA polymerase recognizes the starting and stopping points of a gene came from experiments reported in June, 1970. Leading investigators in this field reported their findings at a meeting at the Cold Spring Harbor Laboratory of Quantitative Biology in New York. Like many other modern biological problems, this one is related to how the information stored in genes is expressed in terms of cell function.

Most biologists accept the concept that the genetic information is recorded as the sequence of chemical units called bases of the deoxyribonucleic acid (DNA) of which genes are made. When a gene is to be expressed—that is, when it is going to initiate production of the protein for which it is responsible—its bases act as a template, or pattern. A complementary base sequence of messenger ribonucleic acid (mRNA) is formed on this template with the aid of RNA polymerase. The genetic information encoded on the RNA is then translated through a complex process into a specific protein. Such proteins, particularly the enzymes, determine the biochemical processes that are characteristic for each specific kind of cell.

Because the DNA in most cells has its millions of bases strung like beads on a necklace, genes are also arranged side by side. The problem discussed at Cold Spring Harbor was how the RNA polymerase knows on which bead (base) along the necklace (DNA) to begin forming mRNA, and how it knows on which bead to stop. In other words, how does it know where the gene it is working on begins and ends?

The questions are of more than theoretical significance. Experiments have led to the conclusion that when, and even whether, a gene will function is controlled by factors affecting mRNA synthesis. This larger area of gene expression is basic to understanding cellular biology, particularly how embryonic cells become the specialized cells of specific organs and how some mature cells become cancerous.

Sigma factor. A team of biochemists led by Wolfram Zillig of the Max

Chloroplasts, bodies containing chlorophyl, were observed and photographed dividing, *below*, outside of the plant cells where they are found. A close-up shows the process in detail, *below right*. The discovery supports a theory that chloroplasts once lived independent from plant cells.

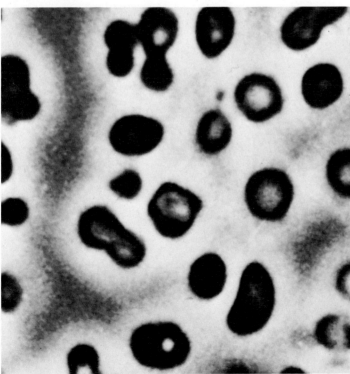

Microbiology

Continued

Planck Institute for Biochemistry in Munich, Germany, told of identifying a substance they called sigma factor, which is responsible for starting mRNA synthesis. The scientists used a combination of chemical and mechanical techniques to isolate RNA polymerase from the bacterium *Escherichia coli*. Next, they separated the enzyme into four subunits by treating it with either urea or lithium chloride under alkaline conditions. Finally, they tested various combinations of the subunits for the ability to bind to the DNA of the bacterium. They found that one subunit, the sigma factor, is required for the enzyme to recognize where to bind onto the DNA, a prerequisite for mRNA synthesis.

Virtually identical results were reported by molecular biologists Joseph Krakow and K. von der Helm from the University of California, Berkeley. They used techniques similar to those of the German group, but worked with the bacterium *Azotobacter vinelandii*.

Rho factor. Molecular biologist Jeffrey Roberts of Harvard University presented evidence that a specific protein, which he named rho factor, in *E. coli* was required to terminate mRNA synthesis. When he mixed DNA and the RNA polymerase of *E. coli* in a test tube without the rho factor, pieces of mRNA too large to be functional were synthesized. This indicated that the polymerase was copying DNA beyond the end points of genes. When the purified rho factor was added, normal-sized segments of mRNA were synthesized.

Cyclic AMP. Also at the Cold Spring Harbor meeting, two groups of microbiologists reported experiments that clarified how the cell chemical 3',5' adenosine monophosphate, or cyclic AMP, helps control gene expression. The question of why some genes are expressed in a cell while others are not has long intrigued biologists.

The expression of certain genes can be experimentally controlled in bacteria. These genes usually carry information for the synthesis of enzymes used in catabolism, the breakdown of foodstuffs into both energy and the molecules that are to be utilized in the synthesis of new cell materials.

"Good morning. My, how you've grown!"

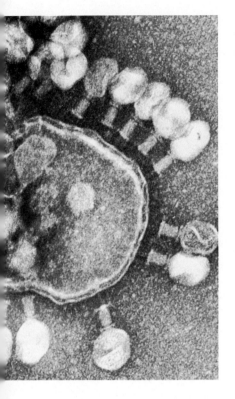

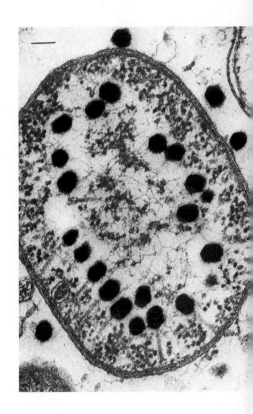

Microbiology

Continued

When viruses infect a cell, they attach to its outer surface, *above*, and inject their strand of nucleic acid into the cell, *center*. Each strand contains all the genes of the virus, and is capable of making many copies of the virus, *right*. This ultimately will destroy the cell.

The enzymes are called inducible. That is, they are synthesized (the gene that produces them is expressed) only when the material upon which the enzyme acts is present. However, the production of most inducible enzymes is repressed (the gene that produces them is repressed) when the sugar glucose is present. This occurs even when the material upon which the enzyme works is present.

In 1965, biologists Richard Makman and Earl Sutherland at Case Western Reserve University reported that repression by glucose in *E. coli* was accompanied by a reduction in cyclic AMP. Since then, it has become increasingly clear that cyclic AMP is involved somehow in both gene repression by glucose and in gene induction.

One of the two teams reporting at the Cold Spring Harbor meeting was led by Robert Perlman and Ira Pastan of the National Institutes of Health, Bethesda, Md. They designed an experimental procedure for measuring the mRNA produced by only one gene. The gene specifies production of the enzyme beta-galactosidase, which helps in the first step in the catalysis of the sugar lactose.

Cells of *E. coli* were induced to form beta-galactosidase by adding lactose to their growth medium. Next, the experimenters added radioactive uridine to the medium. This compound is incorporated only into the newly forming RNA of a cell. After one minute, which is enough time for synthesis of mRNA, the cells were broken open by treating them with a detergent. The mRNA of the resulting cell extract, including that made radioactive during the one-minute period, was isolated and purified. Now the problem was to measure the quantity of radioactive beta-galactosidase mRNA produced during the one-minute period. The experimenters took advantage of the fact that mRNA will combine with DNA having a base structure that is complementary to its own base structure. In this case, the complementary DNA was the beta-galactosidase gene. But, they had to separate the gene for the enzyme from all the other DNA of the cell. The scientists knew of an unusual tool they could use to do this.

There are viruses that infect bacteria and, in the process of destroying them, incorporate one or several bacterial genes into their own DNA. One such virus, lambda H80, carries the needed bacterial gene for beta-galactosidase. The scientists mixed the radioactive mRNA preparation with purified DNA from the virus. The mRNA for beta-galactosidase combined with the corresponding gene in the virus DNA. The uncombined radioactive mRNA was then separated from the viral DNA and its attached mRNA. The researchers then used the amount of radioactivity in the combined DNA and mRNA material to determine the amount of mRNA that had been formed by the beta-galactosidase gene during the one-minute period.

With this technique, the scientists proved conclusively that the mRNA was produced only if lactose was present, and that adding glucose to the cells dramatically reduced the amount of mRNA synthesized even if lactose was present. When cyclic AMP was added, however, the normal amount of the mRNA was synthesized despite the presence of glucose.

The second team to report on cyclic AMP consisted of Geoffrey Zubay of Columbia University and Danielle Schwartz and Jonathan Beckwith of the Harvard Medical School. They discovered that cyclic AMP must be bound to a specific protein molecule before it can activate a repressed gene. They called this protein CAP, which stands for catabolite-gene activating protein.

First, they isolated E. coli that, because of gene mutation, produced only very low levels of many of the enzymes whose production is repressed by glucose. These mutant bacteria had some general defect in the activating machinery for the glucose-sensitive genes. In the case of most of the mutants, this defect was apparently the inability to synthesize cyclic AMP, for they quickly began producing normal amounts of the glucose-sensitive enzymes when cyclic AMP was added.

One mutant, however, did not increase its enzyme production. Testing specifically for this mutant's ability to synthesize beta-galactosidase, the scientists found that none of the enzyme was made with or without cyclic AMP present. But, they discovered that normal amounts of the enzyme were made if cell extract from nonmutated E. coli was added along with cyclic AMP. They found that the material needed along with cyclic AMP was CAP, which they isolated and purified from the nonmutant bacteria. They further showed that the CAP-cyclic AMP complex reacts with DNA but not with mRNA. This indicates that the site of activity of the complex is on the gene itself.

Food from gas and electricity. One of the greatest problems facing the world is feeding its burgeoning population. Food high in protein is particularly scarce. Scientists have tried to convert readily available and cheap inedible materials into edible protein for many years. Their attempts have centered around feeding the raw materials to bacteria. The bacteria, which are high in protein, could then be processed and fed to domestic animals, or, as a direct supplement, to human beings.

Some success has been reported in raising bacteria on sawdust, wastepaper, and methane gas and other hydrocarbons. But, a truly novel process was developed in 1970. Microbiologists Hans Schlegel and Robert Lafferty at the University of Göttingen in Germany worked out a procedure for converting electricity and simple gases into food. A specialized group of bacteria are able to grow rapidly in a simple mineral-salts medium when fed the gases hydrogen, oxygen, and carbon dioxide. Electrolysis, or electrical decomposition, of water yields the hydrogen and oxygen, and carbon dioxide is also a very readily available gas.

Using enzymes, the bacteria harness the energy released when the hydrogen and oxygen recombine to form water. They use this energy to convert carbon dioxide into cell material. Schlegel and Lafferty reported that the hydrogen and oxygen obtained from one kilowatt-hour of electricity will form 33 grams (about 1.2 ounces) of bacterial protein.

The bacteria product could be processed in a number of ways to give it virtually any consistency and texture that is desired by the processor. Flavoring and other additives could then complete its disguise as any of a number of more traditional and acceptable foods. Gasburger anyone? [JERALD C. ENSIGN]

Oceanography

After years of debate, congressional proposals, and studies, the Administration of President Richard M. Nixon created, in 1970, a new environment-oriented agency responsible for ocean activities.

The National Oceanic and Atmospheric Administration (NOAA) was established as part of the U.S. Department of Commerce on Oct. 2, 1970. NOAA now brings together: the National Oceanographic Data Center and the National Oceanographic Instrumentation Center from the Department of the Navy, the National Data Buoy Development Project from the Coast Guard and Transportation departments, the Environmental Science Services Administration from the Department of Commerce, most of the Bureau of Commercial Fisheries programs, most of the marine-mining programs of the Bureau of Mines, the marine sports fish program of the Bureau of Sport Fisheries and Wildlife from the Department of the Interior, the Office of Sea Grant Program from the National Science Foundation, and the U.S. Lake Survey program of the Army Corps of Engineers. Robert M. White was approved as administrator of NOAA on Feb. 19, 1971.

NOAA ocean programs include: creating systems to assess, develop, and conserve ocean resources; accelerated surveys of inshore, continental shelf, and deep-ocean regions; and studies of the circulation of estuarine and coastal-zone waters to better predict tides, currents, and weather. Other programs are: studies of the effects of contaminants, particularly mercury and other heavy metals, on ocean life; further work on weather modification, particularly hurricane modification such as the seeding of Hurricane Debbie in 1969, which brought reductions in wind speed; use of small research submersibles and underwater habitats for the direct observation and study of ocean processes, and increased use of ships, aircraft, buoys, and satellites for the collection of oceanographic data throughout the world.

Tektite II. Ten aquanaut teams, each with four scientists and one engineer, carried out the Tektite II research program from April 20 to Nov. 10, 1970. In the series of dives, the researchers tested

Lady scientist-aquanauts from the Tektite II program check a fish trap about 50 feet under the surface near the Virgin Islands. They are wearing a new bubbleless breathing apparatus that does not disturb surroundings.

Oceanography

Continued

The *Researcher*, a new scientific survey ship designed for science on the high seas and the continental shelves, was commissioned in 1970.

their ability to work underwater for long intervals as they studied ocean life. Their habitat was a two-tank, four-room combination of laboratory and living quarters 50 feet below the surface of the sea near St. John in the Virgin Islands.

Most of Tektite II's scientific work involved direct observation at close range of marine life and its reaction to various stimuli. The scientists found that lobsters can travel up to 1,000 feet in total darkness to feed at night and then return to their daytime burrows. This demonstrates an unexplained ability of lobsters to navigate. The scientists also discovered that DDT and other man-made chemicals significantly lower the metabolism of coral colonies. In addition, they learned that alternate day-night sharing of space by different species of fish – for example, moray eels and queen triggerfish – greatly increases the number of species that a given volume of coral reef can sustain.

The scientists also showed that the geological record can provide much information about the marine life of long ago. For example, some fish and other

sea creatures eat only certain shellfish and deposit their shells in unique collections and concentrations. Other creatures break shells and bore into them in certain ways. Thus, shell deposits in ancient sediments can reveal which animals lived in an area at the time the sediments formed.

The Deep Sea Drilling Project, sponsored by the National Science Foundation, confirmed the theory of continental drift. It also filled in many details of how and when the continents wandered, uncovered new mysteries, and achieved drill-string re-entry – a major advance in man's ability to look backward in geological time. See DRILLING INTO THE OCEAN FLOOR.

Nodules recovered. Deep Sea Ventures, Inc., a division of Tenneco Corporation, Gloucester Point, Va., demonstrated a "vacuum cleaning" method that could bring up sea-floor nodules. Such nodules are stonelike concentrations of various minerals, including manganese oxide. The vacuum cleaner is actually a hydraulic pump designed to lift as much as 400 tons of material from

Oceanography

Continued

the sea floor daily. A demonstration of the pump took place in July, 1970, at a 3,000-foot depth of the Blake Plateau in the Atlantic Ocean, about 100 miles east of Jacksonville, Fla. This area is only one of many sites known to have many of the nodules.

The undersea nodules recovered contained less manganese than can be found in commercial dry-land deposits—25 to 35 per cent rather than 46 to 50 per cent. But the nodules also contain other valuable metals, including as much as 2.3 per cent cobalt, 1.6 per cent copper, and 2 per cent nickel. If ores from deep-sea sources can be recovered and processed at a cost competitive to land ores, commercial exploitation of this virtually unlimited natural resource could soon become important.

Artificial upwelling. Columbia University scientists on Aug. 7, 1970, opened a mile-long pipeline to pump deep, nutrient-rich ocean water to the surface north of St. Croix in the Virgin Islands. The experiment produced artificial "upwelling," and was based on suggestions made by Columbia University oceanographers Robert Girard and J. Lamar Worzel, in 1967. See LIVESTOCK FROM THE SEA.

Currents measured. In December, 1970, oceanographers William J. Schmitz, Jr., and Frederick C. Fuglister, of Woods Hole (Mass.) Oceanographic Institution and Alan R. Robinson of Harvard University reported that they had measured ocean currents in the Gulf Stream as fast as 11 centimeters (about 17 inches) per second. The measurements were taken 200 meters (about 660 feet) above the ocean bottom, using metering devices moored 35 kilometers (about 23 miles) apart.

An analysis of two months of metering showed large variations in these deep currents over a period of about 30 days. The mean velocity—the general rate of flow—was 2 centimeters (less than 1 inch) per second eastward, and 10 centimeters (less than 4 inches) per second northward.

Marine pollution attracted a great deal of attention during the year. A report of the President's Council on Environmental Quality proposed extensive

Recording the Earth's Magnetic Field

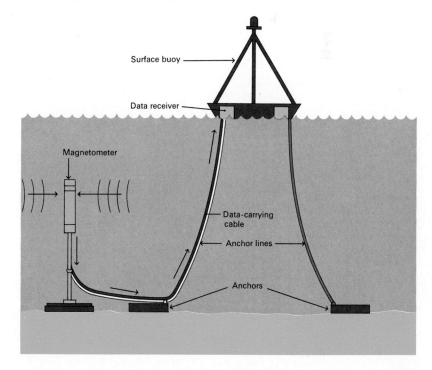

An unmanned device moored in the ocean can measure the earth's magnetic field more precisely than can magnetometers towed by ships. Its data will aid in finding offshore oil and drawing accurate magnetic maps for sea and air navigation.

Surface buoy

Data receiver

Magnetometer

Data-carrying cable

Anchor lines

Anchors

Mopping Up
Oil Spills

When two of its tankers collided in San Francisco Bay, on Jan. 18, 1971, Standard Oil Company of California worked fast. Seven oil-skimming barges quickly attacked the resulting spill – 20,000 barrels of fuel oil. The barges pumped floating oil and some water aboard, separated them by gravity, discharged the water, and stored the oil. Other boats laid down 10,000 feet of floating barriers, called booms, to protect the shoreline. Oil that eluded the booms was soaked up ashore by 9,000 bales of hay. The spill did little permanent damage.

This event was a far cry from what happened when the sandy shores of France and England were heavily fouled by oil from the stricken tanker *Torrey Canyon* in March, 1967. The grounded vessel dumped 700,000 barrels of crude oil into the English Channel. The incident put oil companies as well as governments on guard, and ways were sought to combat such destruction.

There are now a number of ways to cope with oil that has been spilled into the world's waters from tanker collisions or offshore oil-well accidents. For example, the U.S. Coast Guard plans to be stockpiling its Air Delivered Anti-Pollution Transfer System (ADAPTS) in 1972. This system uses huge nylon bags coated with nitrile or polyurethane rubber. The bags can hold up to 140,000 gallons, and are used with pumps that suck oil from the holds of stricken tankers. ADAPTS can be airdropped. It can handle 5.6 million gallons of oil in 24 hours in winds of up to 45 knots and in waves up to 12 feet high.

The Coast Guard tested another new device in the Gulf of Mexico in June, 1971. It is a barrier designed to keep oil from spreading in moderate seas with no more than 5-foot waves, a 2-knot current, and a 20-knot wind. Previous barriers worked only in calm water with little current and wind. Johns-Manville Company of New York City developed the new device – a flexible, floating, nylon curtain that extends about 2 feet above and below the surface. The curtain is impregnated with rubber and supported by inflatable floats.

Reynolds Submarine Services of Miami, Fla., delivered its first oil-removal system to a major oil company in 1971. The floating, bowl-shaped device, known as Medusa, is made largely of aluminum, and has a skirt of neoprene-impregnated nylon. The largest model, 18 feet in diameter, handles 10,000 gallons of oil and water per minute in waves up to 5 feet high and winds of up to a maximum of 30 knots.

Medusa's central pumping unit draws in spilled oil over a circular, flexible barrier. It uses a swirling motion that concentrates the oil in the center where it is sucked into a tube. The concentrated oil is then pumped to floating storage tanks. Medusa can be towed or airdropped.

Another oil-collection system, for use in harbors and rivers, was developed by American Oil Company of Chicago. The system has a rotating drum covered with polyurethane foam that absorbs more oil than water. Mounted on the bow of a vessel, the drum rotates in a slick and absorbs the oil, which is then squeezed by rollers into containers. Large models can recover 350 barrels, almost 15,000 gallons, of oil an hour.

Dispersants are the main chemicals used on oil spills. They speed the dispersion of oil into water and keep spills from moving intact into shorelines. Dispersants also expose more of the oil's surface to microorganisms, probably speeding its natural degradation. However, dispersants tend to harm marine organisms and, also, make oil more poisonous.

Scientists also are experimenting with bacteria that break oil down chemically and feed upon it. The bacteria could be sprayed on an oil slick. Bacteria cultures are especially promising for use on oil-soaked beaches, which are extremely difficult and expensive to clean in any other way.

Another way to combat the effects of oil spills is to detect them as soon as possible. Federal agencies are testing airborne ultraviolet and infrared sensors that can detect and identify floating oil during daylight hours. Microwave sensors have also been tested. They can be used to detect floating foreign material day or night, but cannot identify it.

The great activity in the search for ways to clean up oil spills has been generated by a public no longer complacent about the environment-damaging errors of industry. Both industry and government must continue their efforts if we are ever to be sure that our seas and shores are fully protected from the ravages of spilled oil.   [KENNETH M. REESE]

Oceanography

legislation to regulate all ocean dumping, and ban or strictly limit dumping of harmful materials. Several chemicals, especially persistent pesticides such as DDT and other chlorinated hydrocarbons used to control insects on farms, were found to be concentrated in various sea organisms, including commercial fish. Mercury, another serious pollutant, was found in swordfish in such high concentrations that the sale of the fish was temporarily banned.

Catastrophic accidents, including oil spills from tankers and discarding of industrial chemicals, increased concern over man's polluting the ocean. Measurements taken near New York Harbor–one of the world's most polluted areas, where unrestricted dumping has gone on for years–show the oxygen content of the water has dropped 50 per cent, and the amount of poisonous metals present has increased.

Seabed treaty proposed. On May 23, 1970, President Nixon proposed that the nations of the world renounce all claims to the seabed resources beyond a depth of 200 meters (650 feet). Such resources, he suggested, could be regarded as the "common heritage of mankind." This controversial proposal would establish an international regime for the exploitation of seabed resources beyond the 200-meter limit. Coastal nations could act as international trustees for those deep-sea resources off their coasts.

In another proposal to encourage international scientific cooperation, the Council of the National Academy of Sciences recommended that the United States consider opening ocean waters subject to U.S. jurisdiction to foreign scientific research. Such a step might encourage other countries to relax their own restrictions.

United Nations acts. Serious international concern for man's future use of the sea was shown in the General Assembly of the United Nations (UN). In a 108-to-0 vote, the body declared the seabed and sea floor beyond the limits of national jurisdiction to be the "common heritage of mankind." This resolution evolved from a special UN committee under Ambassador Hamilton Amerasinghe of Ceylon. [RICHARD C. VETTER]

Physics

Atomic and Molecular Physics. During 1970 and 1971, atomic physicists helped to devise new lasers and to make existing laser systems more versatile. And they used lasers to extend their knowledge of atoms and molecules.

In the past, a variety of laser devices have produced light in the visible and infrared regions of the spectrum. But until 1970, no one had succeeded in breaking into the far-ultraviolet range. The very nature of the laser process stipulates that stimulated emission, the mechanism of laser action, becomes increasingly difficult at short wave lengths.

The break into the far ultraviolet was achieved almost simultaneously in 1970 at both the Naval Research Laboratory in Washington, D.C., and the International Business Machines (IBM) laboratory at Yorktown Heights, N.Y.

In the ultraviolet lasers developed at both laboratories, hydrogen molecules in the ground state are pumped to excited vibration-rotation energy levels by a fast, intense electric spark. The electrons of the spark collide with many of the hydrogen molecules, raising them to the excited levels. The excited molecules can release energy by emitting ultraviolet radiation as they fall to an upper vibrational level of the ground state.

The electric spark is so intense that far more molecules will be elevated to the higher energy states than to the upper ground state vibrational levels, and laser action can take place. The stimulated emission is so strong that no resonant laser cavity is needed to produce the necessary amplification. The two laboratories have each reported 10 lasing wave lengths. Both reported six of these in the wave length range from 1,567 to 1,613 angstrom units.

The high-energy photons that are produced by these ultraviolet lasers cannot pass through the atmosphere. Operating in a vacuum apparatus, however, physicists can use the lasers to make more precise studies in higher energy ranges of such phenomena as photoelectric effects and properties such as optical constants.

Tunable lasers have made new experiments in atomic physics feasible by providing a source of high-energy density in a chosen narrow wave length interval. An outstanding example is the work of W. C. Lineberger and B. W. Woodward at the Joint Institute for Laboratory Astrophysics at Boulder, Colo., in 1970. These men used a tunable dye laser—which operates as dye molecules dissolved in water fluoresce—to study exactly how the laser radiation removes an electron from negative ions of sulfur (S^-).

In the experiment, negative sulfur ions are formed by an electrical discharge through hot sulfur gas. As they form, the ions are pulled away from the discharge by an electric field and then passed through a magnetic field to separate the S^- ions from other ions present.

This beam of pure S^- is then passed through a region where it is bombarded by photons from a tunable dye laser. The photons, if energetic enough, will remove the extra electron from the S^- ion, leaving neutral sulfur atoms. The scientists studied in detail the efficiency of the light pulse in removing the electron—called photodetachment—as the energy of the photons is increased—that is, as their wave length is decreased.

Success of experiments such as that on S^- has added impetus to the search for better tunable lasers, and several new systems are now competing with the dye laser. One such system is a solid state diode laser made from lead (Pb), tin (Sn), and Tellurium (Te). It operates at wave lengths near 10 microns, in the infrared. The wave length at which this crystal will *lase* (generate coherent light) can be adjusted to a great degree, however, by changing the ratio of lead to tin. In this manner, the wave length can range from 6.5 to 32 microns.

The output wave length of the laser can then be fine tuned about that approximate value by changing the temperature, which changes the size of the resonant cavity. Such temperature tuning can vary the wave length nearly half a micron.

The resonant cavity in which the diode must operate to obtain enough amplification restricts the output of the laser. These diode lasers jump from one range of lasing wave lengths to the next as the temperature changes. In so doing, they skip over segments in the spectrum.

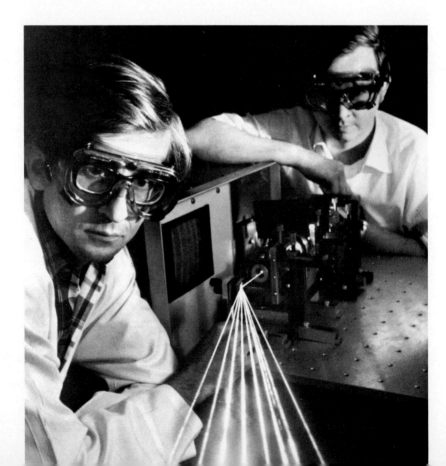

A new dye laser, the exciplex laser, was developed at Bell Labs in 1970. It produces a variety of colors that can range from near ultraviolet to yellow.

As a result, about half of the wave lengths in the tuning range cannot be produced at all.

Also, temperature tuning is not easy to control, nor is the diode temperature easy to measure. Despite its limited applicability, this diode laser was used to produce very high resolution absorption spectra of molecules such as sulfur hexafluoride (SF_6). Its use to analyze atmospheric gases for air-pollution monitoring has been proposed.

The spin-flip laser. The most exciting development in the field of tunable lasers came when Bell Telephone Laboratory scientists announced a tunable Raman spin-flip laser in 1970. In their experiment, they placed a crystal of indium antimonide (InSb) in a magnetic field and irradiated it with 10.6 micron wave length photons from a carbon dioxide (CO_2) laser.

The InSb crystal emits a small amount of stimulated radiation, called Raman scattered radiation, at a different wave length. The new wave length is equal to 10.6 microns plus an amount proportional to the magnetic field. By increasing the magnetic field it is possible to produce a laser output from the InSb that has a longer wave length (or lower frequency) than the CO_2 pumping radiation.

The output of the InSb laser is at a single wave length, and this wave length can be changed in a more nearly continuous and measurable manner than can the output of a temperature tuned diode. Such devices have produced output wave lengths from 30 to 100 watts when the crystal was irradiated by a 1,500-watt pulsed CO_2 laser. The spin-flip laser is far superior to any prism or grating spectrometer for extremely high resolution absorption spectroscopy. Its application to atomic and molecular physics will give physicists more precise and detailed information on the structure of atoms and molecules.

To achieve a maximum tuning range, high field superconducting magnets must be used. The available tuning range can be further extended by using lasers other than the CO_2 laser for pumping, or by using a semiconductor other than InSb.   [KARL G. KESSLER]

Elementary Particles. A ray of hope appeared in the particle beams in 1971, as a number of experiments gave hints of exciting times to come. And good news was indeed welcome in a field beset with economic problems and the slow progress of fundamental theory.

The easiest triumph to recognize came from the European Center for Nuclear Research (CERN) in Geneva, Switzerland. A five-year effort to build huge storage rings there that would produce head-on collisions of high-energy protons finally succeeded. What makes this achievement important is the curious velocity-addition rules of relativistic particles. A head-on collision of two particles moving almost at the speed of light is far more than twice as violent as a collision of either particle with a stationary object. The head-on collision of two proton beams at 25 billion electron volts (GeV), the upper limit of the CERN facility, is equivalent to that of a single proton beam of more than 1,400 GeV with stationary protons. Thus, the CERN experiment gives an advance peek into an energy region that extends far beyond the 500-GeV limit of the new accelerator at the U.S. National Accelerator Laboratory (NAL) near Batavia, Ill.

Head-on into asymptopia. Why are such high energies important? Because many particle physicists believe that the answer to the complexities of particle behavior will be found in collisions of very energetic particles. Physicists hope that in such collisions the forces between particles will stop changing as the energy increases, and reveal their simplest possible form. The region in which this happens is dubbed "asymptopia," combining the words asymptotic and utopia.

But there was no way to predict how much energy it would take to reach asymptopia. In April, 1971, Carlo Rubbia of CERN and Harvard University, acting as spokesman for an international team, announced a successful landing in what appeared to be asymptopia. The team had studied collisions at two different energies somewhat below the CERN peak energy. The forces acting in these collisions had apparently ceased to change with energy. Thus, the 500-GeV

NAL giant may be able to reach well into the promised land, and with the advantage of beams of particles other than protons as well as the convenience of a nonmoving target.

Lumps in the jam. Meanwhile, on the West Coast of the United States, another long-term effort has begun to pay dividends. One of the primary selling points of the 2-mile-long linear electron accelerator at Stanford University, which began service in 1968, was the possibility of using it as a super electron microscope capable of looking at the inner structure of protons and neutrons. From early studies, it now appears that the form of these basic constituents of matter may be lumpy rather than smooth. In the words of Stanford theorist Sidney D. Drell, "We are beginning to look at the seeds in the jam."

The picture is far from clear, however. The Stanford microscope is not quite fine enough to give a clear view of the lumps. Moreover, the interaction of an energetic electron with a proton or neutron produces a fearful disruption of the target particle. Until the details of this disruption are better understood, no one can be quite sure what the Stanford results mean.

The results, however, cannot help but encourage new speculation about the "quarks" of theorist Murray Gell-Mann of the California Institute of Technology or the "partons" of his colleague Richard P. Feynman. Gell-Mann proposes that all particles are ultimately composed of three subparticles called quarks. Feynman's parton theory examines the consequences of subparticle structure without specifying the exact nature of the subparticles.

The field seems ripe for new experiments. The NAL accelerator will produce intense beams of mu-mesons, which are essentially heavy electrons. Mu-mesons are even more suitable than electrons for experiments of this type. If there is any significant structure in the universe below the level of the so-called elementary particles, physicists should get a good look at it during the next few years.

A frozen quark? The quark theory received a boost from another Stanford laboratory, that of low-temperature expert William M. Fairbank. The most obvious characteristic of one of Gell-Mann's quarks is that it must carry an electric charge of either one-third or two-thirds that of an electron. All known charged particles carry a charge that is a whole multiple of that of the electron. Although free quarks should be very rare in ordinary matter, an occasional one might turn up.

In 1971, Fairbank and co-worker Arthur Hebard suspended a tiny sphere of superconducting niobium metal about 1/100 inch in diameter in a magnetic field in a vacuum. The motion of such an object is completely free of friction, and thus is tremendously sensitive to the presence of an electric charge. By squirting positive or negative electrons on the niobium sphere, Fairbank and Hebard tried to reduce the charge on the sphere to zero, and found they could not. Furthermore, the residual charge on the sphere appeared to be about one-third that of an electron!

Fairbank hesitates to claim the discovery of the quark on the basis of an oddball sphere of niobium. The experiment is being refined, repeated, and checked for possible errors.

Moreover, particle physicists are unlikely to believe firmly in quarks until they catch one in a free state, where they can measure its properties in detail. Dozens of attempts to find a free quark have failed, but several tries are early on the agenda at the NAL.

On the other hand, to encourage skeptics, "last year's quark" seems to have evaporated. The 1970 experiment of Charles B. A. McCusker, the Australian physicist who reported quark tracks in cloud chamber photographs of cosmic rays, has now been repeated by H. Faissner and co-workers in Switzerland, Wayne E. Hazen at the University of Michigan, and Wilson M. Powell at the University of California, Berkeley. With apparatus even more sensitive to quarks than McCusker's, all three report negative results.

Something peculiar. As if Stanford had not already created enough excitement for one year, a "shot in the dark" experiment there by Melvin Schwartz may have turned up an even greater surprise. A decade ago, Schwartz was the co-discoverer of two kinds of neutrinos. Neutrinos are devoid of electric charge, and immune to nuclear forces. Thus they can penetrate miles of dense matter

The Physicists' Dilemma

Geoffrey Chew of the University of California, Berkeley, believes that the physicists searching for truly elementary particles will find only frustration. In 1970, he poked this cartoon fun at the quest.

10,000 B.C. The inhabitants of the paper square have no conception of the true nature of the universe they inhabit.

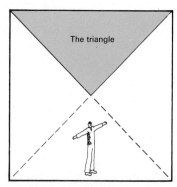

A.D. 1900. Physicists of the square discover a basic subdivision of their universe. They call it the "triangle" and consider it to be the fundamental building block of the universe.

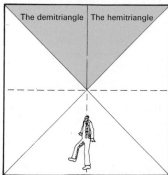

A.D. 1930. Physicists discover that the triangle can be split. Its parts are termed the "hemitriangle" and the "demitriangle." These are thought to be the fundamental building blocks of the universe.

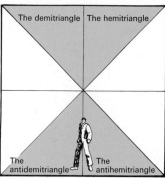

A.D. 1950. Mirror images of the demitriangle and the hemitriangle are discovered. These are termed "antidemitriangle" and "antihemitriangle."

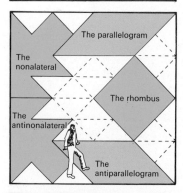

A.D. 1960. Physicists' conception of their universe is further clouded by new discoveries: the rhombus, the parallelogram, the antiparallelogram, the nonalateral, and many others. It is unclear what these discoveries signify.

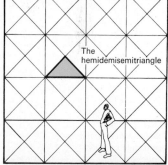

A.D. 1970. A new configuration, the "hemidemisemitriangle," is theorized. From it, all known configurations can be built. The hemidemisemitriangle is thought to be the fundamental building block of the universe.

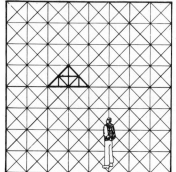

A.D. 1975. Hemidemisemitriangle is discovered. The following year the hemidemisemitriangle is split.

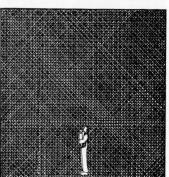

A.D. 2000. The inhabitants of this paper square have no conception of the true nature of the universe they inhabit.

with little chance of being absorbed. But if enough of them pass through a detector, a few will collide with nuclei and signal their presence.

Schwartz wondered whether there might be other types of very penetrating particles. To check out the possibility, he had a pit dug into a hill behind the target end of the 2-mile-long Stanford accelerator. He put his detector, a large spark chamber, into the pit. Then he slammed the accelerator's electron beam into a dense block of metal.

The unstable particles such a collision creates require a long flight path before they decay into neutrinos. The "quick death" of the beam thus ruled out neutrino production. No other known particle could penetrate all the way through the hill. To his surprise, Schwartz found that about once a day something registered in his detector, which is too crude to say much more about the identity of the interlopers than give a rough estimate of their energy. What Schwartz is seeing is anybody's guess. There is no empty place in the current particle roster for anything of this type.

A year of frustration. Despite the exciting developments, 1971 was a year of bitterness and frustration for many particle physicists in the United States. Government agencies, forced to fund new programs at the unfinished NAL out of a dwindling research budget, ended their support of several existing laboratories. Young physicists at the peak of their productive years suddenly found themselves thrown on a job market already glutted with fresh Ph.D.'s. A lucky few managed to land jobs abroad or in other fields.

The picture is somewhat rosier in Europe, where agreement was reached in 1971 to build a European companion to NAL. This 300-GeV machine will be built at CERN's present site straddling the French-Swiss border, and will use CERN's existing 28-GeV accelerator as a booster to start the particles on their journey. By segregating funds for this project from the normal CERN budget, the multinational combine backing the project assured it will not be built at the cost of starving the science it is intended to help.   [ROBERT H. MARCH]

A Supercooled Quark Detector

Supercold quark-sensing device was developed at Stanford University. It has a niobium sphere that hovers between two oppositely charged capacitor plates, and its electrical charge can be measured by its movement up or down. Electrons and positrons are added to the sphere to neutralize its electrical charge. But, if a quark is present in the sphere, its fractional charge would prevent neutralization.

Magnetometer

Cup-arm mechanism

Superconducting coils

Superconducting coils

Capacitor plate

Niobium sphere

Electron

Radioactive positron source

Capacitor plate

Radioactive electron source

Nuclear Physics. A team of British, Israeli, and American scientists under the leadership of John Batty of the British Atomic Energy Commission at Harwell announced in February, 1971, the possible production of the first supertransuranic element – with atomic number 112. The announcement created major excitement among physicists.

During the past 30 years, physicists had systematically extended the known system of the elements from uranium, number 92, to a yet unnamed transuranic species, number 105, and possibly even to numbers 106 and 107.

As the atomic number increases, however, the lifetime against decay through either alpha-particle radioactivity or spontaneous fission decreases to milliseconds or less. At the same time, the difficulty of producing and identifying them greatly increases. At Oak Ridge National Laboratory, Curt Bemis and his collaborators have, for the first time, produced an absolute signature for these short-lived transuranic species. Only about 2,500 atoms are now required to definitely identify a new element.

Bemis detects the alpha particle produced by the radioactive decay of an atom and, simultaneously, a characteristic X ray from the resulting atom. The energy of this X ray is a unique signature for any atom. It seems clear that with much effort we can continue to produce and identify heavier species that have extremely short lifetimes.

Yet, element number 112 is exciting for a different reason. Theorists predict that there should be islands of stability, where nuclei have relatively long lifetimes, in moving to heavier species. When produced, several of these new nuclei should live a very long time, indeed. The Harwell group announced that element 112 had an apparent lifetime measured in months.

Moreover, these new species should release perhaps 10 neutrons when they fission as compared to the 3 neutrons characteristic of each uranium fission. This is the number that determines the critical mass for nuclear-energy release; the critical mass of such new species thus would be much smaller than for uranium. The possibility of using these totally new nuclear species for portable energy sources has provided new impetus to work in this field.

The Harwell group used an unusual production technique. In the European Center for Nuclear Research (CERN) 30-GeV proton synchrotron in Geneva, Switzerland, the proton beam traverses the targets that experimentalists have inserted as part of their apparatus. Then, it is finally stopped in a large block of tungsten. After roughly a year of use, this block was removed, dissolved chemically, and searched for supertransuranic species.

It is expected that the supertransuranic numbers 110, 112, and 114 will be chemically similar to platinum, mercury, and lead, respectively. The Harwell group chemically separated the tiny amount of mercury in the dissolved beam stop and examined it for the alpha radioactivity and fission products that would be expected for the supertransuranics. They found both!

There is far from general agreement that element 112 has actually been found, however. Contaminants might possibly simulate supertransuranics. Also, scientists do not understand exactly how such a heavy species could be produced by the 30-GeV proton bombardment of tungsten. One theory is that in very rare collisions, a recoiling tungsten nucleus gains enough energy to interact with another tungsten nucleus and make a very heavy compound nuclear system. This system fissions, and the new supertransuranics appear among the fragments.

Despite these uncertainties, every laboratory in the world that has the necessary facilities has taken up the search. Although this first report involved use of very high-energy protons, the consensus in the scientific community is that reactions involving heavy nuclear projectiles – heavy ions – are the best route to the transuranics. Russia, West Germany, France, Denmark, and the United States are all trying to develop larger heavy-ion accelerators.

Nuclear molecules. New configurations of highly excited light nuclei were found in 1971. Nuclear physicists have speculated for a long time about what happens when energy is pumped into a nucleus. In the simplest picture, the individual nucleons – the neutrons and protons – speed up, much like water molecules in a heated droplet, until some are actually evaporated, carrying

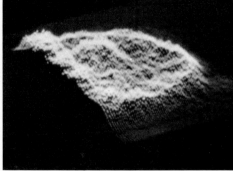

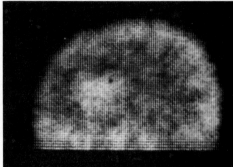

Physics

Continued

The hybrid positron detector, *above*, can be used to locate brain tumors. Scanners move along the patient's head, sensing radiation produced when positrons from a radioactive substance added to the blood are changed into gamma rays inside the brain. A computer maps the radiations in several ways, *right*, to locate the tumor.

away some of the system's excess energy.

The discovery in 1961 of nuclear molecular states in the magnesium nucleus of mass 24 showed, however, that under some circumstances the 24 excited nucleons could cluster into distinct groups. Most of the excess energy is tied up in the internal binding energy of these groups. Relatively little of it remains for simple motions–rotations or vibrations–of the groups relative to one another. In the excited magnesium nucleus, the 24 nucleons separated into identical groups of 12–in other words, there were two carbon nuclei in the nuclear molecule.

During 1970 and 1971, many laboratories found evidence that suggests that more complex configurations exist. In each configuration, the alpha particle, a tightly bound combination of two protons and two neutrons, is one of the dominant groups. The nuclear analogs of the H_2O and CH_3 molecular complexes have been identified. In each case, the hydrogen atom of the chemical is represented by an alpha particle in the nuclear molecule.

Nuclear energy. It is often incorrectly assumed that nuclear physics has little remaining contact with the nuclear energy field. The input data for even the oldest and most widely used boiling or pressurized water nuclear reactors often may be as much as 10 per cent inaccurate. Costly design allowances must be made for this uncertainty. Equivalent nuclear data for fast-breeder reactors is often far less certain. Now, new nuclear instrumentation has improved the accuracy of some of the numbers.

For example, Alan Smith and his collaborators at the Argonne National Laboratory made a series of careful measurements in 1971 on the ratio of neutron capture to fission cross sections for the plutonium isotope of mass 239. They showed that an entire family of water-cooled, plutonium-fueled breeder reactors, to which the utilities, reactor manufacturers, and the Atomic Energy Commission had looked for the future, simply would not breed. Hundreds of millions of tax dollars would have been committed to the development of these reactors.　　　　[D. ALLAN BROMLEY]

Plasma Physics. The year 1971 was the year of the tokamak. A tokamak is a doughnut-shaped device that produces a hot plasma. It holds the plasma together in a magnetic field that is produced partly by an electric current flowing through coils surrounding the doughnut, and partly by the heating current flowing in the plasma itself.

The tokamak era began in 1969, after years of only slight interest in the design by plasma physicists. Lev A. Artsimovich of the Kurchatov Atomic Energy Institute in Moscow announced then that the third tokamak device, T-3, had held a hotter and denser plasma together longer than any other device. The skepticism that greeted Artsimovich's claim turned to enthusiasm after N. J. Peacock of Great Britain's Culham Laboratory took laser light-scattering equipment to Moscow and confirmed the Russian measurements. Soon after, many laboratories, especially in the United States, began building tokamaks. See CURBING THE ENERGY CRISIS.

By 1971, two American tokamaks were in operation: Princeton University's ST Tokamak and the Oak Ridge (Tenn.) National Laboratory's Ormak. Two others were near completion: the Massachusetts Institute of Technology's Alcator and the University of Texas' Texas Turbulent Tokamak. Results have been most encouraging. The Princeton device made a particularly important contribution by verifying the scaling laws that were used to predict such plasma properties as temperature and density of the larger new tokamaks from measurements on the smaller T-3.

Another vital experiment on the ST Tokamak partly answered the question of why tokamaks produce a plasma that lasts longer and is denser and hotter than that produced in stellarators, a class of similar devices that were studied intensively for 15 years at Princeton University. Stellarators have a thinner doughnut than the tokamaks and have no heating current in the plasma. They use external coils to produce the so-called poloidal magnetic field produced by the plasma current in the tokamak.

Physicists had suspected that stellarators were ineffective as plasma traps because their plasma doughnuts, unlike the circular tokamak doughnuts, must be stretched into an oblong shape. They feared that this configuration was somehow disastrous.

Thinner bananas. Experiments in 1971 on the Princeton ST Tokamak disproved this notion. Instead, researchers found that the essential difference lies partly in the skinnier stellarator doughnut and partly in the stellarator's weaker poloidal field. Plasma particles tend to spiral around magnetic lines of force. When these lines are closed on themselves, as they are in doughnut-shaped devices, plasma can escape from the trap only through collisions. A collision causes a charged particle to start spiraling about a new line that is separated from the old line by approximately the radius of the spiral. As time goes on, each particle meanders from line to line and it may eventually escape. Unfortunately, the measured loss rate was thousands of times the predicted rate.

In both stellarators and tokamaks, each magnetic line of force repeatedly passes from a strong field region through a weak field, and back to a strong field. Those particles spiraling with enough velocity perpendicular to that line will be reflected by strong field regions and will move back-and-forth between them. However, a more careful investigation of the particle motion under these conditions reveals that the particle's spiraling and oscillatory motion does not take place precisely in a narrow tubelike area surrounding the line of force. It happens on a much wider, banana-shaped surface surrounding the line.

When one particle collides with another, it starts spiraling on a new banana-shaped surface much farther from the initial banana than the spiral radius. Thus, it takes far fewer collisions for a particle to escape from the confining magnetic field than researchers had thought. This explains why tokamaks are better plasma traps than stellarators. Their stronger poloidal fields hold particles to thinner bananas. And their fatter doughnuts place the walls at a greater distance. This means that it takes far more collisions and course changes for a particle to escape.

"Bootstrap" tokamak, a new concept for a continuously operating device, has recently been proposed independently by B. B. Kadomtsev and V. D. Shafranov at the Kurchatov Institute, and by R. J. Bickerton, J. W. Conner, and J. B. Taylor of the Culham Laboratory. In conventional tokamaks, the plasma is actually the secondary "coil" of a very large transformer. The plasma current is induced in it. This makes continuous operation very difficult because of heat dissipation problems in the transformer equipment.

The new idea reduces or eliminates the need for a transformer. Neutral gas diffuses from the outside of the doughnut to the center, and is ionized there and turned into a plasma. The subsequent outward motion of this plasma, together with the existing poloidal field sets up an electromotive force that produces a plasma current. This current strengthens the existing poloidal field. Such bootstrapping avoids the heat dissipation problems of transformer-produced currents and would permit continuous operation.

Electron beams and sheets. Another area where progress has occurred is in the use of the extremely intense, highly energetic electron beams, which have recently become available. Such beams could be used to vaporize, ionize, and rapidly heat a deuterium-tritium fuel pellet to thermonuclear temperatures. Electron beams are more promising for this purpose than the previously used beams of photons from high-power lasers because they can be produced much more efficiently (35 per cent as against 0.1 per cent).

Intense electron beams can also be used to generate a circulating electron sheet that can ionize, heat, and trap neutral gas molecules. Large-scale experiments on such a device, called Astron, have been conducted by Nicholas C. Christofilos for more than 10 years at the Livermore Research Laboratory of the University of California. The required density of the electron sheet has not been achieved. Recently, however, a group at Cornell University led by H. H. Fleischmann used an intense electron beam to produce an electron sheet dense enough to prove the validity of the Astron concept. This device is much smaller than Christofilos' Astron, but the achievement has been a much-needed shot in the arm to the Astron research.   [ERNEST P. GRAY]

The Oak Ridge National Laboratory's Ormak was one of several new U.S. machines built along the lines of the successful Russian tokamak device.

Solid State Physics. In 1969, while working at the Rensselaer Polytechnic Institute's East Hartford (Conn.) Graduate Center, Helmut Schwarz and Heinrich Hora first reported their observation of a remarkable new interaction between light waves and electrons. In 1971, their findings stimulated extensive experimental and theoretical research.

In the experiment, carried out in a vacuum, they caused an electron beam that had an energy of about 50,000 electron volts to collide with a light beam produced by a high intensity argon ion laser. The two beams cross at right angles to each other as they pass through a solid crystal of aluminum oxide (Al_2O_3) or silicon oxide (SiO_2). The crystal, transparent to the laser light, is exceedingly thin (about 1,000 A thick), to also permit transmission of the electron beam. The laser beam is polarized so that its electric field oscillates parallel to the moving electrons.

The physicists first viewed the electron beam on a luminescent screen, somewhat like watching a television test pattern. They observed the diffraction pattern of the beam that is produced by the regular arrangement of atoms in the crystal.

Then they allowed the electron beam to fall on a nonluminescent screen. This screen would not glow when struck by the electrons, yet they still saw the same diffraction pattern. It had the color of the laser light.

Schwarz and Hora believe that the laser light modulated the electron beam – that the information of the light beam was transferred to the electrons during their collision in the crystal. It was the first time such an effect had been observed. There are reasons to believe that the collision of the two beams must take place in a nonconducting solid for the effect to occur.

Schwarz described further work on this newly discovered effect in March, 1971, at the Cleveland Meeting of the American Physical Society. He noted that when the observation screen is moved farther from the collision crystal, the intensity of the light pattern increases and decreases regularly. The distance between peaks is about 1 centimeter (cm).

Since the original publications of Schwarz and Hora, several laboratories have set up experiments to reproduce this very interesting effect. To date, however, no other laboratories have reported success. Schwarz believes that all of the electrons in the beam must have almost identical energy in order to see this coupling.

The experimental findings have also stimulated much theoretical work in Europe, Russia, and the United States as physicists seek to describe the process involved in this coupling of electrons and light waves. The 1-cm separation of the peaks has been explained theoretically. However, it is difficult to understand the large amount of light observed at the viewing screen. Schwarz and Hora note that the effect they have observed must involve quantum mechanical processes; it cannot be understood by a classical mechanism. In addition, Hora has reported that the electrons in both the beam and the crystal must be *coherent* (moving in a fixed relationship to each other) for very short distances within the crystal.

Crystal bonds. Atoms crystallize in a variety of structures. A large solid is built up from many identical unit cells, which may be cubic, hexagonal, or rhombic in shape. One of the goals of theoretical solid state physicists has been to predict the specific type of unit cell that a group of atoms will assume. This would allow physicists to classify the large variety of known solids in as simple and fundamental a fashion as possible.

James C. Phillips of Bell Telephone Laboratories, working independently and also with J. A. van Vechten, reviewed recent theoretical work in 1970 and proposed a new approach to studying chemical bonds in crystals. The type of bonding that holds the atoms of the unit cell together and the number of neighbor atoms that participate in the bond are of primary importance in understanding a specific structure. For example, sodium chloride (NaCl) may be considered to be a regular arrangement of Na^+ and Cl^- ions. The electrons in each ion are in very stable, closed shell arrangements. Each ion has six nearest neighbors, all of the other species. These ions exert electronic forces upon each other, which keep the solid together at ordinary temperatures.

But in a diamond, for instance, covalent bonds link the carbon atoms. Each

carbon shares two electrons with each of its four nearest neighbors. There are no ions present. In most solids, however, every bond has both ionic and covalent effects present.

Many of our present ideas about the degree of ionicity of the atoms in solids are based on the pioneering concepts introduced by Linus Pauling in the late 1920s and early 1930s. The notion of electronegativity – the power of an atom in a molecule to attract electrons to itself – was of key importance. A measure of the electronegativity was the heat of formation of the bond. The greater the difference in electronegativity between two bonded atoms, the more ionic is the bond.

Phillips proposed a new measure of ionicity based on optical measurements of solids. From these measurements, he determines the energy separations between the electron states that bond and those that do not bond the crystal together. Phillips believes that his approach is in clear contrast to the philosophy of classical physical chemistry, in which the chemist starts from individual atoms and then builds up more complex molecules by adding atoms in a step-by-step fashion.

So far, Phillips has considered 68 compounds having only two chemical elements. His group includes NaCl, gallium arsenide (GaAs) – a well-known laser material, silicon carbide (SiC), used in light-emitting diodes, and cadmium sulfide (CdS) – a sensitive light detector. In each case, he calculated the ionicity from the energy gaps produced separately by the covalent and ionic parts of the bond.

Phillips compared his results to the earlier results of Pauling and also to those of Charles A. Coulson of the University of Oxford in England, whose calculations are based on a quantum mechanical approach that directly uses information in the calculation about the electronic orbits involved in the bonding. He found the differences are not very large even though the starting points and approaches of the theories are quite different.

Furthermore, using his covalent and ionic energy gap parameters, Phillips can sort the 68 compounds into two very clear groups in which an atom has either four or six nearest neighbors. A single straight line on a graph relating co-valent and ionic energy gaps separates all the compounds having fourfold structures from the sixfold ones.

He concluded that the ionicity of the bond between the atoms, which is directly determined by these two parameters, is the critical factor that determines the arrangement of the atoms in the unit cells for these compounds.

Phillips' work stimulated discussion with Pauling in several journals during 1971. This exchange has resulted in a clarification of the relationship between Pauling's approach and results and those discussed by Phillips.

Ion implantation. Solid state electronic devices, such as transistors, are made by placing atoms into the surface layer of a solid or solid substrate. Much better control over the placement of the atoms would offer unusual opportunities for basic research experiments as well as for the production of other small surface-layer devices. The atoms can be implanted by bombardment of the solid with ions in the energy range from 1,000 up to 1 million electron volts.

Only in the past several years have scientists used such ion implanted specimens to study the nuclear properties as well as the electronic and magnetic properties of solids. From these studies, we have learned how to produce a specific impurity distribution in a region under the surface. The atoms can be implanted in a solid at low temperatures and the number of ions and their location can be varied by controlling the external bombarding conditions. The beam current determines the number of impurity atoms, and the energy determines the distribution below the surface.

Ions can also channel into crystals. They can penetrate deeply if they enter the crystal in a direction in which atoms or planes of atoms can regularly steer the ions through the crystal. Because of this, the depth of penetration can also be adjusted by changing the direction of the beam relative to the crystal planes.

Ion implantation in metal oxide semiconductors would allow higher switching speeds, reduction in size, and will allow easier fabrication of three-dimensional devices.   [JOSEPH I. BUDNICK]

Psychology

The results of several new lines of investigation make it clear that the human infant has a wide range of perceptual abilities immediately after birth. In addition, the newborn child "learns" from his environment–that is, he can store and use information from his own experience to influence his behavior and the environment itself. The infant's limited behavioral capacities had previously led psychologists interested in child development to conclude that perception is a slowly developing process, and not generally complete until a year or two after the child is born.

Until the past few years, most tests of the abilities of the very young child had been based upon the development of his ability to move through the environment and interact with objects in that environment. The ages at which he grasps and holds objects, sits up, crawls, and walks had been carefully noted, recorded, and compared with standards.

All the major child-development theories of our time–for example, those of the Swiss psychologist Jean Piaget–have been based largely upon the sequence in which these abilities develop. This approach has been maintained despite lack of evidence that the time it takes various motor skills to develop is related to later mental abilities, except perhaps in cases of extreme forms of mental retardation.

The new lines of child-development research have grown out of advances in methods of studying the infant's perception and behavior. The techniques take advantage of simple responses almost every infant can make from the time of birth, such as sucking, moving an arm, or looking at an object.

Looking at patterns. In the early 1960s, Professor Robert L. Fantz of the Perceptual Development Laboratory of Case Western Reserve University, Cleveland, demonstrated that the newborn baby is much more than a perceptual blank slate. He found that shortly after birth the child looks at checkerboard designs much longer than he looks at indistinct areas in his environment. By the time the child is 4 months old, the contours of the human face become much more interesting to him.

Psychology

Continued

This is measured by how much longer he gazes at facelike patterns than at nonsense designs made up of similar contours. The fact that the infant looks longer at one of two figures indicates that he can discriminate between them and that he has consistent preferences among certain kinds of stimuli.

One of the new techniques psychologists use to study visual perception in the newborn child was developed by Fantz and S. Nevis in 1967. They place two objects, such as pictures or geometric designs, about 12 inches in front of the infant. These objects are presented against a plain background. Usually each pair of objects is shown to the infant twice, each time in a different location. This is done to eliminate contaminating the experimental results with any preference the child might have for either location. Observers whom the child cannot see note how much time he spends gazing at each object.

In other situations, a single object is shown to the infant and observers note the length of time he examines it. Sometimes the number of babbles and coos the infant makes while looking at objects is also recorded as an additional measure of their interest value. Results from studies using these techniques were among the first to show that the mechanisms of perception exist at birth–even in children born prematurely.

Head-turning. Professor Lewis P. Lipsitt, director of the Child Study Center at Brown University, Providence, R.I., and his colleague Einar R. Siqueland have used a simple mechanical device that enables them to record automatically the movements of a newborn baby's head as he lies in a crib. In this way they can determine the number of times that stimuli placed on one side or the other of the infant can cause him to turn his head toward them. The head-turning response can also be used to evaluate the infant's interest in other sensory events, such as odors and sounds.

For example, researchers touch the infant's cheek to make him turn his head. (Such a response is congenital in the newborn, that is, the newborn child does this about 25 per cent of the time without learning.) Every time the baby

A 3-month-old baby learns to keep movie in focus by sucking on a specially wired pacifier at a selected speed. The Harvard University tests are designed to determine perceptual ability in infants.

Psychology

Continued

responds with a head turn, the researchers put a bottle in his mouth for a few seconds. Under these conditions the newborn learns, within a half hour, to turn his head increasingly frequently. The rise in response rate rises from 25 per cent to 80 per cent.

Arm movements. Additional procedures have been developed by many investigators to allow infants of a few months of age to make responses that influence their environment. One of the most interesting was developed by Carolyn K. Rovee and David T. Rovee of Trenton State College, Trenton, N.J. It capitalizes on the high attention value that moving objects have for infants. A hanging mobile, when gently moved by air currents, for example, is unusually effective in capturing the attention of a young baby.

If the child could initiate movements of the mobile, would he learn to do so? According to the Rovees the answer is clearly "yes." When they arranged the mobile so that it moved only when the baby moved his arm, he quickly learned how to activate the mobile. He just as quickly learned not to bother to move his arm when the mobile was held fast. Thus, even in the first few months of life, the child can learn to make responses that will produce changes in his visual world.

Sucking responses. In 1971, Professor Siqueland of Brown University was using another technique to study learning capacities during the first few months of life. He arranged a slide projector to show color pictures to a baby lying in a comfortable position. A small device, much like a pacifier, was put into the baby's mouth. The device contained sensitive switches that recorded the sucking responses of the baby and also controlled the brightness of the picture. The stronger or more frequently the baby sucked, the brighter the projected image became. In this way, the infant directly controlled the intensity of the pictures by his sucking behavior.

By 4 months of age, babies can learn how to increase the intensity of pictures presented to them in this way. Siqueland's technique can also be used to evaluate the attractiveness of stimuli for infants at other ages. In a similar technique at Harvard University, the focus, rather than brightness, of the pictures is regulated by the infant's sucking. On his own initiative, he quickly learns to produce a clear focus.

Babbles and coos. Not only are infants capable of complicated visual perceptions early in life, but they also make responses that indicate they categorize speech sounds in the same way that adults do. In January, 1971, Peter D. Eimas, Siqueland, Peter Jusczyk, and James Vigorito reported in *Science* that infants at both 1 month and 4 months of age discriminate among speechlike sounds on the basis of speech rhythms. These characteristic rhythms are found in most, if not all, languages. This suggests that the discriminative capacity important for recognizing speech sounds may have its roots in the child's heredity. Or perhaps learning to recognize the sounds of speech occurs within the first month after birth. In either case, perceiving this aspect of speech also demonstrates perceptual capacities in the young infant well beyond what researchers had previously imagined existed at such an early age.

Underestimated powers. The results from these new methods of studying the perceptual and learning capacities of the child, as well as the evidence from physiology of the readiness of vision and other senses early in life, indicate that the capacities of the young child have been grossly underestimated.

Furthermore, while it is clear that the infant begins to process information early in life, this should not be interpreted to indicate that hereditary factors have an exclusive or more important effect upon perceptual development. The very fact that children begin to explore, recognize, and profit from their environment early in life makes the effect of this environment all the more important.

Through understanding the processes in which perception develops, we may come to understand the effects of abnormal perceptual development in which visual events do not have their usual motivational or attention-capturing characteristics. This may arise from cultural and physiological deprivation. More important, knowledge of the factors that underlie perceptual development in human beings may show us how to overcome the debilitating effects of such deprivation.   [ROBERT L. ISAACSON]

Science Support

The public's attitude toward supporting high-technology projects changed dramatically in 1971. This was most evident in May when the United States was forced to abandon development of its costly supersonic transport plane (SST).

Sentiment against the SST had been building for several years as more people came to suspect that sometimes technology does more harm than good. The widespread concern built up into political pressure, and Congress took the final, almost unprecedented, step of scuttling the SST.

The federal government and leading aerospace firms had launched the SST program in the 1960s. The SST was designed to fly at about 1,800 mph, thus allowing travelers to reach such distant points as Japan, Australia, and Europe in far less time than is now needed. The initial goal of the project was to produce two prototype planes for testing by 1973. Actual commercial models were to be ready for delivery to airlines by 1978. The Boeing Company and General Electric Company were the prime contractors working on the aircraft.

Protests mounted. There was no major opposition to the SST in its early years, but in 1970 and 1971 a huge ground swell of protest developed against continuing the project. Opponents charged that the aircraft would be uneconomical. They said it was using up money—$1.6 billion in federal and private funds would be needed to produce the two prototypes—that could be better used on pollution control, urban mass transit, and other pressing needs. They also maintained that the SST would cause grave environmental problems ranging from bothersome airport noise and sonic booms to potentially catastrophic changes taking place in the upper atmosphere.

Many critics agreed with Murray Gell-Mann, the Nobel prize-winning physicist from the California Institute of Technology, who complained publicly in 1969 that too often something is built just because it can be built. Gell-Mann called for some landmark acts of "technological renunciation" to protect the environment. He suggested that the SST might be the project best rejected.

"Science for the People" was the theme of those who were protesting the uses of science and technology at the annual meeting of the American Association for the Advancement of Science.

Those who supported the SST – and that included President Richard M. Nixon – argued that it was needed so the U.S. aircraft industry could maintain its technological lead over foreign competitors. Administration officials noted that both Russia and a British-French partnership were building SST's, and they expressed fear that these foreign-built planes would dominate the world market. This, they said, might cause balance of payments problems for the United States and could bring more unemployment in the already hard-hit aerospace industry. The Administration also said that the environmental hazards of the SST had been exaggerated. It maintained that these hazards could be eliminated before huge fleets of SST's began crisscrossing the skies.

The issue reached a political climax in 1971 when Congress voted on whether to continue funding for the two prototype planes. Although about $1 billion in federal and private funds had by then been spent for the SST, the House of Representatives surprised everyone on March 18 by voting against continuing the project. On March 24, the Senate also rejected government support for the SST. A last-ditch effort to revive the project later won approval in the House, but the Senate definitely ended the SST program on May 20.

The significance of the SST rejection went far beyond the simple question of whether a particular aircraft should be built. As an editorial in *Nature*, the prestigious British scientific journal, observed: "The end of the SST project is an important happening. Of that there can be no doubt. It is not often that people, voters, and their elected representatives, play such a decisive part in strictly technical decisions. Who will now say that modern society is powerless to put a rein on technical development?"

Nature went on to lament that the sudden halting of the project before prototypes could even be completed and tested might bring "unwelcome" side effects. "One danger in the situation, which has now emerged, is that it will encourage the confused belief in modern society that technology is not merely suspect but evil."

Florida canal halted. The defeat of the SST was not the only rejection of technology in 1971. Under pressure from environmentalists, President Nixon halted work on the Cross-Florida Barge Canal on January 19. About $50 million in federal funds, plus more than $12 million in state and local funds, had already been committed to this major engineering project. The 107-mile canal, which was to run across northern Florida between the Atlantic Ocean and the Gulf of Mexico, had originally been authorized in 1942 in a move designed to protect U.S. ships from German submarines. The project was not finally funded until two decades later, when its announced purpose was to cut transportation costs for shippers.

Environmental groups strenuously opposed the Florida canal, arguing that it would destroy vegetation and wildlife, especially along the Oklawaha River. The concerns of conservationists were so compelling that the Nixon Administration's own Council on Environmental Quality recommended that the project be halted. President Nixon then announced that he was stopping work on the canal "to prevent potentially serious environmental damages." He explained that he would no longer approve projects solely on the basis of their technological and economic merit.

"The purpose of the canal was to reduce transportation costs for barge shipping," he said. "It was conceived and designed at a time when the focus of federal concern in such matters was still almost completely on maximizing economic return. In calculating that return, the destruction of natural ecological values was not counted as a cost, nor was credit allowed for actions preserving the environment."

Examining other projects. Many other technological projects were subjected to close scrutiny and, in some cases, work was stopped. Nuclear reactors scheduled to be built in several localities were delayed because of citizen opposition to possible radioactive hazards. The large-scale defoliation program carried out by the U.S. military in South Vietnam was halted after scientists complained that it was injuring civilians and seriously damaging the land (see THE WORLD SERIES OF SCIENCE). And several proposed advanced weapons systems – notably the antiballistic missile, the F-14 fighter plane, the B-1 bomber, the Main Battle Tank 70, and proposed new aircraft carriers – aroused intense opposition in the Senate.

Research questioned. There were even suggestions by eminent scientists that basic research along certain potentially dangerous lines should be cut back or abandoned. James D. Watson, a Nobel prize-winning biologist, told a House science subcommittee on Jan. 28, 1971, that "all hell will break loose," politically and morally, when scientists succeed in conceiving a baby in a test tube and then place it successfully inside a woman who will bear the child. Watson suggested that the United States should take the lead in forming an international commission to ask, "Do we really want this?" and perhaps "take steps quickly" to make such research illegal.

Similarly, geneticists Walter Bodmer of the University of Oxford in England and Luigi Cavalli-Sforza of the University of Pavia in Italy argued in the October, 1970, issue of *Scientific American* that scientists should delay research on whether observed differences between American whites and blacks in IQ scores are genetic in origin. "In the pres-

ent racial climate of the United States," they said, "studies on racial differences of IQ, however well intentioned, could easily be misinterpreted as a form of racism and lead to an unnecessary accentuation of racial tension."

The National Academy of Sciences refused to put its prestige behind a demand for such racial studies. At its annual meeting in late April, 1971, the academy voted down a proposal that it seek funds to study genetic differences. Instead, it approved a resolution calling for the academic community to put behavioral genetics into a broader scientific context.

President's adviser disagrees. Although the new effort to curb science and technology brought cheers from many, the trend was deplored by some leaders of the scientific community. Edward E. David, Jr., a communications expert at Bell Telephone Laboratories who became the President's science adviser in 1970, told a science writers' seminar on Feb. 24, 1971, that the American public is becoming increasingly alienated from rational ways of thought. He specifically deplored the

A grim joke to the many unemployed workers in the aircraft industry was this sign erected when a mass exodus followed cancellation of the supersonic transport and cutbacks in the space program.

rejection of the SST before prototypes could be built, James Watson's warning on biological experimentation, and the suggestion that research on racial differences should be delayed.

"Society is losing its courage to experiment," David said. "Already we see timidity in new undertakings. We require overanalysis before we are willing to find out what are the real possibilities. If these trends progress, our society will become dull, stodgy, and altogether stagnant."

Science spending up. Despite the death of the SST and the cries of alarm from a few scientific leaders, the outlook for government support of science and technology seemed brighter in 1971 than it had been for several years. This brighter outlook could be seen partly in the U.S. budget, and partly in improved relations between the White House and the President's science adviser.

Earlier, total federal spending for research and development had dropped—from a high of $17 billion in fiscal 1968 (July 1, 1967, to June 30, 1968) to $15.6-billion in fiscal 1970. The decline, caused largely by the government's efforts to finance the Vietnam War and curb inflation, meant that some laboratories and universities had to reduce their scientific staffs and salaries.

Thus, there was mild jubilation in the scientific community when the Nixon Administration boosted spending to an estimated $15.9 billion in fiscal 1971 and proposed to boost it still further (to almost $16.3 billion) in the budget recommended for fiscal 1972. The 1972 budget proposal was almost certain to be reduced somewhat by Congress, but scientists still took heart in the seeming upward trend after several years of financial decline. There had actually been no specific desire to increase science spending—instead, the budget boost resulted largely from the Nixon Administration's desire to pump money into the economy to head off a recession.

Fighting cancer. The most dramatic scientific initiative proposed in the 1972 budget was a $100-million program to combat cancer, one of mankind's major killers. The President's proposal was made in his State of the Union message on Jan. 22, 1971. "The time has come," the President said, "when the same sort of concentrated effort that split the atom and took man to the moon should be turned toward conquering this dread disease. Let us make a total national commitment to achieve this goal." His proposal was believed to be an effort to head off private health lobbyists and Senator Edward M. Kennedy (D., Mass.), believed to be planning an even more ambitious attack on cancer.

The health lobby, headed by philanthropist Mary Lasker, pushed hard for creation of a new anticancer agency that would be highly visible and have substantial funds. A bill to create such a National Cancer Authority and to provide it with $1.2 billion over the next three years was proposed by Kennedy and Senator Jacob K. Javits (R., N.Y.).

The idea of a new cancer agency was endorsed by the American Cancer Society and the American Heart Association. But it was opposed by the chairmen of 77 departments of medicine at the nation's medical schools and by several medical groups.

Opponents argued that progress against cancer could best be made by funding programs through the existing National Institutes of Health, rather than by creating a whole new bureaucracy of uncertain value. They particularly feared that a separate cancer agency would quickly absorb a disproportionate share of the nation's health dollars. They said a large new agency might waste funds in a fruitless search for a quick cure for cancer while ignoring the need for basic scientific research into the nature of the complex disease.

As presidential science adviser David expressed it in a speech on Feb. 13, 1971: "To isolate the cancer effort would prejudice the very outcome we seek. The problem of cancer straddles virtually all the life sciences—molecular biology, biochemistry, virology, pharmacology, toxicology, genetics—any of these or all of them, will contribute to the final solution."

On June 10, 1971, after months of quarreling, the Nixon Administration and Kennedy announced agreement on a compromise bill that would set up a semi-independent cancer agency that would, nevertheless, be part of the National Institutes of Health. The bill seemed likely to pass.

The U.S. budget figures were not the only cause for moderate optimism

Federal Spending for R&D

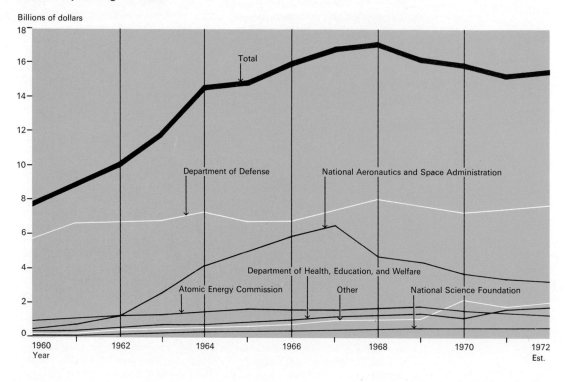

Billions of dollars

Total

Department of Defense

National Aeronautics and Space Administration

Atomic Energy Commission

Department of Health, Education, and Welfare

Other

National Science Foundation

1960 Year 1962 1964 1966 1968 1970 1972 Est.

Science Support
Continued

among scientists. Equally important was the enhanced stature of the President's new science adviser. President Nixon's first science adviser, Lee A. DuBridge, for reasons never made public, seemed to fall out of favor with the White House. As a result, he was frozen out of the key stages of budget making, and he reportedly lost much of his influence in important decisions involving science and technology. DuBridge announced his resignation on Aug. 19, 1970. David seemed successful in re-establishing close ties with the White House, and participated in final budget reviews.

Joblessness. Institutional support for science and for protection against damaging by-products of technology was strengthened by the creation of two new federal agencies in 1970. One was the Environmental Protection Agency, an independent unit given the responsibility for pollution control. The other was the National Oceanic and Atmospheric Administration, a part of the U.S. Department of Commerce, which was given responsibility for developing marine resources. See OCEANOGRAPHY.

The most troubling problem for science in 1971 was a sharp rise in unemployment among scientists, particularly those who had been working in the defense and aerospace industries. As the government cut back its spending for the space program and for certain military projects, many leading aerospace companies began laying off their technical and blue-collar workers. The U.S. Department of Labor estimated on March 12 that as many as 65,000 scientists, engineers, and technicians who had worked on defense or aerospace jobs were out of work. Other Administration officials privately estimated that the total unemployed might ultimately reach 100,000.

The joblessness was considered especially alarming because the unemployed were skilled professionals whose talents could presumably be used in solving many crucial problems confronting society. In an effort to help, President Nixon on April 2, 1971, announced the establishment of a $42-million fund to assist the unemployed professionals with job retraining, counseling, and travel expenses associated with finding and

Science Support

moving to new employment. Most manpower experts, however, agreed that the problem would not really be solved until the economy turned upward and more jobs were created.

Overseas competitors. Another weak spot involved the ability of U.S. companies to hold their own against technically sophisticated foreign industrial competitors. A number of pessimistic experts predicted that Japan, West Germany, and other industrial powers might overtake the United States and claim a major share of the world market in high-technology products, thus threatening future economic growth in the United States and causing severe balance of payments problems as well.

"Our technological superiority is slipping," Secretary of Commerce Maurice H. Stans told a joint committee of Congress on Feb. 17, 1971. Stans cited figures indicating that American imports of sophisticated "technology-intensive" manufactured products have been increasing twice as fast as exports.

Not all experts agreed that there was cause for alarm. Some pointed out that the United States was still the world's leading exporter of technology-intensive products. They suggested that the reduction in the American lead was rather small, and was due to temporary economic factors related to the Vietnam War. Still, experts in some of the nation's leading scientific groups – including the National Academy of Engineering and the President's Science Advisory Committee – expressed fears that American technological superiority was being challenged in a fundamental way.

As John R. Pierce, an executive at Bell Telephone Laboratories, expressed it at a symposium in October, 1970: "Today we are facing a technological challenge far more important to us and far more difficult to meet than the challenge of Sputnik."

By mid-1971, there was both cautious optimism and continuing concern over unemployment, world trade, and public attacks on technology. The hope remained that the bad side effects of new technologies could be eliminated without abandoning technological advance entirely.   [PHILIP M. BOFFEY]

Space Exploration

Russia achieved one of its long-term goals in space exploration in 1971 by establishing a manned orbiting space station. But the achievement ended in tragedy. The three cosmonauts who made the flight were found dead in their vehicle when it returned to earth on June 30. Apparently, a leaky seal caused a sudden loss of cabin pressure during the space capsule's descent.

The project – known as *Cosmodom* (home in space) – had begun seven weeks earlier on April 19 when the unmanned laboratory, called Salyut, was launched. Three days later, Soyuz 10, a spacecraft also manned by three men, was orbited. The two ships met and docked, but apparently something went wrong because the crew did not transfer to Salyut. On April 24, the cosmonauts returned to earth, leaving Salyut in orbit.

A successful transfer took place on June 7, however, as the three-man Soyuz 11 crew – Lieutenant Colonel Georgi Dobrovolsky, commander, and civilians Viktor I. Patsayev and Vladislav N. Volkov – moved from the Soyuz spacecraft into the more spacious Salyut.

Tass, the Russian news agency, listed five main objectives of the flight: (1) checking out the space station, (2) testing its navigational instruments, (3) beginning geographical and earth-resources studies, (4) investigating the atmosphere, and (5) studying human biological reactions.

The Soyuz 11 crew used television far more than did cosmonauts on earlier Russian space flights so that ground controllers could watch their activities. People all over the world saw taped excerpts of these transmissions. In one episode, Dobrovolsky engaged in weightless acrobatics while wearing an electric exercise garment. The cosmonauts nicknamed it the "penguin suit" because on earth it forced a man to waddle like a penguin. In the state of zero gravity, however, the penguin suit gave its wearer something to work against – like an isometric exercise device – which is necessary to keep muscles in tone on long orbital missions.

Skylab. The Russian Salyut space station was about two years ahead of the somewhat larger United States orbiting

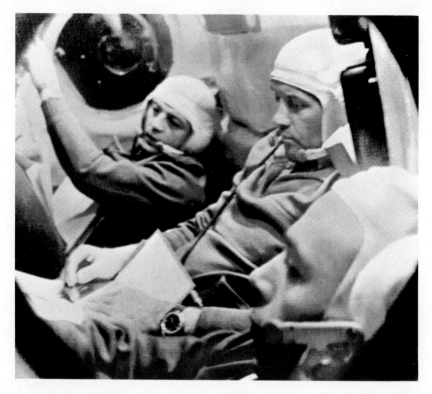

Russian cosmonauts Vladislav Volkov, flight engineer; Georgi Dobrovolsky, commander; Viktor Patsayev, test engineer; left to right, rode Soyuz 11 first to a space achievement and then to a tragic conclusion in June, 1971.

Space Exploration

Continued

laboratory called "Skylab," scheduled for flight in 1973 and 1974. The Russian and American approaches to space stations are similar. Hardware already tested in manned space flight is being used in both programs. Current size limitations, however, rule out large-scale scientific experiments for either of the stations.

The United States has plans – not yet officially approved or funded – for a larger orbiting platform called Space Station, to be supplanted in the 1980s by a permanent space base. It is reasonable to assume that Russia has parallel plans, and Russia's timetable for second- and third-generation stations may be ahead of those of the United States.

Project Apollo. U.S. activities in space continued to concentrate on lunar exploration. On Jan. 31, 1971, the Apollo program was resumed, after nearly a year, with the lift-off of Apollo 14's Saturn V launching rocket from Cape Kennedy, Fla. Captain Alan B. Shepard, Jr., U.S. Navy, the first American in space, was in the commander's seat for his first flight since the brief but

historic suborbital Mercury-Redstone 3 mission of May, 1961. Shortly after the Apollo 14 mission, Shepard was selected for promotion to rear admiral, the second astronaut to achieve flag rank. Brigadier General James A. McDivitt, U.S. Air Force, was the first.

Shepard was accompanied on the mission by Commander Edgar D. Mitchell, U.S. Navy, as lunar module pilot and by Major Stuart A. Roosa, U.S. Air Force, as command module pilot. Apollo 14 was the most ambitious flight ever attempted. It was far less comprehensive, however, than the three flights scheduled to follow it in the shortened Apollo program.

Unlike the landing sites chosen for Apollo 11 and 12 in 1969, which were on flat and almost featureless plains, the Apollo 14 lunar module landed in the Fra Mauro highlands, a more rugged and geologically more interesting area. Shepard and Mitchell, who walked on the moon, were far more active, and for longer periods, than either of the pairs of astronauts who preceded them. Improved suits and life-support equipment, and a lightweight two-wheeled cart that they pulled while making an exploratory trip, gave Shepard and Mitchell unprecedented mobility. The cart conveniently held equipment that would have been cumbersome to carry.

The Apollo 14 mission was acclaimed as the most productive yet undertaken. It brought the first phase of Apollo moon exploration to a satisfying close.

Apollo 15 and beyond. The second and final phase of Project Apollo opened in July, 1971, with a 12-day mission to an area of prime scientific interest in the northern hemisphere of the moon, a plain called Hadley-Apennine. This region is named for the nearby Apennines mountain range and the long, narrow canyon called Hadley Rille.

Apollo 15 was first of three missions to transport heavier payloads than the earlier flights and to carry an electric-powered vehicle resembling a golf cart, called Lunar Rover. With the added mobility afforded by the Rover, crews of these missions can make extensive trips away from the lunar module – up to a theoretical maximum of 35 miles.

The mission originally numbered Apollo 15 was to be one much like that undertaken by Shepard, Mitchell, and Roosa, with no Lunar Rover aboard. But a curtailment of the Apollo program in September, 1970, brought about by budgetary restrictions, forced elimination of two of the six remaining flights.

Thomas O. Paine, who was NASA administrator at the time, announced on September 2 that the missions then designated Apollo 15 and 19 would be canceled and those numbered 16, 17, and 18 would be redesignated. Apollo 20, the final flight in the program, had already been eliminated in February, 1969, because of budget restrictions.

Apollo 15 was commanded by Air Force Colonel David R. Scott, who flew in space in March, 1966, with Neil A. Armstrong in Gemini 8. Scott also flew in March, 1969, as command module pilot of Apollo 9. His lunar surface partner was Lieutenant Colonel James B. Irwin, a newcomer in space, and the command module pilot was Major Alfred M. Worden, also making his first flight. The Apollo 15 crew was the first

Over a rugged landscape in southern California, Apollo 15 astronauts James Irwin and David Scott test ride a vehicle that is similar to a lunar vehicle designed for trips across the surface of the moon.

Space Exploration

Continued

A Saturn S-4B rocket stage is being converted into an orbital workshop by McDonnell Douglas Astronautics Company. The vehicle will provide living and working space for three astronauts for up to 56 days.

in the moon program made up exclusively of Air Force officers.

Unmanned space projects. During the last half of 1970, Russia showed renewed interest in the planets, and stepped up the pace of its unmanned exploration of the moon. As its lunar program developed, Russian spokesmen repeatedly asserted that their way was considerably cheaper than the American way, and involved no risk of human life.

Russian moon shots. In the last half of 1970 Russia took two significant steps in lunar exploration, one of which continued well into 1971. On September 12, an unmanned payload, Lunar 16, was launched toward the moon, and on September 16 went into lunar orbit. Four days later, on September 20, the spacecraft landed, as other Soviet moonships had done before it.

This time, however, there was a difference. The unmanned craft scooped up a small sample of lunar material and then, on September 21, launched itself from the moon on command from earth. It re-entered the earth's atmosphere and was recovered on September 24. The

soil sample was turned over to Russian scientists for study and analysis. In June, 1971, U.S. and Russian space officials agreed to exchange Luna 16 and Apollo 11 and 12 lunar samples.

The amount of material brought back by Luna 16 was almost insignificant alongside the amount recovered by Apollo crews, but Soviet scientists defended unmanned lunar explorations on economic grounds. The Russian news agency Tass quoted Boris N. Petrov, a "Soviet space expert," as saying, "The flight of an unmanned craft compares in cost to a manned one by a factor of 1 to 20 or 50." Reckoning the cost of an Apollo mission at $385 million (as officially set by NASA), this would mean that a shot such as Luna 16 cost Russia between $8 million and $14 million, a figure Western experts consider impossibly low.

There was no dispute, however, over the fact that unmanned flights are cheaper than manned ones, and some space authorities in the West expressed the view that, in terms of cost-effectiveness, unmanned flights are probably

Unmanned self-propelled Lunokhod-1, *right*, sent to the moon by the Russians late in 1970, transmitted pictures to earth, including one of its own tracks, *below*.

more justifiable than manned ones. James A. Van Allen, discoverer of the globe-girdling radiation belts that bear his name, expressed this opinion after the flight of Luna 16.

Luna 17, the second Russian unmanned lunar spectacular, was launched Nov. 10, 1970. This soft-landing spacecraft contained a roving vehicle equipped with television – a sort of robot version of the U.S. Lunar Rover. The Russian vehicle, called "Lunokhod," carried a French-built laser reflector somewhat simpler than reflective devices placed on the moon by Apollo crews. The reflectors enable scientists on earth to measure accurately the earth-moon distance by determining precisely the time it takes for a laser beam to travel from the earth to the moon and back.

Russian spokesmen stressed the "international" character of the Luna 17 mission, but this was a Russian show all the way, and a genuine "first" in the sense that Russian-made wheels were the first ever to roll across lunar soil.

Powered by electricity generated from solar energy, the Lunokhod functioned only during the two-week-long lunar day. Each sunset the vehicle was "put to sleep." Two weeks later, at each sunrise, it was "awakened." Commanded by shirt-sleeved men in a Russian control center, the vehicle rolled across the lunar surface, photographing "targets of opportunity" as it went.

On Oct. 20, 1970, about three weeks before Luna 17 was launched, Russians sent up No. 8 in the *Zond* (probe) series of round-the-moon spacecraft. Experts in the United States speculated that it was an unmanned predecessor to manned flights around the moon. Previous Zonds had made direct, high-deceleration reentries over the Southern Hemisphere with splashdown in the Indian Ocean. But Zond 8 came in over the Northern Hemisphere and made a more gentle reentry before achieving an Indian Ocean landing on October 27. This, U.S. experts reason, is the way Russia's future manned missions will return, taking advantage of Russian land-based tracking stations and holding the gravitational forces of deceleration to levels acceptable to man.

The Russians' Luna 16, *right*, made round-trip flight from the earth to the moon, returning soil samples collected in the capsule, *above*.

Space Exploration

Continued

U.S. space experts believe that manned missions of the Zond type will try to establish a lunar orbit space station, perhaps similar to the Salyut station sent into earth orbit in April, 1971. Russian spokesmen have dropped hints of such intentions, and have said repeatedly that landings on the moon are not part of the Russian manned space program.

Planetary probes. Four instrumented spacecraft, three of them Russian, were successfully started on interplanetary paths. A fifth, the U.S. television probe Mariner 8, failed to achieve escape velocity when launched May 8, 1971.

The first of the planetary explorers to go was Venera 7, the fourth in a series of progressively more sophisticated Russian atmospheric entry vehicles to be sent to Venus, the earth's nearest planetary neighbor. Venera 7, carrying a 2,596-pound scientific payload, was launched Aug. 17, 1970. It entered Venus' dense, hot atmosphere just four months later on December 15. The craft continued to send signals for 20 minutes after it reached the surface. The signals

confirmed previous indications that no human being could survive on Venus because of the high temperatures and crushing atmospheric pressures. See ASTRONOMY, PLANETARY.

The "window" for flights to Mars in 1971 – the period when planetary orbits are favorable – opened in May. For the first time, Russia made a double-barreled attempt to intercept the red planet. The first attempt in 1962 with a single spacecraft, Mars 1, had failed. Mars 2 and Mars 3 were successfully launched on May 19 and May 28, respectively, and set on courses that would get them to Mars in November.

As is their custom, Russian spokesmen were vague about the missions of the two spacecraft. American analysts inferred from some available data that the probes were at the very least Mars orbiters and probably both could orbit and land. One of the revealing items was their announced weight – 4,650 kilograms (10,230 pounds).

Some space experts in the United States believe that Russia will attempt unmanned landings, as NASA hopes to

Space Exploration

Continued

do in 1976 with two craft in the Viking series. Russian scientists are obviously interested in the surface characteristics of heavenly bodies. They have been successful in landing unmanned craft on the moon and even more inhospitable Venus. And a planetary landing would give them an opportunity to score another impressive first in the continuing competition with the United States.

Mariner 9, a television-equipped spacecraft with a capability of going into orbit around Mars, was the sole successful U.S. planetary launching in 1971. Space officials had hoped that Mariners 8 and 9 would complement each other. With two spacecraft orbiting the planet, scanning assignments could be divided up and together compile a complete photographic atlas of Mars. With the failure of Mariner 8, a new program had to be written for Mariner 9.

Other nations in space. China, for the second year in a row, launched an instrumented earth-orbiting spacecraft on March 3, 1971. Its announced weight was 221 kilograms (486 pounds) – notice-

ably more than the 173 kilograms (386 pounds) of the first Chinese satellite launched on April 24, 1970. This makes experts in the United States believe that a new launching rocket, or at least a new upper stage, was used.

Peking gave details of neither launcher nor spacecraft and did not name the spacecraft. The United States designated it "China 2."

Japan, which entered the space club early in 1970 with the successful launching of an orbital payload on a rocket of the Lambda class, tried in September, 1970, to send up another satellite using a larger Mu-class rocket. This attempt failed. But on Feb. 16, 1971, a spacecraft was successfully launched into orbit with a Mu rocket from Japan's Kagoshima Space Center. The spacecraft was named Tansei, which roughly translates into English as light-blue sea.

France continued its modest space program with the launching of a spacecraft from Kourou, near Devil's Island in French Guiana, on Dec. 12, 1970.

The British attempted to place a spacecraft in orbit with a made-in-Britain

The Apollo 14 command service module comes up to the lunar module in one of a series of unsuccessful attempts to dock on the way to the moon. The connection was finally made.

launching rocket of the Black Arrow class from Woomera, Australia. The attempt failed.

Other nations were involved in space on a cooperative basis. Italy sent up two payloads—Uhuru, on Dec. 12, 1970, and San Marco C, on April 24, 1971, using U.S.-built Scout rockets. Canada built the ISIS-B satellite that was launched by American technicians on March 31, 1971. ISIS—International Satellites for Ionospheric Studies—is a follow-up program to the Alouette project instituted in the 1960s to study the upper atmosphere from above.

Russian cooperative efforts, in addition to Luna 17's French-built laser reflector, included an Intercosmos satellite that carried scientific instruments from other eastern European countries.

The United States, principal moving force of the International Telecommunications Satellite (Intelsat) Corporation, provided launching facilities for a large communications payload. On Jan. 25, 1971, Intelsat IV-F-2, was placed in synchronous orbit over the equator at an altitude of about 23,000 miles (see

COMMUNICATIONS). Another internationally oriented U.S. launching was that of Natosat, a communications satellite for the North Atlantic Treaty Organization, on February 2.

Russian-U.S. box score. Although each major space power has impressive "firsts" to its credit, it was clear by 1971 that American initiative had peaked and that Russia was forging ahead.

From 1958—the first full year of the space age—until 1966, the United States led Russia in the number of successful launchings to earth orbit or beyond. In 1966, the margin was 73 to 44.

Then, in 1967, Russia took the lead, 66 to 57. Since then, the gap has steadily widened as Russian activities accelerated and American ones slowed. In 1970, the score was 81 to 19, and the indications for 1971 were about the same.

At the end of 1970, the grand total box score in the space race between Russia and the United States stood at 592 launchings for the United States and 457 for Russia. In total number of payloads sent aloft, the United States led, 748 to 485.   [WILLIAM HINES]

Transportation

The quality of the environment was once considered a relatively minor issue in transportation, confined mostly to efforts to beautify highways or to avoid their construction in scenic areas or parklands. In 1971, environmental concerns created a new wave of social action that hampered, stopped, or seriously altered some of the largest transportation projects in history.

As a result of the discovery of one of the world's largest oil reserves at Prudhoe Bay, along the desolate North Slope of Alaska, a group of seven oil companies had planned to build and operate an 800-mile pipeline to the ice-free port of Valdez to the south. From Valdez, the oil was to be transported by giant tankers to the United States' West Coast.

The $2-billion project had progressed to the point where 48-inch steel pipe had been delivered and construction camps built. Then, work was stopped in 1970, by court injunctions brought by conservationists. The hot oil in the pipes, they argued, would melt the permafrost, seriously erode the fragile tundra, increase the danger of spillage due to

ruptured pipe, and interfere with migrating herds of caribou. Furthermore, the proposed pipeline route and the port of Valdez were located along Alaska's most active earthquake regions.

The results of public hearings held in early 1971 in Anchorage, Alaska, and Washington, D.C., portend a considerable delay in the pipeline project. Serious consideration is being given to alternate routes through Canada.

SST trouble. A similarly ambitious project in air transport was also in serious trouble. The supersonic transport (SST) was the center of controversy among U.S. scientists, economists, and politicians throughout 1970 and 1971. The primary thrust of the attack against further federal or private support for the project was the issue of the environment. One fear was that of the sonic boom at supersonic speeds, and the high sideline noise at take-off. The introduction of water vapor in the stratosphere at about 45,000 feet was viewed as a serious threat to the climate.

The SST is also one of the least efficient forms of passenger air transport.

As such, it is increasingly considered to be a questionable economic venture. It is deemed certain that the *Concorde*, an SST being developed by a British-French combine, will require a subsidy to operate. According to Richard A. Rice, a visiting professor at Carnegie-Mellon University in Pittsburgh, the SST's approximately 13 passenger-miles per gallon of fuel is exceeded in inefficiency only by that of the helicopter. In contrast, a 747 jumbo jet will deliver about 22 passenger-miles per gallon. Although this difference does not appear significant at first glance, it means that the extra fuel used per day by a fleet of 500 SST's, in comparison to the fuel used by subsonic jet aircraft to deliver the same total number of passenger-miles, would equal the daily output of the Alaskan North Slope oil fields—about 2 million barrels.

Plans to continue federal funding to develop a prototype SST were finally ended by a negative vote in the Senate in May, 1971. Without government support the aircraft seemed doomed. The decision is considered to be a major victory for conservationists.

Canal ditched. As if to round out the environmentalists' victories over large-scale transportation projects, construction of the Cross-Florida Barge Canal was permanently stopped on Jan. 19, 1971, by order of President Richard M. Nixon on the advice of his Council on Environmental Quality. The notable feature of this victory is that about one-third of the $180-million, 107-mile canal had been completed.

Congress had authorized the building of the canal in 1942 to protect American shipping from German attack, but construction did not begin until the mid-1960s. Conservationists had long opposed the project on the grounds that it would destroy the Oklawaha River and endanger fish and wildlife in the Florida Everglades.

Emission standards. Environmental pressures of the pipeline, airplane, and barge, however, are relatively minor in their long-range impact when compared with the recently instituted federal standards for automobile-engine emissions. Ever since biochemist Arie J. Haagen-Smit of the California Institute of Technology discovered in the 1950s that the Los Angeles smog is caused by the emissions of gasoline-powered internal-combustion engines, public concern has mounted over the effect of automotive exhaust on our atmosphere.

Unlike many environmental issues, which to some extent are subjective or speculative, the automobile exhaust problem is real and documentable. Accordingly, there was widespread support for the 1970 Clean Air Act. It requires that by 1975 all automobiles sold in the United States must meet standards so rigid that harmful emissions will be reduced to 10 per cent of the 1971 levels.

The new standards announced April 30, 1971, speed the development of alternative engines such as the closed cycle, external-combustion engine (the steam engine). Automobile manufacturers, however, appear confident that they can meet the standards with the internal-combustion engine through new designs, add-on devices, and strict maintenance and inspection.

Amtrak. The first sign of hope that the steady decline of rail passenger service in the United States might be brought to a halt or reversed—came in October, 1970, when Congress created the National Railroad Passenger Corporation, which was later nicknamed Amtrak. The goal of this new government-backed passenger system is to unify and improve service and to operate at a profit.

Early in 1971, Amtrak announced its final routes, which went into effect on May 1. Although this network represents another drop in the total number of trains in operation, the new approach is a positive step to revitalize this important mode of travel. Amtrak announced that it was dropping about half of the 350 passenger trains then operating, down from about 6,000 that were running at the end of World War II. The drop is less drastic than the numbers indicate, however, because the discontinued units were little used.

In other countries, however, plans are being pushed for expanded rail corridors. The Japanese National Railways is extending the highly successful New Takaido Line and is looking forward to the development of a high-speed rail

"*That* ought to satisfy Ralph Nader!"

A Magnetic Train

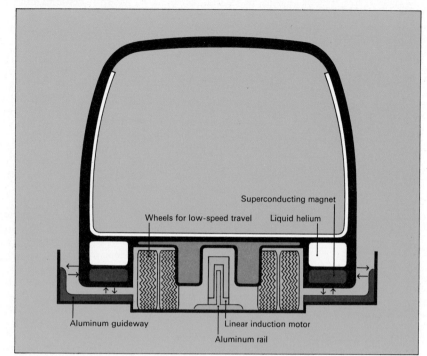

Proposed 300-mph train using electromagnetism for suspension and propulsion will float silently above its aluminum guideway. The superconducting magnets lift the train and induction motors provide the propulsion.

Superconducting magnet

Wheels for low-speed travel Liquid helium

Aluminum guideway

Linear induction motor

Aluminum rail

Transportation

Continued

network linking all major Japanese cities. England, France, Germany, and Italy are all building new lines and upgrading service – up to 130 miles per hour (mph) in some instances. Puerto Rico has announced plans to girdle the island with a modern ground transport system, of yet undetermined technology, to link its major towns and cities, and to help develop new towns as a means of halting its urban sprawl.

Air-cushion vehicles. The U.S. Department of Transportation (DOT) made its first effort to construct radically new systems in actual service situations in 1971. The most innovative and futuristic will be a high-speed link between the Los Angeles International Airport and the San Fernando Valley. The system will consist of 80-passenger vehicles capable of speeds of up to 150 mph. They will ride on a layer of air forced under the vehicle by on-board blowers, and will be powered along a rail by linear electric motors.

This air-cushion system, similar in many respects to the French Aerotrain, will be the most advanced form of

ground transportation to be constructed in the United States. All systems, including propulsion and suspension, will be electrically powered from a wayside pickup, thus providing a high-speed system that is extremely quiet and smooth.

People movers. The Urban Mass Transportation Administration of DOT awarded the Jet Propulsion Laboratory of the California Institute of Technology a contract to develop designs and specifications for a "people mover," a vehicle smaller than a streetcar, for West Virginia University in Morgantown. The system will provide continuous service between the old campus in town and the new campus in the suburbs of this small mountain city. The type of suspension, guidance, and propulsion are as yet undetermined, but candidates for such small-scale, urban transit systems are numerous.

Efforts by private concerns to market new systems of urban transportation are increasing. A wide range of systems, varying from monorail to small-scale systems for major activity centers, are available.   [JAMES P. ROMUALDI]

Zoology

Two intrepid chemists from the University of Hawaii, Richard E. Moore and Paul J. Scheuer, located and collected specimens of a rare and dangerous sea anemone, or polyp, called limu-make-o-Hana. The scientists, who reported their work in 1971, sought to analyze the chemistry and determine the toxicity of a poisonous substance the polyps produce.

The polyps were known to live in only one tidepool whose location was carefully guarded by superstitious persons in the Hana district on the island of Maui. According to legend, disaster will strike anyone who attempts to gather the poisonous polyps. However, the chemists reported that, with the help of "an elaborate chain of informers, the tidepool was located at the end of a lava flow at Muolea (Kanewai), south of Hana, Maui . . ."

The scientists collected some of the polyps, ground them up, and extracted some of the poison, which is called palytoxin. They learned several things about the palytoxin, but its incredible potency was the most interesting.

A mouse dies from a dose of about one ten-billionth of its own weight—about one ten-billionth of an ounce will kill a 1-ounce mouse. Thus, this seemingly harmless little polyp produces one of the most poisonous substances ever discovered. But why it needs such a lethal weapon is still unknown.

A desirable hitchhiker. Another polyp, *Calliactis*, lives on the empty snail shells occupied by the hermit crab. Not only does the polyp get a free ride in this way, but it also feeds on the scraps discarded by the crab. When the hermit crab changes shells, he transfers his passenger to the new residence. This suggests that the crab may receive some benefit from the polyp. What this benefit is has long eluded zoologists.

Working at the Stazione Zoologica in Naples, D. M. Ross has finally discovered why the hermit crab likes the polyp's company. Hermit crabs are considered a delicacy by the octopus. In 1971, Ross reported that when he put hermit crabs without polyps on their shells in a tank with an octopus, the octopus immediately pulled the crabs

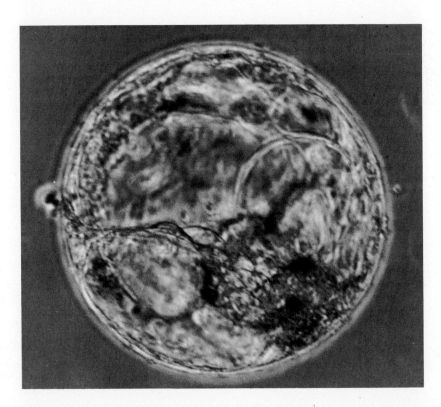

This embryo, containing more than 100 cells, was raised from a human egg that was fertilized outside of the body.

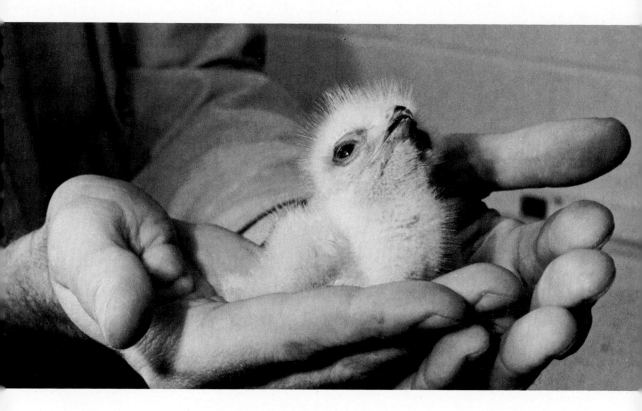

Zoology

Continued

A baby red-tailed hawk became the first wild bird of prey successfully hatched from an artificially inseminated egg. The technique could help save several endangered bird species.

from their protective shell and ate them. However, when hermit crabs with polyps living on their shells were put into the tank, the octopus attacked as before, but quickly withdrew when his tentacles touched the polyps. Judging from his behavior, the octopus had been stung. And after several more such encounters, the octopus carefully avoided the hermit crab and his rider.

Thus, zoologists have learned what generations of octopuses have found out the hard way: The polyps produce a poisonous substance that not only protects them but also protects their sources of food and transportation.

Worm turns–against cancer. According to Polynesian accounts, patients with cancer can be cured by consuming the cooked tentacles of the kaunaoa worm, which lives buried in the sea bottom off Hawaii. Frank L. Tabrah, Midori Kashiwagi, and Ted R. Norton of the Department of Pharmacology of the University of Hawaii tested the validity of this legend.

They collected some of the worms, ground their tentacles, and put them

in alcohol. They then injected extracts of the resulting material into mice that were also inoculated with tumor cells. In control animals injected only with the tumor cells, tumors grew and killed all of the mice within 25 days. However, more than half the animals treated with the kaunaoa-tentacle extracts developed no tumors. Preliminary tests indicate that the active anticancer ingredient in the worms is a large, nonpoisonous molecule that is not a protein.

Regeneration. It has been known for over 200 years that salamanders can regenerate, or grow back, their legs following amputation. It is also well established that the regeneration will not occur if the nerves that led to the legs are also cut. What is not known is how the nerve fibers stimulate the regeneration.

In an ingenious experiment, Ferdinand Hui and Alfred Smith of the New York University Medical College, tried to find out if it is the presence or the functioning of the nerves that is required for limb regeneration. They partially paralyzed salamanders for up to 11

Zoology

Continued

weeks with repeated injections of hemicholinium, a drug that interferes with neuromuscular transmission. To avoid asphyxiation because of paralysis of the salamanders' breathing mechanisms, the scientists used larval salamanders, which can breathe passively under water through gills.

In animals not treated with the drug, the legs, which had been amputated at the knee, regenerated about four-fifths of an inch in 10 weeks. The animals treated with hemicholinium, however, regenerated less than half as much leg tissue. Thus, although paralysis did not completely inhibit the outgrowth of new legs, it definitely retarded regeneration.

Such regeneration has long been considered impossible in higher vertebrates such as mammals. However, zoologists L. Nichols Grimes and Richard J. Goss of Brown University discovered that rabbits can regenerate new tissue to fill in holes about the size of a dime that have been punched through their ears.

To find out exactly which tissues of the ear are responsible for this phenomenon, belly skin was grafted in place of the inner and outer skins of some rabbits' ears, and holes were cut through the grafted regions. These holes grew in, too, suggesting that any kind of skin at all will do.

Next, the cartilage was removed from the ears of some rabbits, leaving the inner and outer skin layers back to back. When holes were punched in these ears, the tissue did not grow back. This proved that the cartilage produces the unusual capacity for rabbit ears to regenerate.

The task for the future will be to discover what is so special about this cartilage that it can promote such exceptional outgrowths of new tissue. The answer may give scientists a clue as to why most of the rest of the mammalian body is totally incapable of growing back any lost parts.

Mystery mouse birth. One of the "facts" of reproduction is that an animal produces only one litter at a time. It came as a considerable surprise to S. A. Barnett and Kathleen Munro at the University of Glasgow, Scotland, therefore, when some of their laboratory

Fossil of Archaeopteryx, *below,* earliest known bird, may help solve mystery of how birds developed flight. Restoration, *below right,* shows details of three claws on each wing. This suggests that wings developed on reptiles that leaped after their prey rather than from their jumping from trees to avoid foes.

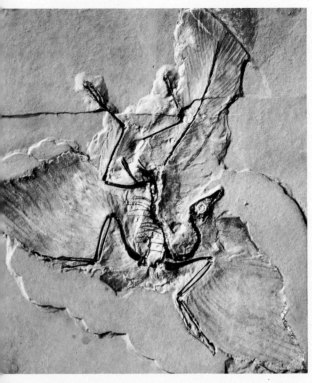

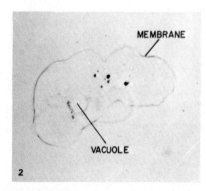
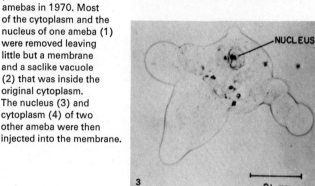
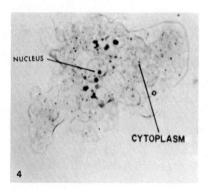

New one-celled animals called amebas were assembled from the parts of several natural amebas in 1970. Most of the cytoplasm and the nucleus of one ameba (1) were removed leaving little but a membrane and a saclike vacuole (2) that was inside the original cytoplasm. The nucleus (3) and cytoplasm (4) of two other ameba were then injected into the membrane.

Zoology

Continued

mice gave birth to second litters in about two weeks rather than the usual 21 days after previous offspring had been born. To add to the mystery, no male mice had been with the females since the first birth.

Further investigation showed that these animals must have mated during the early stages of their first pregnancy. A second crop of fertilized eggs developed as far as they could without implanting in the uterus. Then, when the previous litter was born, the partly developed eggs moved down into the uterus and resumed their growth.

Test-tube embryos. The early human embryo spends its first week or so floating free in the fluids of the Fallopian tube and uterus. There, it goes through the early stages of cleavage in which it repeatedly doubles its number of cells. Eventually it develops into a mass of cells called a blastocyst, which is characterized by the formation of an internal cavity. These preliminary stages of development go on prior to the embryo's attachment to the mother's uterus. Why, then, could they not be induced to occur outside the body if the right fluids could be provided?

Until recently, scientists could keep early human embryos alive in this way only during the very early cleavage stages. For example, P. C. Steptoe, Robert G. Edwards, and J. M. Purdy of the University of Cambridge in England, had raised human embryos to the 16-cell stage. But in 1971, the same investigators reported the successful culture of early human embryos for as long as six and one-half days after fertilization. During this time, each embryo underwent repeated cell divisions and had more than 100 cells. They also differentiated to the preimplantation stages, including development of the characteristic cavity.

Clearly, if the next step, implantation, is to be achieved, an artificial uterus of some kind will have to be developed. This will be a formidable task, indeed.

Mama's aroma. Many newborn animals have to have some way of recognizing their mothers. Aside from visual means, it is believed that the young often recognize their mother's smell. To test

Recognition by a Cat's Whisker ... Holes

Scientists in Kenya have found that they can identify individual lions by recording whisker-hole patterns. Lions have four or five rows of whisker holes but only the top two rows are needed for identification. Numbers on the drawing show possible sites, and the dots indicate actual whisker holes.

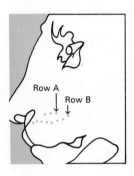

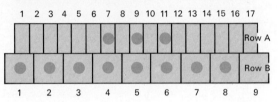

Row A

Row B

Zoology

Continued

this theory, Dietland and Christine Muller-Schwarze at Utah State University took a black-tailed deer fawn from its mother shortly after birth and trained it to accept a four-legged wooden rack set on wheels as a substitute mother.

The artificial mother was equipped with a milk bottle from which the fawn could suckle, and a buzzer that could be sounded to imitate the low bleat of the fawn's real mother. This maternal contraption was swabbed twice daily with the scent of a pronghorn antelope instead of that of a deer.

The fawn was left with this "mother" for two weeks. It reacted as a fawn does to its natural mother. For example, when startled by thunder or other loud noises, the fawn ran to the wooden rack for protection.

Then, for a second two-week period, the fawn was put in with an unscented dummy, after which it was tested to see if it could pick out the antelope smell from among several dummies. Time and again, it picked out the antelope smell from others.

Even when the fawn was 2½ months old, it preferred the company of a live pronghorn antelope to that of a black-tailed deer or a mule deer. These tests show that the odors to which a young deer is exposed make such a deep impression upon him that they are used for recognition thereafter.

Fog blocks hearing. Bats can fly in the dark because they are able to detect and avoid obstacles by virtue of an echo location mechanism. However, zoologist J. D. Pye at the University of London Kings' College reported in 1971 that bats refuse to fly in fog.

When these bats encounter a fog bank, they turn sharply to avoid flying into it. The reason for this, according to Pye, is that each fog droplet absorbs much of the ultrasonic energy emitted by the bat, rather than bouncing it back as a sharp echo useful for navigation.

Unable to "focus" its sensitive hearing, the bat probably perceives only a ringing sound. To fly in such an ultrasonic blackout would be to invite certain disaster, so the creature wisely stays in clearer atmospheres. [RICHARD J. GOSS]

Men of Science

Of increasing concern to many scientists is the proper use of their science. This section, which recognizes outstanding scientists, features two who have not been afraid to question the wisdom of those who determine what are to be the goals of science and technology.

Albert
Szent-Györgyi

By Judith Randal

**His scientific career twice interrupted by war,
this distinguished Nobel laureate has persevered
in exploring some of the fundamental secrets of life**

Seated in the living room of his comfortable old house on Cape
Cod, Massachusetts, Albert Szent-Györgyi poured tea for his visitor.
It was a raw afternoon, and the sea through the windows was as leaden
a gray as Szent-Györgyi's eyes are bright blue. In fluent English with
the flavor of central Europe, he transported himself and his guest to
another country and another era—when both Szent-Györgyi and the
science of biochemistry were young.

"As a schoolboy in Budapest, I was looked upon as retarded, and
had to have private tutors to pass my exams," said the man who won
the 1937 Nobel prize in physiology and medicine, isolated and identified vitamin C, and unraveled much of the mystery that had sur-

rounded the phenomenon of muscle contraction.

"When I told my uncle, a scientist, that I wanted to follow in his footsteps," Szent-Györgyi recalled, "he seriously objected. He had been a bright student, and had become a distinguished anatomist. Obviously, he thought me dull and my aspirations impertinent. It was my older brother Paul who was considered the gifted child in the family.

"When, at last, I began to do well in my studies, my uncle gradually gave in. He suggested first that I might go into producing cosmetics and, later, that I consider dentistry. And when I went to medical school in 1911, he still wanted me to stick to something practical. Since he had hemorrhoids he urged me to specialize in proctology." Szent-Györgyi was drawn more to the laboratory than to the bedside, but to please his uncle, he dutifully wrote his first research paper on the tissues of the anus and rectum.

Albert Szent-Györgyi (pronounced Saint JOR jee) was born in 1893. His mother came from a family that produced four generations of scientists, and his father was a member of landed nobility. In medical school, young Albert had scarcely whetted his appetite for the microscope and the dissecting room when, in 1914, World War I broke out. Hungary, then a part of the Austro-Hungarian Monarchy, fought on the side of the Central Powers—Bulgaria, Germany, and the Ottoman Empire (now Turkey). Szent-Györgyi was to be in a soldier's uniform for most of the next five years.

As the fighting dragged on, it became increasingly apparent that the Central Powers could not win against the Allies—chiefly Belgium, France, Great Britain, Italy, Russia, Serbia, and, later, the United States. Szent-Györgyi was convinced that a small, powerful group of militarists was causing needless slaughter, and, in addition, he was eager to return to science. He concluded that, "In the long run, I could best serve my country by staying alive."

Thus, one day, after a struggle with his conscience, he deliberately shot himself in the arm. The wound sent him back to Budapest. There he finished medical school, receiving his M.D. in 1917. With the war still in progress, he resumed his army service in a bacteriological laboratory. But soon he was in trouble because he objected to the dangerous experiments being performed on prisoners of war. "Since the officer in charge of the experiments outranked me," Szent-Györgyi recalled, "I was ban-

The author:
Judith Randal is medical editor of The Washington Star. She has written many award-winning articles, and is a frequent contributor to Science Year.

At the age of 77, Albert
Szent-Györgyi has turned
his considerable energy
and skill toward trying
to discover some of
the secrets of cancer.

ished to the malaria-ridden swamps of northern Italy as punishment."
The protest was typical of a Szent-Györgyi, who has always willingly
risked his life for principle. "Living for years fingering a trigger instead of a test tube was frustrating," he said, "but the idea of being
killed for my ideas has never frightened me."

After the armistice on Nov. 11, 1918, he went to Pozsony, a town in
western Hungary where his maternal great-grandfather had begun his
scientific career. There Albert became an assistant in the pharmacology department of the university. But not long afterward, Pozsony was
ceded to Czechoslovakia and renamed Bratislava. Hungarians were no
longer welcome in the city. Szent-Györgyi explained, "When I left to
go back to Budapest, I had to disguise myself as a workman and wait
for dark to collect my scientific equipment."

The Austro-Hungarian Monarchy was overthrown in 1919, and
Socialists and Communists briefly took over the newly independent
Hungary. Landowning families like the Szent-Györgyis were stripped
of their property and wealth. Except for 1,000 pounds sterling
(then equal to $4,425) the young scientist—who by this time was mar-
ried and had a daughter—had nothing but his scientific skill and a determination that "My goal as a scientist would be to understand life."

He felt, then as now, "that life is most likely to reveal its secrets in its simpler forms." That is why, for him, pharmacology was wrong. Although the drugs were relatively simple, the experimental animals in which they were studied were not. Bacteriology, too, had seemed overly complex. He would, therefore, use his savings for further education, and if the funds ran out and he could not find work in a laboratory, he could always go to work as a physician for some colonial government in the tropics.

To prepare himself, Szent-Györgyi, with his wife and daughter, went first to Prague, Czechoslovakia, to study electrophysiology. He then went to Berlin to learn about hydrogen ion concentrations, which are a measure of the acidity or alkalinity in the fluids of living organisms. In 1919, with the last of his money and a determination to reach the goal he set himself, he enrolled at the Institute of Tropical Medicine in Hamburg, Germany.

There, his luck suddenly changed. He met a professor from the University of Leiden in the Netherlands who offered him a position as an assistant in pharmacology with the opportunity to study chemistry at night. The salary was modest and the life quiet, but Szent-Györgyi enjoyed the academic atmosphere and he stayed at Leiden for two years. His next post was at the University of Groningen,

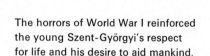

The horrors of World War I reinforced
the young Szent-Györgyi's respect
for life and his desire to aid mankind.

in northern Holland. When he arrived, in 1922, he hoped to earn Dutch medical credentials so that he could go to the Dutch West Indies as a doctor. The head of a laboratory at the university had been repeatedly disappointed because his dogs died after experimental surgery despite stringent efforts to maintain germfree conditions. Szent-Györgyi recalled from earlier experience that dogs could fight infection much better than they could cope with the antiseptics used to control it. Thus, when he was invited to try his hand at solving the problem, he completely eliminated antiseptics. All the animals survived. Thereupon, the head of the laboratory offered him a job as an assistant. Szent-Györgyi accepted, and began to embark on investigations of biological oxidation that ultimately led to his Nobel prize.

In those days, little was known of oxidation, the chemical process by which a substance takes up oxygen. A critical step in many of life's most important processes, oxidation is also the reason that certain fruits and vegetables, such as apples, bananas, and potatoes, turn brown after they are peeled and exposed to the oxygen-laden air.

Turning his attention to this phenomenon, Szent-Györgyi discovered that fruits and vegetables that do not turn brown in air contain a substance not present in those that do. He was not sure then what it was, but he correctly reasoned that this substance must be a reducing agent—a chemical that counteracts oxidation. The reducing agent was later found to be vitamin C, which is known chemically to scientists by the name ascorbic acid.

Szent-Györgyi's further investigation of biological oxidation disclosed information that was to provide German biochemist Hans A. Krebs with a start toward his ultimate explanation of how men and animals convert food into energy. It was primarily for this work that Krebs himself was later chosen by the Caroline Institute in Stockholm, Sweden, to share a Nobel prize in physiology and medicine.

In 1926, the professor who was Szent-Györgyi's sponsor at Groningen died and was succeeded by a psychologist who, said Szent-Györgyi, "took a dislike to me and had no interest in chemistry." Thinking his research career was ended, Szent-Györgyi sent his family back to Budapest. He then decided to bid a farewell to science by attending an international conference in Stockholm. The decision turned out to be a fateful one for Szent-Györgyi.

The main address at the conference was given by one of the most distinguished biochemists of the day—Sir Frederick Gowland Hopkins of Cambridge University in England—who had apparently read some of Szent-Györgyi's papers. Much to the younger scientist's astonishment, he heard his name and work mentioned three times in the speech. Getting up the courage to introduce himself, the man who thought his laboratory days were over found himself invited to study under Hopkins on a Rockefeller Foundation fellowship in England. He would stay there from 1926 to 1930, with a few months out in 1929 for research in the United States.

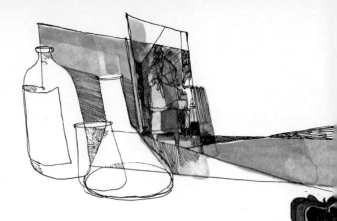

Laboratory tests of red peppers prepared for the dinner table revealed them to be a very rich source of vitamin C.

During the years at Cambridge, Szent-Györgyi isolated the unidentified reducing agent from cabbages, oranges, and adrenal glands, then crystallized it. He still did not know the exact chemical nature of his substance, but suspected it was something like a sugar. The chemical names of sugars end in "ose," and, in his typically humorous fashion, he referred to his substance as "ignose" (unknown sugar) or "Godnose." But the editor of a prominent scientific journal, disapproving of puns, refused to publish Szent-Györgyi's paper on the subject until the flippant names were replaced by something more orthodox. Szent-Györgyi substituted the term "hexuronic acid," and the paper was accepted and ultimately led to his Ph.D. degree in 1927.

One trouble he faced in his research on the reducing agent was that so little of it was available. Adrenal glands, the richest source then known, were hard to obtain in Great Britain. So Szent-Györgyi went to the United States in September, 1929, to work with Edward C. Kendall, then biochemist at the Mayo Foundation in Rochester, Minn. The nearby St. Paul slaughterhouses would solve the problem of supply. In the spring of 1930, Szent-Györgyi returned to England with 25 grams, almost an ounce, of the reducing agent—"An enormous amount in those days," he recalled with understandable pride.

Then his career took another unexpected turn. Hungary's minister of education, who wanted to modernize the nation's science, asked Szent-Györgyi for his help. Szent-Györgyi was reluctant to leave Cambridge, but feeling duty-bound to assist, the scientist returned to his native land to serve as the head of the department of medical chemistry at the University of Szeged in southeast Hungary.

Scientific equipment and supplies were hard to obtain in Hungary. However, Szent-Györgyi's new laboratory was quickly filled with able assistants, one of whom told his chief he knew how to detect whether a substance contained vitamin C. Vitamins—complex organic compounds—are necessary for health and growth, and vitamin C specifically is required for bones, teeth, and body tissue. Szent-Györgyi suspected that his reducing agent was, in fact, vitamin C. But, believing that vitamins were the "concern of the chef, not the scientist," he had not checked the matter out. Now, however, he gave his last few grams of the reducing agent to his young colleague, who proved within a month that "Godnose" and the vitamin were one and the same. Szent-Györgyi, quite naturally, was eager to dig further, but

adrenal glands were as scarce in Hungary as they had been in Great Britain, and citrus fruits, another source, were unavailable. Then, on an evening in 1932, fate and Szent-Györgyi's natural kindness combined to solve his problem in an unusual way. He did not feel like eating the red peppers his wife had planned as a part of supper. To avoid hurting her feelings, he told her he wanted to test the vegetable in his laboratory. "By midnight, I knew that red peppers were a treasure chest of vitamin C, containing 2 milligrams of vitamin per gram of pepper, or 1 part in 500." Red peppers were abundant in Szeged; they are the source of paprika, one of the city's main products. "A few weeks later," Szent-Györgyi recalled "I had many kilograms of crystalline vitamin C that I distributed all over the world among researchers who wanted to work on it."

Quite by accident, he had found one of nature's greatest storehouses of the vitamin. This pioneering work with vitamin C, together with earlier studies, including those on oxidation, brought him the 1937 Nobel prize in physiology and medicine.

When Szent-Györgyi arrived in Stockholm to receive the Nobel prize, he was treated as a hero who had benefited all mankind. The experience made up for the disappointments of the past and would sustain his faith during the harsh times that lay ahead.

In 1938, the Hungarian government fell, and the progressive minister of education died. In the rising tide of fascism and anti-Semitism spurred on by Adolf Hitler, books were burned and Jews persecuted. Szent-Györgyi, who was by then acting president of the University of Szeged, had friends and students who were Jews, and he felt he could not stand idly by. His opposition to anti-Semitism angered many, and one day the entire law faculty came after him with sticks, threatening him for defending "non-Aryans." This and other aspects of the gathering storm that led to World War II began to interfere with his scientific work.

Hungary was quickly falling under total German domination in 1942 when a group of its leading citizens, knowing of Szent-Györgyi's commitment to democratic ideals, came to him secretly. They asked him to help save the nation from becoming Germany's vassal. Turkey was neutral territory, so Szent-Györgyi, using the pretext of a lecture engagement, went to Istanbul, where he planned to contact the British and Americans.

It was a dangerous mission because Nazi-occupied territory had to be crossed, but the visit to Turkey went without mishap. However, trouble lay ahead. The British Secret Service in Istanbul had instructed Szent-Györgyi to set up a secret radio station in Hungary. The Gestapo, the Nazi secret police, who by then were everywhere in Hungary, learned of the plot, and placed Szent-Györgyi under house arrest.

Later, when German troops actually occupied the country in 1944, Hitler personally ordered Szent-Györgyi taken into custody. Warned in time, the scientist and his second wife, Martha—he had remarried in 1941—found sanctuary from the Nazis in the Swedish legation in Budapest. Fearing that he might not survive the war, Szent-Györgyi sent his unpublished papers to Hugo Theorell, a biochemist friend in Stockholm. Not knowing Szent-Györgyi's address, Theorell cabled the Swedish legation in Budapest to say he had received the items. The Nazis intercepted the message and discovered Szent-Györgyi's whereabouts. In the nick of time, the scientist and his wife were smuggled in the trunk of a Swedish diplomat's automobile to a new hiding place.

The couple decided it would be best to separate temporarily. Martha Szent-Györgyi went to her parents' home. The scientist himself fled from one hiding place to another until he found refuge near the advancing Russian lines.

If the Nazis were cruel overlords, so, too, were the Russians. In 1945, after the Russian Army had driven the Nazis from Hungary, it rounded up thousands of Hungarian soldiers. Those who did not die of poor treatment or disease in camps were sent to unknown destinations to become slave laborers. One entire Hungarian regiment, overjoyed

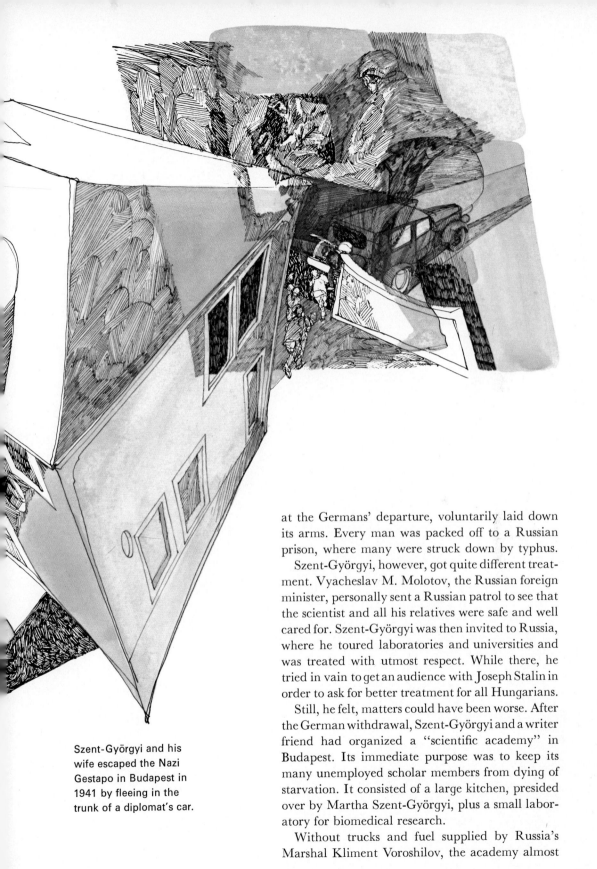

Szent-Györgyi and his
wife escaped the Nazi
Gestapo in Budapest in
1941 by fleeing in the
trunk of a diplomat's car.

at the Germans' departure, voluntarily laid down
its arms. Every man was packed off to a Russian
prison, where many were struck down by typhus.

Szent-Györgyi, however, got quite different treatment. Vyacheslav M. Molotov, the Russian foreign
minister, personally sent a Russian patrol to see that
the scientist and all his relatives were safe and well
cared for. Szent-Györgyi was then invited to Russia,
where he toured laboratories and universities and
was treated with utmost respect. While there, he
tried in vain to get an audience with Joseph Stalin in
order to ask for better treatment for all Hungarians.

Still, he felt, matters could have been worse. After
the German withdrawal, Szent-Györgyi and a writer
friend had organized a "scientific academy" in
Budapest. Its immediate purpose was to keep its
many unemployed scholar members from dying of
starvation. It consisted of a large kitchen, presided
over by Martha Szent-Györgyi, plus a small laboratory for biomedical research.

Without trucks and fuel supplied by Russia's
Marshal Kliment Voroshilov, the academy almost

Szent-Györgyi considers
his achievements in work on
muscle contraction to be a
high point in his career.

surely would have failed. Szent-Györgyi used the vehicles for what he called a "travel agency." Because the city's public transportation was at a complete standstill, and many people were eager to leave the capital, business was brisk. The money collected in fares went to buy much-needed food for the starving scientists and artists. For a time, the arrangement seemed to work well.

The venture, however, was doomed. In 1947, the Russians arrested one of the academy's backers in Budapest and falsely accused him of espionage. Szent-Györgyi eventually obtained the man's release from jail and arranged to have Martha drive him to Switzerland. Learning firsthand of continued Russian brutality to Hungarians behind prison walls completed Szent-Györgyi's disillusionment with the Russians. He decided he could not continue to live in Hungary, and turned his hopes toward the United States.

During his visit to the United States in 1929, Szent-Györgyi had attended a scientific meeting in Boston. While there, he also went to the Marine Biological Laboratory at nearby Woods Hole, Mass., for a clambake where lobsters (which he had never seen before) were served. If it hadn't been for the memory of the delicious lobsters, he probably would have forgotten both the laboratory and its working space, available to scientists at modest cost.

Arriving in the United States in 1947 on a visitor's visa, his memory jogged by the thought of the red crustaceans, he headed for Massachusetts to begin work at Woods Hole. Szent-Györgyi was determined to extend and refine experiments on muscle contraction that he had begun in Hungary. He was interested in the treatment and prevention of diseases of the most important muscle of all—the heart. Accordingly, with this goal in mind, he proceeded to set up the Institute of Muscle Research.

Szent-Györgyi's chief preoccupation as a biologist has always been with the most fundamental problems, what he calls "the symptoms of life." He is bored by endless data, preferring simply to dig in and do a job. He seeks "a fingertip friendship with living matter." He believes this approach gives him the opportunity to notice and pursue seemingly insignificant details that may lead to important basic discoveries. And, unlike many scientists who are afraid to speculate, he boldly proceeds from "the wildest theories" even though they may later turn out to be wrong. Men who have been his students say this boldness, along with incredibly perceptive instincts, are his strong points. As one says, "He has the keenest sense of what is new and what is worth studying."

Szent-Györgyi had first turned his attention to muscular contraction in the 1930s. He regarded motion as one of the most distinguishing characteristics of life. It had long been known that a protein called myosin was involved in contraction, and Szent-Györgyi began to extract this substance from muscle fiber. He often noticed that as he carried on the extraction process, the material became stickier.

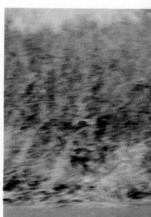

This seemingly trivial observation led him to suspect that he was isolating a second substance. He had discovered another muscle protein. Once one of his students isolated the protein called actin, Szent-Györgyi was able to demonstrate that actomyosin–actin and myosin together–contracted and relaxed when he exposed it to a body chemical called adenosine triphosphate, or ATP.

Looking back to his work with actomyosin, Szent-Györgyi said, "To have reproduced *in vitro* (in the test tube) one of the oldest signs of life, motion, was perhaps the most thrilling moment of my life." The discovery was all the more exhilarating because it related directly to so many important life processes. Birth begins when the muscular uterus contracts to push a baby through the birth canal and into the

Albert Szent-Györgyi's life today is centered around cancer cell research. But he also finds the time to write extensively and to enjoy pleasures such as long strolls and motorbiking. In "The Seven Winds," his comfortable seaside home, he entertains colleagues and relatives. He prefers living by the ocean "because it gives a man a wider perspective and allows him to ignore the trivial things that take up so much valuable time."

An articulate critic, Albert Szent-Györgyi has written many books and articles, and given lectures on campuses throughout the United States. His views reflect sympathy with the ideals of the young and their concern for mankind.

light of the outer world. Life often ends when another muscle, the heart, after many billions of rhythmic motions, at last is still.

As his work at Woods Hole progressed, Szent-Györgyi began to realize it was approaching the point of diminishing returns because technology had not yet produced the tools he needed to answer his scientific questions with suitable precision. The studies had taught him, however, that energy may travel between living cells much as a spark jumps between separated wires, enabling the cells to influence each other without actually touching. This gradually led him, during the 1950s, to such problems as the interplay of electrons when stimulated by magnetic fields, and ultimately to his current scientific interest, the electrochemical behavior of cancer cells.

Just as Szent-Györgyi's spirit of scientific inquiry never dims, neither does his commitment to his own special view of life. He is passionate in his defense of young people, and alert to the hypocrisy of government and other institutions on environmental and racial questions. He is a strong critic of militarism and the U.S. involvement in Southeast Asia. He has written and spoken extensively on all these subjects.

In the past few years, misfortune again marked Szent-Györgyi's life. Cancer took the two people dearest to him—his wife Martha in 1963 and his daughter Nelly in 1969. He remarried in 1966, but the union was not a happy one and lasted only a few months.

He now lives alone in the rambling, shingled house he calls "The Seven Winds" on Penzance Point outside Woods Hole. He has only his niece Csilla for company, except when friends or his three grandchildren from Boston drop in to pay a visit. As it has always been since

childhood, when his mother taught him to love the classics, music remains his companion. Bach and Mozart are special favorites.

There is also a special place in his heart for another composer. "My mother was a talented singer," he explained, "and when she was young she went to see Gustav Mahler to ask his professional advice on whether she should marry my father or pursue a concert career. She was a tiny woman of rather fragile appearance, and Mahler, taking this into account, advised her against the rigors of a career. So she decided to get married. I feel I owe my life to Mahler."

Szent-Györgyi used to keep some of his experimental animals on the grounds, but abandoned this practice after vandals destroyed a delicate cancer experiment by drowning a group of rats late in 1970. "I could redo the experiment," Szent-Györgyi explained, "but it took three months to prepare. Three months are precious to me now."

In many ways, his appearance is that of a much younger man. His 5-foot 7-inch frame is still solidly muscular, kept in condition during summer by a long daily swim in Vineyard Sound and Buzzards Bay. His open, lively face and his smile and expressive hands also belie his age. His spirit, despite a lifetime of struggle, seems stronger than ever. Each morning, he arises at daybreak to spend an hour or two writing before going to work at the Marine Biological Laboratory.

His nephew Andrew Szent-Györgyi, who is a biochemist and professor at Brandeis University, describes his uncle's vigorous nature and outlook on life and work in this way: "He fishes with a big hook, and he always goes after big fish." In a long and inspiring lifetime, Albert Szent-Györgyi has often landed the big ones.

Eugene Shoemaker

By Joseph N. Bell

His geological mapping of the moon began a 10-year struggle to get science in the manned space program

Eugene Merle Shoemaker knows more about the lunar surface than any other man, even those who have walked on it. The first astrogeologist of the space age, he charted the course of this new science of lunar and planetary geology. Long before President John F. Kennedy started the moon race in 1961, Gene Shoemaker had begun to map the moon out of a compelling intellectual interest.

Today, at the age of 43, he heads the Division of Geological and Planetary Sciences at the California Institute of Technology (Caltech). Without the mustache that he grows and shaves so often it seems to wax and wane with the phases of the moon, Gene looks too young, too optimistic, to have weathered the frustrations he encountered during the hectic lunar decade of the 1960s. He could still be the brash young scientist who knew exactly where he was going, and who, even as a child, was always several years ahead of everyone else.

"Like many youngsters, I just started to pick up rocks," Gene recalls fondly, his eyes crinkling as he talks of his first love—geology. "Only I never quit. My first serious collecting was in 1937, when I was 9 years old, on a trip with my father through Wyoming and the Black Hills of South Dakota. Dad had studied enough geology in school to whet my interest with his own. He encouraged me, and so did the school principal in Buffalo, N.Y., where we moved from our farm in Wyoming after that trip. But what determined me most was the program for youngsters at the Buffalo Museum of Science. I was lucky enough to get started in it when I was in the fifth grade."

Soon afterward, Gene was collecting rock specimens in earnest—first at glacial areas around Niagara Falls and on the nearby shores of Lake Erie, then at the classic fossil sites in upstate New York. "I went to special classes in Buffalo where I could take archaeology and other subjects related to geology, and I got steadily more fascinated."

The Shoemakers moved to Los Angeles when Gene finished junior high school. His mother was a schoolteacher and his father, a free soul, was at various times a teacher, coach, farmer, and trucker. He moved the family to California to operate a trucking business. This was during World War II and he could not get the maintenance equipment he needed, so he hired on as a stagehand at a Hollywood movie studio.

"I knew I wanted to go into some aspect of geology even before I started high school," Gene says, "and my parents were already aiming me at Caltech in Pasadena. It's not necessarily something I recommend, but when you know for sure at a very young age what you want to do, it's a tremendous advantage. I never went through the troubles so many kids have, including my own, in trying to find themselves. But it can't be forced. It has to come from inside."

Gene entered Caltech when he was just 16. "I suffered the same ego shock most students experience at Caltech," Gene says. "They've been going through school being top banana without having to work at it, and suddenly, for the first time in their lives, they have to work very hard. Of course, they really aren't working any harder at Caltech than they will be the rest of their lives as productive scientists. But it's all been easy before and now they're surrounded by a bunch of guys just like themselves—and some of them a lot smarter. That's tough on the ego."

In 1948, when Gene was 20, he was awarded his M.S. degree from Caltech and joined the U.S. Geological Survey, a bureau of the Department of the Interior. "That summer," Gene recalls, "I was driving through southern Colorado, thinking about the tremendous advances Germany had made in rocket technology during World War II. I suddenly realized that within my lifetime man was going to go to the moon, and I could be in the middle of it, actually involved." From that moment on, the moon was never far from Gene Shoemaker's thoughts.

The young scientist began work on a Geological Survey search for uranium ore deposits that was sponsored by the Atomic Energy Commission (AEC). The work involved extensive geological mapping of the sedimentary areas in the rugged Southwest, and it was to last for almost a decade. One project Gene started resulted in the discovery of about a million tons of uranium ore. During this period, he met and married a pretty young schoolteacher from Chico, Calif., named Carolyn Jean Spellman, and also managed two years of graduate study at Princeton University, where he eventually received a Ph.D. degree in 1960.

The author:
Joseph N. Bell is a free-lance writer for national magazines and lecturer in English at the University of California, Irvine.

Gene's long search for uranium deposits was to lead him into lunar research. In 1957, his geological mapping of the Navaho and Hopi country of Arizona brought him to the 4,150-foot-wide Meteor Crater, located about 50 miles southeast of Flagstaff. Gene set out to determine whether the crater was in fact an exploded volcano, many of which contain uranium deposits.

Gene Shoemaker's roles in the moon program ranged from teaching geology to astronauts, *above left,* to designing lunar tools, such as the Jacob's staff, *above right.* Studying a TV monitor, *left,* an intent Shoemaker watches first close-up pictures of the moon from Surveyor 5.

"Some volcanoes," explains Shoemaker, "are highly charged with volatiles that erupt violently, and I was interested in the structure and mechanisms of these volcanoes. So was the AEC, because they figured that underground nuclear explosions might produce an effect similar to that of an explosive volcano. They thought plutonium produced in an underground nuclear explosion might be localized enough to be mined. I was concerned about where the plutonium might go during an underground explosion. It seemed to me that by studying Meteor Crater, I could estimate the effects of shock in a nuclear crater and answer that question.

"So I first began examining the deformation of rocks in the crater wall. I soon realized, however, that a big part of the story was in the thrown-out material, so my investigation grew into a broad-gauged study of the whole crater. I must admit I got into that with considerable malice aforethought. By then, the moon was my long-range goal. I could see that studying the mechanics of crater formation would be the logical means for studying the geology of the moon. It offered me a great opportunity to learn how to distinguish craters formed by volcanoes from those formed by meteorites."

An earthbound geologist sits amid tools the astronauts will use to study structure of the surface of the moon.

From a study of rocks collected on the floor and rim of Meteor Crater, Shoemaker and an associate, Edward Chao, discovered the mineral coesite, which is produced when the atoms in quartz are squeezed into a different arrangement. Because this process requires more pressure than an exploding volcano generates, the Chao-Shoemaker studies confirmed that the 570-foot-deep crater had been formed by a giant meteorite crashing into the earth. This work—especially the discovery of coesite—would win Chao and Shoemaker international recognition and, in 1965, they would be awarded the Wetherill Medal of the Franklin Institute.

Meteor Crater put Shoemaker firmly on the moon track. He had proposed a small program to map the geology of lunar craters by telescope to his superiors at the Geological Survey in 1956. They were still pondering his request on Oct. 5, 1957, when word came over the radio that the first Russian Sputnik had gone into orbit. Shoemaker heard the news upon returning to his Meteor Crater base camp in an isolated canyon in the Hopi buttes of Arizona. "I knew right then," he remembers, "that the heat was on. I saw myself on a straight road from Arizona to the moon."

The moon and the earth are essentially a two-planet system, and the two bodies are probably closely related in origin. But the moon's surface seems not to have been subjected to erosion by running water, which destroyed the early geological history on earth. To Gene, the chances seemed very good that the record of very early events is still preserved in the moon's rocks. "I had been forming an image in my mind of what I thought a manned flight to the moon should accomplish for geology," he remembers. "There seemed to me then a magnificent opportunity to make this come to pass."

At the time of the first Sputnik, however, no detailed lunar map had ever been made. No one knew whether the lunar crust would support the astronauts or even if they could land safely, because surface features such as elevations, slopes, and surface texture were unknown. "I threw myself into lunar work after Sputnik," Gene recalls. "When Congress created the National Aeronautics and Space Administration (NASA) in 1958 and it was clear that it would be in control of space exploration, I began to develop a new program of lunar mapping from the earth with the idea that it would be funded by NASA."

But it was a difficult plan to sell. "Lunar geology was pretty far out then," recalls Alfred Chidester, a geologist at the Center of Astrogeology in Flagstaff. "Gene was a bright, precocious guy who hit the old-timers as a child prodigy, but maybe a screwball. He came on strong, even then, and he was resented by some of the traditionalists."

Shoemaker examines a montage of photos taken by a Lunar Orbiter spacecraft, and ponders the enormous amount of information still to be learned from them.

Shoemaker's plan failed to convince his superiors at the Geological Survey. He was debating leaving the survey in hopes of working directly in the space program, when his maps of the moon based on his crater research began to pay off. They convinced NASA officials that Gene had an "interesting program" to offer. Shortly thereafter, in August, 1960, NASA agreed to sponsor a lunar geology program directed by the Geological Survey.

Shoemaker set up shop at the survey's field center at Menlo Park, Calif., and began what was to become a 10-year program of geological lunar mapping. The previous year, Gene had prepared a preliminary map of the great crater of Copernicus. Using only photographs and telescopic measurements, he showed that he could unravel the sequence in which layers of rock had been built up from the debris of repeated meteorite impacts and lava flows. From this beginning, Shoemaker and his unit of 15 geologists and chemists developed improved mapping techniques and by 1962 had geologically charted extensive regions of the moon's surface. By this time, the astrogeologic unit had become a formal branch of the Geological Survey with Gene Shoemaker as its chief.

In September, 1962, Gene was discouraged to learn that the original seven astronauts, all test pilots, were to be joined by nine more test pilots. There were no scientists—thus no geologists—appointed to the flight program. So, at the invitation of Oran W. Nicks, director of NASA's Lunar and Planetary Program Office, Shoemaker took leave from Menlo Park and moved to Washington, D.C., to try and bore from within. "I wasn't completely naïve," he says. "I knew that NASA's primary goal was just to get a man on the moon before the Russians did. I didn't think NASA's manned flight program was just going to jump up and embrace a scientific program. But I felt that I had to get in and push and try to develop a greater science effort within NASA. That's why I went to headquarters, to try and help build some bridges between the science side of the house and the manned flight program, which at the time was on a separate course going its own way."

Poring over maps of Colorado River country, Gene and a Flagstaff scientist pinpoint areas of unknown geology for future survey by earth-orbiting satellite.

During a trip down the Colorado River in 1968, Gene and geologist Bruce Julian compared a section of the canyon walls with photographs taken of them a century earlier by naturalist John Wesley Powell.

He had another motive, however: Gene Shoemaker wanted to go to the moon himself, and he went to Washington to explore his chances. Although Shoemaker and his associates were at first unsuccessful in persuading NASA that the time had come to select some scientist-astronauts, a year later, NASA changed its mind. But Shoemaker was then two years over the maximum age of 34 and he found himself appointed chairman of the selection committee instead. It was like an athlete who tries out for the team and is named coach. Gene would have resigned this job and sought an age waiver, but he had been struck down by an exotic ailment known as Addison's disease.

Carolyn remembers that awful year. "I started noticing the symptoms when we first moved to Washington," she says. "Gene became progressively weaker. There were just a lot of different and seemingly unconnected ailments that doctors couldn't pin down at first. We took a river trip that spring and he came back dragging instead of refreshed. In June, when tests showed he had the disease, he was given cortisone, and his recovery was unbelievable. Within two days he was surfing, and he has had no problems since."

The disease, however, left its mark. Shoemaker's life line is his supply of cortisone tablets that he must take twice a day to correct the adrenal deficiency that causes Addison's disease. As long as he takes the cortisone, he can function normally, but the disease destroyed the last possibility of his becoming an astronaut.

Despite his illness during that year in Washington, Shoemaker set up and served as acting director of NASA's Manned Space Sciences

Division, which was to develop and finance the scientific program on manned space flights. Gene also planned the Geological Survey's new Center of Astrogeology, at Flagstaff, which would continue the lunar mapping program. And he began site studies for the location of a lunar telescope at Flagstaff.

In 1963, Shoemaker and part of his Menlo Park staff moved to Flagstaff. From this new headquarters, Shoemaker and others began work on the television experiment on Project Ranger. The unmanned Ranger spacecraft were to crash into the moon's surface, sending back close-up television pictures during the last minutes before impact. Gene was also the principal investigator on Project Surveyor, a later series of unmanned soft landings. His Flagstaff team designed the television camera techniques for both projects and helped select the impact and landing sites. "Our goal was to learn about the surface properties of the moon," adds Gene. "From data sent by the television cameras, we hoped to explain what the debris layer is, its thickness, roughness, and how it was formed."

After many disappointments, the Ranger spacecraft flew successfully, and moon mapping took on new dimensions. On July 31, 1964, Shoemaker and his fellow scientists gathered at the Jet Propulsion Laboratory of Caltech as Ranger 7 neared the moon. When the voice on the hot line from the Goldstone tracking station reported: "We're receiving pictures to the end . . . impact . . . impact," cheers rang out in the room. "Before that spacecraft hit the moon," recalls Shoemaker, "we'd gone about as far as we could with earth-based optics. But 17 minutes before Ranger impacted the moon, its cameras began taking hundreds of pictures that brought us a thousand times closer than the best telescopes ever could." Ranger 7's cameras showed that the moon's surface, seen for the first time at close range, was not radically different from Shoemaker's predictions.

At the newly built Center of Astrogeology, Shoemaker's staff had now grown to 200 scientists, engineers, and technicians. Most of the group in Flagstaff today recall this time with the warmth and camaraderie of a wartime regiment bloodied in combat. "We were almost a bootstrap operation," remembers Al Chidester. "We did it, then we found out if it could be done." Gordon Swann, now the prime geology investigator on Apollo 14 and 15, remembers Shoemaker as friendly, congenial, and, above all, energetic. "He yelled at people and they yelled back at him. That was the atmosphere here. We didn't work for him; we worked with him—and that's the way he wanted it."

Carolyn Shoemaker remembers her husband being home "about one-third of the time." He was constantly shuttling between Flagstaff and Washington, Houston, Menlo Park, and Pasadena, where he served sporadically as a visiting professor at Caltech. In the winter of 1964, he also resumed teaching a graduate course in astrogeology as a Caltech research associate. There were a few respites from work to fulfill civic duties and to play, however. Shoemaker ran twice for the local school board in Flagstaff and lost both times. And in 1966, he started to build his "dream house" on a pine-forested mesa, 300 feet above a canyon floor.

Shortly after Surveyor 1 soft-landed on the moon in June, 1966, Gene resigned as administrator and took on full-time research responsibilities with the new title of chief scientist with the Geological Survey. He began an intensive study of the lunar surface as principal geological investigator for Apollo 11, 12, and 13. But gradually disappointments and difficulties began to outnumber successes. Typical of the growing frustrations was a stereo camera system for Surveyor, which would have scanned the landing areas, recording surface textures in great detail. NASA turned down the proposal in spite of Gene's strenuous efforts to get it aboard.

The instrument he wanted most, however, was a Jacob's staff that Apollo astronauts could carry on the moon. On earth, a Jacob's staff is used to measure the thickness of sedimentary rocks. The lunar version, first proposed in 1965, was designed to automatically record a geological traverse as the astronaut walks along, giving a detailed record of where he is and what he is seeing. The system would provide Houston with the astronauts' exact location and figuratively place earth-based geologists on the moon. "This was the system," Shoemaker says, "that I would have wanted were I on the moon myself. It was not built or even seriously considered by NASA."

With each passing month in 1966 and 1967, Shoemaker found himself fighting an increasingly rearguard action to keep science in a NASA program that concentrated on engineering feats. Hal Masursky, Shoemaker's successor at Flagstaff and a friend of more than two decades, recalls, "Gene, as usual, was several years ahead of everyone else. He began the fight to save Apollo back in 1963 by asking what we should do to justify the program after the first landing. Science doesn't exist independently of the system, and he was trying to redirect the program toward science while it was still possible."

"The patches I tried to apply didn't stick," says Shoemaker, "but at least a few lasted long enough to get some science effort started. Early in the game, I think Congress would have supported a reasonable marriage of science and engineering. What was needed was a top-level NASA policy to capitalize scientifically on the gains made in the space program. And of course that just didn't happen. I don't think NASA officials were necessarily antiscience. They just never conceived of the manned space program as a scientific effort or as having a scientific goal. They saw it solely in terms of engineering and national prestige."

But even though science was being downplayed, the Project Apollo astronauts still had to be given a crash course in geology so they would have some idea of what they should look for when they set foot on the moon. This was one of Shoemaker's jobs and he was pleased with his students. "Neil Armstrong and Ed Aldrin," he says, "were very sharp guys who realized what an enormous lot there was to observe on the

Gene and Carolyn Shoemaker often visit their still unfinished dream house which was begun during their years in Flagstaff.

At the Shoemaker home in Pasadena, daughter Linda, 14, *above left,* gets a lift to pick oranges from back yard tree. Son Pat, 15, now owns Gene's rock collection, *above right.* Father and daughter Christy, 18, play with the family pets, *left.*

moon's surface. As it turned out, they had enough background to do a lot more geology than they had time for."

The crews of Apollo 11 and Apollo 12 sampled the moon's well-churned debris layer. Some samples had been thrown from depths as great as 50 feet when a meteorite formed a nearby crater. The rocks the astronauts brought back were meticulously studied by many geologists to determine the relationships of the rocks to the structures from which they came. See SOLVING THE MYSTERY OF THE MOON.

Gene Shoemaker had hoped that this preliminary sampling would be followed by a more deliberate geologic inspection of the lunar surface. But even before the first astronauts walked on the moon, he sensed this would not happen. "I tried for years to get a long-term commitment of support for our work," he says. "And it was becoming clear we weren't going to get it." Only one trained geologist, Harrison H. (Jack) Schmitt, had been named to the astronaut program. Schmitt was scheduled on the back-up crew of Apollo 15. This meant that, in the normal sequencing of crews, he would be on the prime crew of Apollo 18, a mission that was later canceled. "He had been my hostage to destiny," says Shoemaker sadly, "my alter ego, a guy I got involved in the program and the only scientist who survived the whole process and actually had a chance to fly to the moon."

For the first time in his life, Gene Shoemaker found real disenchantment setting in. "I'd reached the point of exhaustion. For a long time, I suppose, I was the only one who could carry the ball for astrogeology. Now there was a group of guys equally able to run with it. It was time for me to get away and leave the job to them."

In 1968, Gene took a leave from Flagstaff to do the one thing that is a constant source of renewal to him—river running. It combines his love of the outdoors, his curiosity at the workings of nature, and his

Gene, Carolyn, and their daughters head for the badminton court near their Pasadena house.

The Shoemakers spend afternoon sightseeing in London while awaiting the launch of Apollo 14.

pursuit of adventure. He is very good at it. That summer Shoemaker led a Geological Survey expedition that retraced the historic voyage of geologist John Wesley Powell down the Green and Colorado rivers a century earlier. The group, including Gene's wife and his son Patrick, spent three months navigating rubber boats through the rapids to locate the sites from which Powell's men made a series of photographs. By comparing new photographs with the old they would determine exactly how much the earth had changed at those sites during the intervening 100 years.

Soon after the trip, Gene accepted the chairmanship of Caltech's Division of Geological Sciences. Leaving Flagstaff and his half-built dream house was difficult not only for Gene but for Carolyn and their three children. Arriving in Pasadena on Jan. 1, 1969, Gene, at first, found it difficult to adjust to academic life. Departmental duties weighed heavy on him, particularly when he much preferred to teach or "do science," both of which suffered under administrative loads.

In October, 1969, Shoemaker's problems were compounded by an impromptu speech he made to a university group in Caltech's faculty club. In simple, tough language, Shoemaker charged that little effort had been made to utilize the best capabilities of the human beings sent into space at such tremendous cost to earthly taxpayers. He insisted that unless the goals of the space program were drastically changed the

Gene analyzed Apollo 14 moon landing for the British Broadcasting Corporation (BBC) television audience in February, 1971.

vast expense of interplanetary manned space flights would be as wasteful as the Apollo program had been.

Gene did not know that a local reporter was present, and the next day newspapers throughout the United States carried an account of his talk. The shock waves were sudden and strong.

Shoemaker was dismayed. "It wasn't my intent," he says, "to take a public shot at NASA. I wouldn't retract the merits of the statements, but I wish they could have stayed in that room. The temptation to say those things publicly had been very strong for a long time, but the thing that always restrained me was the possibility that Jack Schmitt might still fly.

"The net effect of that speech now has been to cut me off totally from having any significant internal voice in NASA," Gene adds. "It has also undermined the position I've long held that you have to work within the system to make things happen."

Caltech geologist and geophysicist Gerald Wasserburg says, "Gene tried very hard for very long to work within the system. That's why it surprised us so much when he was the one who blew the lid off—or at least the one who got caught."

And so, Gene Shoemaker burned his lunar bridges. The man who dedicated his life to getting to the moon—as astronaut, or geologist passenger—is not going to make it. He is resigned to this now—but he

is not happy about it. "I still had hopes even two years ago, but not anymore, not the way the Apollo program is going," he says wistfully. "It will be so long before we are prepared to do any serious exploration of the moon that I'll be an old man. As it turned out, it's much better for me that I didn't get involved in the astronaut program. It would have been an enormous frustration."

Today Gene is seeing more of his family, and they all like that. He has time to play badminton with his children and "to behave like a father for the first time in a decade." Patrick, now 15, is an outdoor boy showing indications of exploring his father's interests, although "we're trying hard not to push." Christine, 18, and Linda, 14, are interested in American Indians in the Southwest. Carolyn manages their big, rambling Spanish house in Pasadena with a quiet efficiency, and Gene keeps it full of scientists and students, inviting them on whim as he used to do in Arizona. He finds some recreation in reading current events and biography but cannot remember when he last saw a movie or casually watched television. He has little interest in music, literature, or art, "except," he says, "for southwestern Indian and pre-Columbian Mexican art—on which I'm a bona fide nut!" Even his close associates are unaware of his political convictions because he simply never talks about them.

In the summer of 1970, Gene shifted his interests to a new field. He began sampling rocks again on the Colorado Plateau. "In a way," he says, "it's a return to an old love with a completely new twist." He is using paleomagnetism—the magnetism of rocks—to study stratigraphy here on earth. Gene wants to develop a dating system 1,000 times more accurate than fossil and isotopic dating. Using this, he could work out a very precise correlation of rocks between the continents. Shoemaker's successor at Flagstaff, Hal Masursky, says, "Gene always does the new thing. We knew what to do with the moon rocks because we spent 10 years under Gene studying what to do. Now he's leading the revolution in earth science resulting from the astrogeology program."

Shoemaker is still fascinated with lunar research, however. "Because we have now reached out to the moon," he says, "we can deal more intelligently with how the earth and the solar system were created, and make use of this information in practical ways. Ultimately, knowledge about the origin and early history of the earth will provide new insight into the relationship of various mineral deposits, for example, and why and how they got where they did. Using this insight, we may be able to predict with some degree of accuracy where to find more. And that's going to be vital in the next 50 years, because we're using up the minerals that underpin our civilization at alarming rates.

"Apollo has opened an era that will soon need fresh ideas, fresh approaches to the problems. We may still be debating the origins of the earth and the moon when we bring back the first rocks from Mars. But we must go and get those rocks and go on from there."

His desk cluttered with administrative papers, the department chairman today has less time than he would like for geological research.

Awards
And Prizes

A listing and description of major science
awards and prizes, the men and women
who won them, and their accomplishments

Chemistry. Major awards in the field of chemistry included:

Nobel Prize. Luis F. Leloir, director of the Institute for Biochemical Research in Buenos Aires, Argentina, won the 1970 Nobel prize for chemistry and a $78,400 cash award. Leloir was honored for "his discovery of sugar nucleotides and their role in the biosynthesis of carbohydrates."

In the 1950s the Argentine biochemist discovered and isolated the first of a class of compounds, the sugar nucleotides, which act as a catalyst in the body to transform one type of sugar into another. These compounds were the key to solving a major biochemical problem —how living tissue synthesizes complex carbohydrate polymers.

Born in Paris, France, on Sept. 6, 1906, Leloir has spent most of his life in Argentina. After graduating from medical school at the University of Buenos Aires with an M.D. degree in 1932, he worked at biochemical laboratories in Cambridge, England, and the United States. He has been at the Institute in Buenos Aires since 1946.

Perkin Medal. James F. Hyde, senior scientist at Dow Corning Corporation, Midland, Mich., received the 1971 Perkin Medal of the American Section of the Society of Chemical Industry for outstanding work in applied chemistry. Hyde carried out the first successful research leading to commercial production of organosilicon compounds, especially the organosiloxane polymers. His work helped transform the silicones into a family of widely useful chemical structures. Today, silicones in the form of fluids, resins, and rubbers are used in nearly every industry.

Hyde earned his Ph.D. degree in organic chemistry at the University of Illinois in 1928.

Priestley Medal. Frederick D. Rossini, professor of chemistry and vice-president for research at the University of Notre Dame, Indiana, won the 1971 Priestley Medal of the American Chemical Society.

Rossini obtained a Ph.D. degree in physical chemistry in 1928 from the University of California, Berkeley. He joined the National Bureau of Stand-

Agronomist Norman E. Borlaug won the 1970 Nobel Peace prize for his work in developing new high-yielding, semidwarf varieties of wheat.

Luis F. Leloir

Frederick D. Rossini

Hannes O. G. Alfven

ards in Washington, D.C., in 1928. In his research there he achieved extremely high standards of precision in purifying compounds and measuring their physical and thermochemical properties. In 26 years spent characterizing the principal components of petroleum, Rossini and his co-workers identified 500 compounds in the gasoline fraction alone. He joined the Notre Dame faculty in 1960 as dean of the college of science.

Physics. Awards for important work in the field of physics included:

Nobel Prize. Hannes O. G. Alfven of the Royal Institute of Technology in Stockholm, Sweden, and the University of California, San Diego, and Louis E. F. Néel of the University of Grenoble, France, shared the 1970 Nobel prize for physics and the $78,400 cash award.

Swedish space physicist Alfven was honored for his imaginative and pioneering theories in plasma physics. His main contribution was to initiate the field of magnetohydrodynamics, the study of electrically conducting gases in a magnetic field.

Born in Norrköping, Sweden, on May 30, 1908, Alfven received his Ph.D. degree in 1934 from the University of Uppsala. He has been a professor at the Royal Institute since 1940.

French physicist Néel was cited for "fundamental work and discoveries concerning antiferromagnetism and ferrimagnetism, which have led to important applications in solid state physics." He is especially well-known for investigations of magnetic fields of atoms and groups of atoms in a solid.

Néel was born Nov. 22, 1904, in Lyons, France. From 1928 to 1945 he studied at the École Normale Supérieure in Paris and did research at the University of Strasbourg. At Grenoble he directs the higher national school of electrotecnics and applied mathematics.

Fermi Award. Norris E. Bradbury, director of Los Alamos Scientific Laboratory at Los Alamos, N. Mex., since 1945, received the Enrico Fermi Award for 1970 and a $25,000 cash award from the Atomic Energy Commission. He was cited "for his inspiring leadership and superb direction of the laboratory . . . and for his great contributions to the national security and to the peacetime applications of atomic energy."

Bradbury received a Ph.D. degree in physics in 1932 from the University of California. His association with the atomic energy program began in 1945 when he was assigned to Los Alamos and put in charge of the assembly of nonnuclear components for the first nuclear device which was tested at Alamogordo in July 1945. He became director of the laboratory in 1945. Under his leadership, it grew from a complex of temporary wartime structures to one of the nation's largest institutions dedicated to basic and applied research in virtually all fields of nuclear science. Los Alamos was the leader in nuclear rocket propulsion technology when he retired in September, 1970.

Franklin Medal. Wolfgang K. H. Panofsky of Stanford University was awarded the 1970 Franklin Medal of the Franklin Institute. He was cited "for his contributions to high energy physics and particularly for the solution of many complex engineering problems concerned with the design, construction, and use of the eminently successful high energy linear electron accelerator at Stanford University."

Panofsky received his Ph.D. degree from the California Institute of Technology in 1942. One of the scientists who conceived the accelerator, he has been its director since 1961. He joined Stanford as a professor in 1951 and in 1953 became director of its high energy physics laboratory and accelerator center.

Lawrence Award. Five nuclear physicists, including a three-man team, received the Ernest Orlando Lawrence Memorial Award for 1971 from the U.S. Atomic Energy Commission (AEC). Each received a $5,000 cash prize. The team included Robert L. Fleischer, General Electric Research Laboratory, Schenectady, N.Y.; P. Buford Price, Jr., University of California, Berkeley; and Robert M. Walker, Washington University, St. Louis. The other recipients were Thomas B. Cook, Jr., Sandia Laboratories, Livermore, Calif., and Robert L. Hellens, Combustion Engineering Inc., Windsor, Conn.

Fleischer, Price, and Walker were honored for contributing to the discovery and understanding of the phenomena of etching charged particle tracks in solids and for the detection and study of latent tracks registered in solids.

Earth and Physical Sciences

Continued

Luis E. F. Néel

Cook was cited for his "significant contributions to the study of nuclear weapons effects, for his original work in the translation of this knowledge into advanced technology for peaceful and military uses of atomic energy and for his outstanding contributions to the nation through his service as an adviser to the AEC and the Department of Defense on the effects of nuclear detonations."

Hellens was recognized for his "pioneering contributions to the field of light water reactor physics, for work in reactor statics, and the Fourier transform approach to the solution of the Boltzman transport equation."

Geology. Major awards in the earth sciences were:

Penrose Medal. Ralph A. Bagnold of Edenbridge, England, received the 1970 Penrose Medal from the Geological Society of America in recognition of his work on the formation and movement of sand dunes.

After military service in World War I, Bagnold was awarded an engineering degree at Cambridge University. He re-

joined the army and while abroad explored the desert as a hobby, studying sand dunes and their formation. After his release from the army in the 1940s he began studying the physics of water-transported sediments on seabeds and rivers. This research is used today by geologists and engineers studying flood control, harbor and beach erosion problems, and airborne particulate matter pollution problems.

Vetlesen Prize. Three scientists received the 1970 Vetlesen Prize for geophysics and a $25,000 honorarium. Half of the prize money went to S. Keith Runcorn, director of the school of physics at the University of Newcastle-upon-Tyne in England. Allan V. Cox, professor of geophysics at Stanford University, and Richard R. Doell of the U.S. Geological Survey shared the other half.

Runcorn pioneered in using remnant magnetism in rocks to determine the past positions of the magnetic poles when these rocks were formed. Through rock magnetism, Cox and Doell showed that the earth's magnetic field reverses its polarity at irregular intervals.

Life Sciences

James D. Watson

Biology. Important research in biology resulted in the following awards:

Carty Medal. James D. Watson, who shared the 1962 Nobel prize in physiology and medicine for discovering the double-helical structure of deoxyribonucleic acid, won the National Academy of Sciences' John H. Carty Medal and $3,000 cash award in April, 1971.

Watson has concentrated on the role of ribonucleic acid (RNA) in the synthesis of protein, and with his colleagues has provided the most precise available characterization of the ribosome particles in which protein synthesis occurs. These studies identified the small portion of RNA that carries genetic information. Other significant contributions of Watson include research on heredity in bacteria and on the reproduction of viruses in bacteria.

Born in Chicago in 1928, Watson received his Ph.D. degree at Indiana University in 1950. He joined the faculty of Harvard University in 1955, and became director of the Cold Spring Harbor Laboratory of Quantitative Biology on Long Island in 1968.

Elliot Medal. Richard D. Alexander, professor of zoology and curator of insects at the University of Michigan's Museum of Zoology, was given the Daniel Giraud Elliot Medal of the National Academy of Sciences and a $1,000 cash award in April, 1971. He was recognized for his outstanding work on the classification, evolution, and behavior of crickets and other insects.

By listening to the sounds made by crickets, Alexander refined the classification of these insects, discovered many new species, and identified behavior patterns—mating, fighting, communicating, and the establishing of territories. Through exhaustive studies of tape-recorded cricket songs, he characterized 250 different songs from more than 100 species of crickets. His analysis of cricket sound patterns shows an almost infinite variety of songs for individual species.

Alexander received his Ph.D. degree from Ohio State University in 1956. He joined the faculty of the University of Michigan in 1957.

Lilly Award. David Baltimore, associate professor of microbiology in the

Life Sciences

Continued

department of biology, Massachusetts Institute of Technology (M.I.T.), received the 1971 Eli Lilly and Company Award in Microbiology and Immunology and a $1,000 honorarium. He was recognized for his contribution to the development of a new field of research – the molecular biology of viruses.

Born in New York City in 1938, Baltimore received his Ph.D. degree from Rockefeller University in 1964. In his first work there, he and Richard Franklin discovered the first virus-coded polymerase found in mammalian cells.

U.S. Steel Foundation Award. Masayasu Nomura, co-director of the Institute for Enzyme Research at the University of Wisconsin, received the U.S. Steel Foundation Award in Molecular Biology and a $5,000 honorarium in April, 1971. He was cited for "studies on the structure and function of ribosomes and their molecular components."

Nomura showed that parts of bacterial ribosomes may be broken down into their components – RNA and proteins – and then reassociated to regenerate active particles. This discovery has been crucial to the study of ribosome structure and function.

Born in Hyogo-ken, Japan, Nomura received his Ph.D. from the University of Tokyo in 1957.

Medicine. The highest awards in the field of medicine included:

Nobel Prize. Julius Axelrod, chief of the Section on Pharmacology in the Laboratory of Clinical Science, National Institute of Mental Health (NIMH), Bethesda, Md.; Dr. Ulf S. Von Euler, professor of physiology at the Royal Caroline Institute, Stockholm, Sweden; and Sir Bernard Katz, professor of biophysics at University College in London, shared the 1970 Nobel prize for physiology and medicine and a $76,800 cash prize. The three neurophysiologists worked independently on studies of chemicals that carry messages across the junctions of nerve cells.

Axelrod identified mechanisms involved in regulating the formation and inactivation of the chemical noradrenaline in nerve cells. Dr. Von Euler found that noradrenaline serves as a transmitter at nerve terminals. And Dr. Katz discovered that the chemical acetylcholine is released at nerve-muscle junctions.

Axelrod has been section chief at NIMH since 1955, the year he received his Ph.D. degree from George Washington University. Von Euler was born in Stockholm and has spent most of his scientific career at the Institute. His father won a Nobel prize in chemistry in 1929. Dr. Katz was born in Leipzig, Germany, in 1911. In 1935 he fled to England from the Nazis and began his studies on nerve and muscle function at University College in London.

Gairdner Award. Five doctors received the 1970 Gairdner Foundation International Awards for medical research and $5,000 cash prizes. The recipients were: Dr. Vincent P. Dole, Rockefeller University, New York City; Dr. W. Richard S. Doll, regius professor of medicine, Oxford University; Dr. Robert A. Good, professor and head of the department of pathology, University of Minnesota; Dr. Niels K. Jerne, director of the Basel Institute of Immunology, Basel, Switzerland; and Dr. R. Bruce Merrifield, Rockefeller University.

Dole was honored for developing a treatment for heroin addiction – the use of Methadone to eliminate narcotic drug hunger and block the euphoriant actions of heroin.

Doll helped prove a relationship between the number of cigarettes smoked and the likelihood of developing cancer of the lung. He also showed that exposure to therapeutic radiation increases the risk of developing leukemia.

Good contributed to the knowledge of immunity, particularly in the study of childhood diseases.

Jerne worked on the theory of immune reactions, on explaining why the body makes antibodies against proteins and other substances foreign to it while not doing so in respect to proteins that are part of the body.

Merrifield devised a new approach for synthesizing proteins and protein fractions which allows the study of many such substances in a more pure form than any that could be obtained by isolating them from natural sources.

Lasker Award, Basic Research. Dr. Earl W. Sutherland, Jr., professor of physiology at the Vanderbilt University School of Medicine, Nashville, Tenn., received the 1970 Albert Lasker Award for Basic Medical Research and a $10,000 honorarium.

Julius Axelrod

Bernard Katz

Irvine H. Page

Life Sciences
Continued

He was honored for his discovery in 1960 of cyclic AMP, a compound found inside and outside cells all over the body. He also provided an understanding of its role in regulating the action of hormones at the cellular level.

Dr. Sutherland earned his M.D. degree in 1942 from the Washington University School of Medicine in St. Louis.

Lasker Award, Clinical Research. Dr. Robert A. Good, professor of microbiology and regents' professor of pediatrics and microbiology at the University of Minnesota Medical School, received the 1970 Lasker Award for Clinical Medical Research and an honorarium of $10,000.

Dr. Good, a pediatrician, anatomist, and immunologist, was honored for his contributions to the understanding of the body's defenses against disease. He showed that there are two defense systems—one in blood and tissue fluids, and another in certain white blood cells. He also showed how they develop and function. In addition, he helped discover the key role of the thymus gland in immunobiology, a landmark in research.

Dr. Good received both his M.D. and Ph.D. degrees in 1947 at the University of Minnesota, Minneapolis.

Stouffer Prize. Dr. Irvine H. Page, consultant emeritus of the research division, Cleveland Clinic Foundation, Cleveland, Ohio, and Sir George W. Pickering, master of Pembroke College at the University of Oxford in England, shared the $50,000 Stouffer Prize for research in heart disease.

Dr. Page was cited for recognizing that hypertension, or high blood pressure, is a disease that can be treated, and for producing the first reversal of malignant hypertension in 1937 with the use of drugs. He received his M.D. degree from Cornell University in 1926, and has been associated with the Cleveland Clinic since 1945.

Sir George was recognized "for his highly original contributions" to the understanding and treatment of hypertension and cardiovascular disease. He completed medical studies at St. Thomas's Hospital in London and is one of the few physicians to be elected as a Fellow of the Royal Society.

George W. Pickering

Space Sciences

Aerospace. A selected list of awards in aeronautics and astronautics were:

Founders Medal. Clarence L. Johnson, senior vice-president of Lockheed Aircraft Corporation, Burbank, Calif., was awarded the 1971 Founders Medal of the National Academy of Engineering. He was cited for his experimental and theoretical investigations in aerospace sciences, and is best known for his work in designing the Hudson bomber, Constellation and Super Constellation transports, P-38, T-33 trainer, the Jet-Star, and the Warning Star.

Born in Ishpeming, Mich., Johnson received an M.S. degree in aeronautical engineering in 1933 from the University of Michigan, Ann Arbor.

Guggenheim Award. Cosmonauts Andrian G. Nikolayev and Vitali I. Sevastyanov received the Daniel and Florence Guggenheim International Astronautics Award in 1970. They were cited for experiments performed during the flight of Soyuz 9 in June, 1970. The flight lasted about 17 days and 17 hours, providing valuable data on the effects of space conditions on human beings.

Nikolayev and Sevastyanov spent several years at the "City of Stars," the U.S.S.R. cosmonaut training center. The Soyuz 9 flight was Nikolayev's second in outer space. Sevastyanov, a research engineer, has been a lecturer on aerodynamics at the training center.

Hill Space Transportation Award. Christopher C. Kraft, Jr., deputy director of the National Aeronautics and Space Administration's (NASA's) Manned Spacecraft Center near Houston, received the Louis W. Hill Space Transportation Award and $5,000 honorarium in October, 1970. He was cited for "his outstanding management and leadership in directing and planning and operational control of all United States manned space flight missions from the first Mercury suborbital mission through the first Apollo lunar landing."

Kraft received a B.S. degree in aeronautical engineering from the Virginia Polytechnic Institute in 1944. He has been associated with NASA and its predecessor—the National Advisory Committee for Aeronautics—since 1945.

Clarence L. Johnson

Cosmonauts Vitali I. Sevastyanov and Andrian G. Nikolayev received the 1970 Guggenheim International Astronautics Award.

Space Sciences

Continued

Astronomy. Major awards in astronomy cited fundamental contributions:

Draper Medal. Subrahmanyan Chandrasekhar of the Laboratory for Astrophysics and Space Research, University of Chicago, was awarded the Henry Draper Medal and $1,000 honorarium by the National Academy of Sciences in April, 1971. He was cited for his major contributions to theoretical astrophysics, and particularly for his studies of the structure, evolution, and dynamics of stars.

This scientist was the first to set the correct upper limit (now called "Chandrasekhar's limit") to the mass of white dwarf stars–stars of enormous density. They are believed to be the end products of stellar evolution in which nuclear reactions convert hydrogen into helium, helium into carbon, and so on, to still heavier elements. In this process, stellar matter finally collapses into a state from which electrons have been squeezed, allowing millions of atoms to fill space originally occupied by one.

Gould Prize. Elizabeth Roemer, professor of astronomy at the department of astronomy and lunar planetary laboratory, University of Arizona, was awarded the Benjamin Apthorp Gould Prize and $5,000 honorarium of the National Academy of Sciences in April, 1971. She was honored for her contributions to cometary astronomy.

Over the past 18 years, Elizabeth Roemer and her associates have been responsible for the initial sighting of 51 periodic comets as they journey from outer space toward and around the sun. This has given astronomers valuable time to study these objects.

Smith Medal. Edward Anders of the Enrico Fermi Institute, University of Chicago, was awarded the J. Lawrence Smith Medal and $2,000 cash prize of the National Academy of Sciences in April, 1971. The chemist was cited for his basic studies on the origin and history of meteorites.

His probing of space objects has resulted in some original hypotheses–including that meteorites have remained essentially unchanged since their origin; and that gold, silver, and iridium found in moon rocks came from meteorites.

General Awards

Science and Man. Two awards for outstanding contributions to science and mankind were:

Nobel Peace Prize. Norman E. Borlaug, director of the International Maize and Wheat Improvement Center of Mexico, won the 1970 Nobel Peace prize and an $80,000 cash award. He was cited "as the prime mover in the 'Green Revolution' [and] for his great contribution toward creating a new world situation with regard to nutrition."

The Iowa-born crop expert received his Ph.D. degree in plant pathology in 1942 from the University of Minnesota. In 1944 The Rockefeller Foundation appointed him to a program to improve varieties of wheat for Mexico. Since then, he has spent most of his time there working on agricultural projects to increase the production of food.

Borlaug is best known for developing high-yield dwarf wheats by crossing Mexican varieties with Japanese dwarf strains. These new wheats eventually included strains from Africa, Australia, South America, and several states in the western United States. The dwarf wheats, which are much shorter than normal wheat and can support a heavier head of grain, revolutionized wheat production in Mexico.

In the early 1960s, Borlaug began to test his new varieties in India. They were quickly adopted, and India and Pakistan have now more than doubled their wheat production.

National Medal of Science. The highest award of the U.S. government for outstanding contributions to scientific and engineering development, the National Medal of Science, was presented to nine of the nation's most distinguished scientists. Each was judged as having profoundly changed the whole field of science or engineering in which he works. The recipients for 1970 included the following scientists:

Richard D. Brauer, professor of mathematics, Harvard University, Cambridge, Mass., "for his work on conjectures of Dickson, Cartan, Maschke, and Artin, his introduction of the Brauer group, and his development of the theory of modular representations."

Robert H. Dicke, Cyrus Fogg Brackett professor of physics, Princeton University, Princeton, N.J., "for fashioning radio and light waves into tools of extraordinary accuracy and for decisive studies of cosmology and of the nature of gravitation."

Barbara McClintock, distinguished service member, Carnegie Institution of Washington, Cold Spring Harbor, N.Y., "for establishing the relations between inherited characters in plants and the detailed shapes of their chromosomes, and for showing that some genes are controlled by other genes within chromosomes."

George E. Mueller, senior vice-president, General Dynamics Corporation, New York City, "for his many individual contributions to the design of the Apollo system, including the planning and interpretation of a large array of advanced experiments necessary to ensure the success of this venture into a new and little-known environment."

Albert B. Sabin, president of the Weizmann Institute of Science, Rehovot, Israel, "for numerous fundamental contributions to the understanding of viruses and viral diseases, culminating in the development of the vaccine which has eliminated poliomyelitis as a major threat to human health."

Allan R. Sandage, staff member, Hale Observatories, Carnegie Institution of Washington, California Institute of Technology, Pasadena, Calif., "for bringing the very limits of the universe within the reach of man's awareness and unraveling the evolution of stars and galaxies—their origins and ages, distances and destinies."

John C. Slater, professor of physics and chemistry, University of Florida, Gainesville, Fla., "for wide-ranging contributions to the basic theory of atoms, molecules, and matter in the solid form."

John A. Wheeler, Joseph Henry professor of physics, Princeton University, Princeton, N.J., "for his basic contributions to our understanding of the nuclei of atoms, exemplified by his theory of nuclear fission, and his own work and stimulus to others on basic questions of gravitational and electromagnetic phenomena."

Saul Winstein (posthumously), former professor of chemistry, University of California, Los Angeles, "in recognition of his many innovative and perceptive contributions to the study of mechanism in organic chemical reactions."

Barbara McClintock

Allan R. Sandage

Major Awards and Prizes

Award winners treated more fully in the first portion of this section are indicated by an asterisk (*)

Adams Award (organic chemistry): Herbert C. Brown

American Cancer Society Award: Sidney Farber

American Chemical Society (ACS) Award in
Analytical Chemistry: George H. Morrison

ACS Biological Chemistry Award: David F. Wilson

ACS Inorganic Chemistry Award: Jack Lewis

ACS Pure Chemistry Award: R. Bruce King

American Heart Association Research Achievement
Award: Robert W. Berliner

American Institute of Physics: U.S. Steel Foundation
Award (science writing): Jeremy Bernstein

American Medical Association (AMA) Distinguished
Service Award: George R. Herrmann

AMA Scientific Achievement Award: Robert B. Woodward

AMA Sheen Award: Maxwell Finland

Arches of Science Award (science writing):
Margaret Mead

Bonner Prize (physics): Maurice Goldhaber

Bowie Medal (geophysics): Inge Lehmann

Bronfman Prizes (public health): Karl Evang
and Alan F. Guttmacher

Bruce Medal (astronomy): Jesse L. Greenstein

Buckley Solid State Physics Prize: Erwin L. Hahn

*Carty Medal (biology): James D. Watson

Cleveland Prize (general): Cornelia P. Channing

Collier Trophy (aerospace): William M. Allen

Day Medal (geology): George J. Wasserburg

Debye Award (physical chemistry): Norman Davidson

*Draper Medal (astrophysics): Subrahmanyan
Chandrasekhar

*Elliot Medal (zoology): Richard D. Alexander

*Fermi Award (physics): Norris E. Bradbury

Fleming Award (geophysics): Walter M. Elsasser

*Founders Medal (engineering): Clarence L. Johnson

*Franklin Medal (physics): Wolfgang K. H. Panofsky

*Gairdner Awards (medicine): Vincent P. Dole,
W. Richard S. Doll, Robert A. Good, Niels K.
Jerne, and R. Bruce Merrifield

Gibbs Brothers Medal (engineering): Henry A. Schade

Goldberger Award (nutrition): John E. Canham

*Gould Prize (astronomy): Elizabeth Roemer

*Guggenheim Award (aerospace): Andrian G.
Nikolayev and Vitali I. Sevastyanov

Haley Astronautics Award: Fred W. Haise, Jr.;
John L. Swigert; and James A. Lovell, Jr.

Heineman Prize (physics): Roger Penrose

High Polymer Physics Prize: John D. Hoffman
and John J. Lauritzen

*Hill Space Transportation Award (astronautics):
Christopher C. Kraft, Jr.

Hofheimer Prize (psychiatry): Joseph J. Schildkraut;
William E. Bunney, Frederick K. Goodwin, and
Dennis L. Murphy

Horwitz Prize (biology): Albert Claude, George E.
Palade, and Keith R. Porter

Ives Medal (optics): Robert E. Hopkins

Jacobi Award (pediatrics): Amos Christie

Kalinga Prize (science writing): Konrad Z. Lorenz

Langmuir Prize (physics): Michael E. Fisher

*Lasker Awards (basic research): Earl W. Sutherland, Jr.;
(clinical research): Robert A. Good

*Lawrence Award (atomic energy): Robert L. Fleischer;
P. Buford Price, Jr.; Robert M. Walker; Thomas B.
Cook, Jr.; and Robert L. Hellens

*Lilly Award (microbiology): David Baltimore

Macelwane Award (geophysics): Carl I. Wunsch

Michelson-Morley Award (physics): Charles H. Townes

*National Medal of Science (general): Richard D. Brauer,
Robert H. Dicke, Barbara McClintock, George E.
Mueller, Albert B. Sabin, Allan R. Sandage, John C.
Slater, John A. Wheeler, and Saul Winstein

*Nobel Prize: chemistry, Luis F. Leloir; peace, Norman
E. Borlaug; physics, Hannes O. G. Alfven and Louis
E. F. Néel; physiology and medicine, Julius
Axelrod, Ulf S. Von Euler, and Bernard Katz

Passano Foundation Award (medical science):
Stephen W. Kuffler

*Penrose Medal (geology): Ralph A. Bagnold

*Perkin Medal (chemistry): James F. Hyde

*Priestley Medal (chemistry): Frederick D. Rossini

Reed Award (aerospace): Ira G. Hedrick

Ricketts Award (medical science): Solomon A. Berson
and Rosalyn S. Yalow

Rockefeller Public Service Award: Robert J. Huebner

Roebling Medal (mineralogy): George W. Brindley

Royal Aeronautical Society Gold Medal: P. A. Hufton;
Silver Medal: M. J. Brennan

Royal Astronomical Society Gold Medal: Frank Press

*Smith Medal (astronomy): Edward Anders

*Stouffer Prize (medicine): Irvine H. Page and
George W. Pickering

*U.S. Steel Foundation Award (molecular biology):
Masayasu Nomura

*Vetlesen Prize (geophysics): S. Keith Runcorn;
Allan V. Cox and Richard R. Doell

Warner Astronomy Prize: John N. Bahcall

Deaths of Notable Scientists

Deaths of notable scientists, from June 1, 1970, to June 1, 1971, included those listed below. An asterisk (*) indicates that the person has a biography in *The World Book Encyclopedia*.

Alikhanov, Abram I. (1904-Dec. 9, 1970), Russian nuclear physicist, was director of the Institute of Theoretical and Experimental Physics, Moscow, from 1945 to 1965, and one of the designers of Russia's first atomic bomb.

Barabashov, Nikolai P. (1894-April 29, 1971), Russian astrophysicist and director of the Kharkov observatory since 1930. He conducted studies of planets and mapped the far side of the moon.

Beberman, Max (1925-Jan. 24, 1971), mathematician, was called the "father of the new mathematics." The science makes use of guided discovery to help pupils learn basic mathematics.

Boss, Benjamin (1880-Oct. 17, 1970), astronomer, director of Dudley Observatory, Albany, N.Y., from 1912 to 1956. In 1937 he published *General Catalogue of 33,342 Stars*, which he had worked on for more than 30 years.

Carnap, Rudolf P. (1891-Sept. 14, 1970), German-born philosopher, founded the school of logical empiricism, or logical positivism, a branch of modern philosophy that accepts logical and mathematical propositions only if they can be verified by experiment or observation. Carnap's major work was *Philosophy and Logical Syntax* (1935).

Casman, Ezra P. (1904-Oct. 16, 1970), bacteriologist, received the distinguished superior service award of the U.S. Food and Drug Administration in 1962 for isolating organisms that cause food poisoning.

Chaney, Ralph W. (1890-March 3, 1971), geologist and paleobotanist, discovered the dawn redwood in China and arranged to have it planted in the United States and elsewhere to ensure its survival.

Chapman, Sydney (1888-June 16, 1970), English-born geophysicist and mathematician, led scientists from 66 countries who supervised the International Geophysical Year of 1957-1958.

Chisholm, G. Brock (1896-Feb. 2, 1971), Canadian psychiatrist, founding director-general of the World Health Organization, from 1948 to 1953.

***DeKruif, Paul** (1890-Feb. 28, 1971), bacteriologist and science writer who reached a large audience with such best-selling books as *Microbe Hunters* (1926) and *Men Against Death* (1932). DeKruif also wrote many popular articles on science.

Evans, Herbert M. (1882-March 6, 1971), anatomist and embryologist who discovered vitamin E in 1922. Evans was head of the Institute of Experimental Biology at the University of California, Berkeley, from 1930 to 1952, specializing in the growth and reproduction of animals.

Fabing, Howard D. (1907-July 29, 1970), doctor, was president of the American Academy of Neurology from 1953 to 1955. He was one of the first to study the effects of hallucinogenic drugs, such as LSD.

***Fairchild, Sherman M.** (1896-March 28, 1971), inventor and businessman, pioneered in aerial mapping photography. Fairchild developed an aerial camera for the U.S. War Department in World War I, and later made many refinements in aerial photography. He also designed special hydraulic landing gear for aircraft, and a radio compass.

***Farnsworth, Philo T.** (1906-March 11, 1971), television pioneer who, as a 16-year-old high school student in 1922 devised basic principles of electronic television in use today. He held more than 165 patents.

Farrar, Clarence B. (1874-June 3, 1970), Canadian-born psychiatrist and editor of *The American Journal of Psychiatry* from 1931 to 1965.

Ferree, Gertrude Rand (1886-June 30, 1970), psychologist and researcher on human vision. With her husband Clarence E. Ferree, she obtained patents for lighting devices, and optical and ophthalmological instruments. She also helped develop the shaded pictures of numbers to detect color blindness.

Gerasimov, Mikhail M. (1907-July 22, 1970), Russian anthropologist and archaeologist who scientifically reconstructed the faces of prehistoric men on the basis of skull formation.

Gershon-Cohen, Jacob (1899-Feb. 8, 1971), doctor and radiologist, devised a way to detect breast cancer in the early stages.

Gey, George O. (1899-Nov. 8, 1970), cell biologist and cancer researcher, was director of the Finney-Howell Cancer

Max Beberman

Paul DeKruif

Philo T. Farnsworth

Deaths of Notable Scientists

Continued

Sir Chandrasekhara
Venkata Raman

Theodor H. E. Svedberg

Igor Y. Tamm

Research Laboratory at Johns Hopkins medical school and hospital, Baltimore.

Glueck, Nelson (1900-Feb. 12, 1971), an archaeologist, used the Old Testament to identify hundreds of ancient Biblical sites, including King Solomon's mines at Khirbet Nahas, Palestine, in 1934.

Gregory, William K. (1876-Dec. 29, 1970), paleontologist and morphologist, American Museum of Natural History and Columbia University, New York City. Gregory studied the development of teeth in fish and mammals.

Iselin, Columbus O'Donnell (1904-Jan. 6, 1971), oceanographer, was director of the Woods Hole Oceanographic Institution, Massachusetts, from 1940 to 1950. He charted the Gulf Stream in 1936 and received the Medal of Merit from President Harry S. Truman in 1948 for oceanographic research.

Lewis, Oscar (1914-Dec. 16, 1970), anthropologist who studied family life among the poor. He wrote *The Children of Sanchez* (1961), dealing with Mexico, and *La Vida: A Puerto Rican Family in the Culture of Poverty—San Juan and New York* (1966).

Long, C. N. Hugh (1901-July 6, 1970), English-born physiologist and biochemist who pioneered in the study of adrenal cortical hormones and how they affect health.

Maslow, Abraham H. (1908-June 8, 1970), psychologist on the staff of Brandeis University, founder of the study of humanistic psychology.

Mikoyan, Artem I. (1905-Dec. 9, 1970), Russian designer who along with mathematician Mikhail I. Gurevich created Russia's MIG jet airplane, widely used during hostilities in Korea in the 1950s and elsewhere.

Powdermaker, Hortense (1900-June 15, 1970), anthropologist who wrote about differing patterns of life. Her observations dealt with life in such places as the southwest Pacific islands, Rhodesia, Mississippi, and Hollywood.

Radot, Louis Pasteur-Vallery (1886-Oct. 9, 1970), a French doctor, pioneered in anaphylaxis—the increased sensitivity to medicines—and also in allergies and kidney ailments.

*Raman, Sir Chandrasekhara Venkata (1888-Nov. 21, 1970), Indian physicist, won the Nobel prize in physics in 1930 for his discovery that light pass-

ing through a liquid, gas, or crystal is scattered and its wave length and color changed. This transformation came to be called the Raman effect.

Sampson, Milo B. (1909-April 10, 1971), physicist at Indiana University since 1946 was an authority on the cyclotron or atom smasher.

Schwarzkopf, Paul (1886-Dec. 27, 1970), Austrian-born industrialist, was known as the "father of powder metallurgy." He produced the first drawn tungsten wire in 1911, later created tungsten carbide for use in drill bits.

Shemyakin, Mikhail M. (1908-June 27, 1970), Russian organic chemist, had been director of the Soviet Academy of Sciences' Institute for the Chemistry of Natural Compounds since 1960. Shemyakin developed ways to synthesize various kinds of organic compounds.

*Svedberg, Theodor H. E. (1884-Feb. 26, 1971), Swedish chemist, won the 1926 Nobel prize in chemistry for his work in colloidal chemistry. Svedberg developed the ultracentrifuge used to study molecular weights of proteins.

*Tamm, Igor Y. (1895-April 12, 1971), a Russian physicist, shared the 1958 Nobel prize for physics. Tamm and fellow prizewinner Ilya M. Frank explained how radiation, or blue light, originates. The phenomenon was named for Pavel A. Cherenkov, who also shared in the prize.

Van Slyke, Donald D. (1883-May 4, 1971), biological chemist, had been associated with Rockefeller University from 1907 to 1948. He gained prominence for applying chemistry to medicine.

Warburg, Otto H. (1883-Aug. 1, 1970), German biochemist, won the Nobel prize in physiology and medicine in 1931 for his discoveries in connection with the respiratory enzyme. Warburg had headed the Max Planck Institute for Cell Physiology in Berlin since 1931.

Wood, W. Barry, Jr. (1910-March 9, 1971), doctor and bacteriologist, pioneered in reporting on the effects of penicillin in 1943. Wood was also the first to describe how white blood cells combat bacteria, and helped isolate pyrogen, a substance that produces fever.

Zenkevich, Lev A. (1889-June 20, 1970), Russian oceanographer and professor at Moscow University since 1930, was the author of well-known studies on ocean fauna.   [CHARLES LEKBERG]

Index

This easy-to-use index covers all sections of the 1969, 1971, and 1972 editions of *Science Year,* The World Book Science Annual.

An index entry which is the title of an article appearing in *Science Year* is printed in boldface italic letters as: ***Archaeology.*** An entry which is not an article title, but a subject discussed in an article of some other title is printed: **Amino acids.**

The various "See" and "See also" cross references in the index list are to other entries within the index. Clue words or phrases are used when the entry needs further definition and when two or more references to the same subject appear in *Science Year*. These make it easy to locate the material on the page, since they refer to an article title or article subsection in which the reference appears, as:

    Heart surgery: coronary artery by-pass, 72-178, *71*-325; endarterectomy, *72*-185, 330, *71*-324; infarctectomy, emergency, *69*-324; revascularization, *72*-182, 330; saphenous vein by-pass graft, *72*-188, 330. See also **Transplantation**

The indication "*il.*" means that the reference is to an illustration only.

Index

A

Index

Index

Index

Index

Index

Index

Index

Index

Index

Index

Index

Index

core studies, *71*-346; Coriolis force, *71*-345; fission, *69*-340; heavy ion accelerators, need for, *69*-342; heavy nuclei, structure of, *71*-345; isotope studies, *71*-346; nuclei, new configurations, *72*-351; pulsar (Crab Nebula), unknown bulk properties, *69*-342; reactors, fast-breeder, *72*-352; shell models, *69*-340; spectroscopy, nuclear, *69*-340; supertransuranic element (at. no. 112), *72*-351; tandem Van de Graaff accelerator, *71*-347; transuranic elements, *72*-351; triton, reactions involving, *69*-340; vibration states, pairing, *69*-340.

Physics, Plasma: 72-353, *71*-348, *69*-343; Bohm diffusion, *69*-343; dynamic stabilization, *71*-348; feedback stabilization, *71*-348; fusion, progress in controlling, *71*-348, *69*-343; loffe's magnetic well principle, *69*-344; lasers, plasma ions and, *71*-349; loss-cone instability, *69*-345; M-S stabilization scheme, *71*-348; octupole, *71*-348; open systems, successful use of *69*-344; plasma containment, *71*-348, *69*-344; reactors, thermonuclear, *71*-349; shear, Bohm diffusion reduction, *69*-344; apherator, *71*-348; tokamak, *71*-340; Tokamak 3 and 10, *71*-348, *69*-343.

Physics, Solid State: 72-355, *71*-349, *69*-345; amorphous semiconductors, *69*-346, 347; atoms, pictures of, *71*-351; electrical resistance, *69*-345; electrocardiograms, *71*-351; ferromagnetism, *69*-348; Kondo effect, *69*-361; magnetic fields, *71*-351; magnetometer, superconducting, *71*-351; metal-nonmetal transition, *69*-346; organometallic superconductors, *71*-350; semiconductors, *69*-345; superconductivity, *71*-349

Piaget, Jean, *72*-357
Pickering, Sir George W., *72*-419, 422
Pickering, William H., *69*-416, 418
Picosecond pulses, *72*-288
Picturephone, *72*-293, *71*-288
Pierantoni, Ruggero, *71*-375
Pimples, *72*-158
Pineal gland, *71*-376
Pitman, Walter, *72*-323
Plagioclase, lunar soil, *72*-42
Plancus, Janus, *71*-112
Plankton, *72*-112
Plants. See *Botany*
Plasma. See *Physics, Plasma*
Plato, *69*-145
Platt, Benjamin S., *71*-200
Platt, Joseph B., *69*-27
Plimmer, J. R., *71*-276
Plowshare project, *71*-307, *69*-300
Pluto, *71*-51, *69*-259
Plutonium 239, *72*-219
Pochi, Peter E., *72*-158, 172
Pohm, Arthur V., *69*-298
Poincaré, Henri, *71*-32
Polyester fibers, *71*-280
Polyps: *Calliactis, 72*-376; limu-make-o-Hana, *72*-376
Pomeranchuk, Isaak Y., *71*-343
Polyinosinic-polycytidylic acid, *69*-96
Popp, Richard L., *71*-322
Population: *Special Report, 71*-80, *69*-125, 181, 291, 296
Population Council, *71*-90
Porpoise, *69*-54
Porter, Wayne P., *71*-295
Post, Richard F., *71*-349
Potassium-argon technique, *72*-257

Potato, frost-resistant, *71*-254
Potter, L. T., *72*-274
Pottery, neutron-activation analysis, *71*-256
Potts, John T., *71*-271
Poverty, *69*-354
Powdermaker, Hortense, *72*-424
Powell, Cecil F., *71*-422
Powell, Wilson M., *72*-348
Pratt, Perry W., *69*-416, 418
Precambrian rocks, *72*-320
Prednisolone, *69*-85
Premack, David, *72*-151, *71*-66
Premoli-Silva, Isabella, *72*-123
Price, Buford, Jr., *72*-416
Priestley Medal (chemistry), *72*-414, 422, *71*-413, 420, *69*-412, 418
Primate behavior: *Special Report, 71*-64
Prince, Alfred, *71*-320
Priori, Elizabeth S., *72*-326
Proctor, Charles, *72*-85
Proctor, Richard J., *72*-319
Promethium, stellar radioactivity, *72*-265
Prostaglandins, *69*-289
Proton beam, *72*-208
Protons (high-energy), head-on collisions, *72*-347
Psychiatry, orthomolecular, *69*-382
Psychobiology, molecular, *72*-137
Psychology: 72-357, *71*-352, *69*-348; antibiotics, as amnesic agents, *71*-354; arm movements, infant, *72*-359; autonomic nervous system, conditioned response and, *69*-348; babbles and coos, *72*-359; *Books of Science, 72*-280, *69*-272; child development theories and research, *72*-357; electroconvulsive shock (ECS), as amnesic agent, *72*-352; head-turning, infant, *72*-358; hypothalamus, stimulation and behavior, *69*-349; learning capacities underestimated, *72*-359; memory, brain mechanisms, *71*-352; sucking response, *72*-359; visual perception, infant, *72*-357. See also **Brain**
Ptashne, Mark, *69*-78
Public Health: 71-354, *69*-351; birth control, *71*-356, *69*-353; blood banks, *69*-353; *Books of Science, 69*-271; caloric intake, recommended, *69*-351; cigarette smoking, lung cancer and, *69*-351; contraceptives, oral, *71*-356, *69*-353; food additives, *71*-356; general practitioners, specialty licensing board, *69*-353; gonorrhea, *69*-353; Medicaid, *71*-354, *69*-353; Medicare, *71*-354, *69*-353; mental health, *71*-355; National Institute of Health (NIH), *71*-355; Neighborhood Health Center Plan, *71*-354; neighborhood health centers, *71*-354; poverty, *69*-354; regional medical programs, *69*-353; Secretary of Health, Education, and Welfare, *69*-354; vaccines, *71*-356. See also **Cancer; Medicine**
Puck, Theodore T., *72*-196
Pulsars: *Special Report, 69*-37; Arecibo Ionospheric Observatory, *il., 69*-39, 43; cosmic rays, source of, *69*-50; Crab Nebula pulsar, *69*-46, *il.,* 47; definition, *69*-37; discovery of, *69*-37; light pulses, *69*-49; neutron star, *71*-263, *il., 69*-42, 47; pulsation intervals, *69*-38, *ils.,* 40, 43; pulse shape, *69*-45; radio emissions theories, *69*-47, 49; size, relation to pulses' duration, *69*-41; source of, *69*-41, 42; subpulses, *il., 69*-41, 46; X rays, *69*-49; white dwarf stars, *69*-50. See also *Astronomy, High Energy*
Purcell, John R., *72*-65
Purchase, H. Graham, *72*-255

Purdy, J. M., *72*-379
Puromycin, protein synthesis inhibition, *72*-143
Purser, Douglas B., *71*-252
Pye, J. D., *72*-380

Q

Quantasomes, leaf-cell particles, *69*-273, *il.,* 273
Quantum electrodynamics (QED), *69*-336
Quantum mechanics, *71*-209, *69*-158, 384
Quarks, *72*-348, *71*-345
Quasars, *72*-268, *71*-42, *69*-265
Quinacrine hydrochloride: chromosome-stain technique, *72*-315, *71*-309

R

Rademacher, Hans Adolph, *69*-420
Radiation: hazards of, *71*-305; Raman scattered, *72*-347; vitamin D-producing, *72*-256. See also **Molecules, interstellar**
Radiation Control for Health and Safety Act, *71*-303
Radiocarbon dating, *72*-98, 228, *71*-257
Radio interferometers, *72*-77
Radot, Louis Pasteur-Vallery, *72*-424
Rainfall, measurement of, *72*-88
Raleigh, Barry, *72*-323
Ralph, Elizabeth, *69*-145
Raman scattered radiation, *72*-347
Raman, Sir Chandrasekhara Venkata, *72*-424
Ramsay, Sir William, *69*-167
Randal, Judith, *72*-382, *71*-195, *69*-100
Ranger Project, *72*-40, 406
Read, Dwight W., *71*-256
Read, Kenneth R. H., *71*-115
Recycling, *71*-145
Red Paint Indians, *72*-261
Red-shift distance relation, *71*-268, *69*-265, 407
Reed Award (aeronautics), *72*-422, *69*-418
Reed, Michael, *71*-11
Rees, Martin J., *72*-64
Regim-8 (2, 3, 5-triiodobenzoic acid), *69*-250
Reid, Brian K., *72*-65
Reisner, Ronald M., *72*-172
Relativity theory, *72*-55, *69*-263
Renfrew, Colin, *72*-323
Rentzepis, Peter M., *72*-288
Research Corporation Award, *69*-418
Retinoblastoma, *71*-100
Retinoic acid, *72*-172
Revascularization, *72*-182
Reveal, James L., *72*-281
Revelle, Roger, *71*-80, 405
Rhizotron, *71*-253, *ils.,* 253
Rho factor, mRNA synthesis termination, *72*-338
Rice: IR-5 and IR-8 strains, *69*-120; IR-20, *71*-254
Rice, Stuart Arthur, *71*-422
Rich, Saul, *72*-282
Richardson, Stephen, *71*-310
Rickets, in Neanderthal Man, *72*-256
Ricketts Award (medical science), *72*-422, *69*-418
Riedel, William, *72*-123
Riedl, Rupert J., *71*-377
Rifampin (rifampicin), antituberculosis remedy, *72*-299
Riley, Gordon, *71*-401
Ringwood, Alfred E., *71*-310

438

Index

Index

Index

2, 4, 5-T, weed and brush killer, 71-253

U

Uhuru, astronomy satellite, 72-269
Ultrasound, il., 69-109, 110
Undersea cables, 72-293, 71-288
Unemployment, scientists and
  engineers, 72-350, 364, 71-359
Ungar, Georges, 72-148, 247
Unger, William, 72-12
Unidentified flying objects, 69-366
United Nations Food and Agriculture
  Organization, 72-119
United States Department of
  Defense (DOD), 71-357, 358, 69-359
United States Department of
  Health, Education, and Welfare
  (HEW), 71-170, 69-354, 358
United States Department of
  Justice, 72-299
United States House of
  Representatives, Subcommittee on
  Science, Research, and
  Development, 71-357
United States Office of Economic
  Opportunity, 71-359
United States Steel Foundation
  Award, 72-418, 422, 71-415, 69-413
Universities Research Association,
  72-205
University Group Diabetes Program
  (UGDP), 72-298
Upper Mantle Project (UMP),
  71-315, 69-314
Upwelling, artificial, 72-343
Uranium-235, 72-214, 69-184
Uranus, 72-263, 71-51
Urbanization, population growth and,
  71-86
Urea, treatment of sickle cell anemia,
  72-329
Urey, Harold C., 71-271, 69-239, 383
Urschel, Harold C., 72-189, 71-325
Uvarov, Sir Boris P., 71-422

V

Vaccine. See Viral vaccines
Van Allen, James A., 72-370
Vandermeer, John H., 71-296
Van Dyne, George M., 72-82, 69-293
Van Slyke, Donald D., 72-424
Van Tamelen, Eugene E., 71-285
Venketswaren, Subramaniaan,
  72-255
Venus, 72-264, 368, 71-261, 69-257,
  368
Vesalius, Andreas, 72-11
Vetlesen Prize, 72-417, 422
Video picture storage, 72-295
Vietnamese war: chemical and
  biological warfare (CBW), 72-247,
  69-359; impact on American science,
  71-357
Viets, Frank G., Jr., 71-252
Vigorito, James, 72-359
Villegas, Evangelina, 72-255
Vinci, Leonardo da, 72-11
Vincristine, 69-85
Vine, Allyn, 71-402
Vine, Frederick J., 71-316
Vineberg, Arthur M., 72-184
Viral vaccines: chicken (cancer),
  72-255; German measles (rubella),
  71-356, 69-319; influenza, 71-356,
  69-352
Virchow, Rudolf L. K., 72-256
Viruses: breast cancer, 72-326; C-type,

sarcoma, 72-326, 71-319; Epstein-Barr
  (EB), 71-319, 69-317; herpes simplex,
  69-318; A₂-Hong Kong-68, 71-356,
  69-352; inactivated Sendai virus (ISV),
  72-192; kuru, 69-252; lambda H80,
  72-339; Lassa fever, 71-321; Phi X 174,
  69-69; Q-beta, 71-271; R 17, 71-271;
  rubella, 71-356; SV-40, 72-201, 69-329;
  tobacco mosaic, 69-198. See also
  Cancer; Leukemia; *Microbiology*
Visceral responses, 71-183
Vital statistics: heart attacks, 72-182
Vitamin A acid, 72-172
Vitamin C, 72-382, 387, 389
Vitamin D: 69-268; Neanderthal
  Man's skeleton, 72-256
Vitamins, mental illness and, 69-382
Volcanoes, 72-318, 71-312
Volkov, Vladislav N., 72-365
Von Euler, Ulf S., 72-419, 422
Voroshilov, Kliment, 72-391
Vosa, C. G., 71-309

W

Wald, George, 69-361
Walker, Robert M., 72-416, 422
Walkinshaw, Charles H., 72-255
Warburg, Otto H., 72-424
Warner Astronomy Prize, 72-422,
  69-418
Washburn, Sherwood L., 71-69
Wasps, parasitic, 71-109
Wasserburg, Gerald, 72-38, 71-311
Waste recycling, community, 72-308
Watanabe, Tsutomu, 71-334
Water desalination, 69-188
Water pollution, 72-119, 285, 302,
  306, 344, 71-125, 126, 138, 146, 304,
  69-294, 306, 361
Water Quality Improvement Act of
  1970, 71-304
Water reclamation, 72-286
Water testing: *Special Report*, 71-146;
  bacteriology, 71-148, 154; dissolved
  oxygen, 71-147, 153; organic sludge,
  71-148, 152; plankton tow, 71-148,
  150; species diversity, 71-147, 151;
  turbidity, 71-148, 149
Watkins, J. F., 72-194
Watson James D., 72-362, 417,
  69-68
Watson Medal (astronomy), 69-417,
  418
Watson, John B., 72-150
Wattonberg, Hermann, 69-204
Weather. See *Meteorology*
Weber, Joseph, 71-39, 342, 69-263
Weeks, Kent, 72-260
Wegener, Alfred A., 71-314
Weinberg, Alvin M., 72-214, 69-184
Wells, Herbert, 72-258
Westall, Fred C., 72-274
Westheimer, Frank, 71-166
Westphal, James, 71-59
Wheat: agriculture, Shabarti Sonora,
  71-254
Wheeler, John A., 72-421,
  69-412, 418
White House Conference on Food,
  Nutrition, and Health, 71-205
Whiteheads, 72-158
Whooping cranes, 71-377
Wiesel, Thorsten, 72-138
Wiesner, Jerome B., 71-358, 69-359
Wigner, Eugene, 71-209, 69-185,
  204, 359
Wilderness Society, 72-308
Wilkes, Charles, 71-30
Willens, H., 71-349
Williams, Carroll M., 72-245
Willows, Dennis, 72-142
Wilshire, Howard G., 69-306

Wilson, Allan C., 71-256
Wilson, Kent R., 71-283
Wilson, Robert Rathbun, 72-211
Wilson, Robert W., 72-73, 71-264
Winick, Myron, 71-198
Winstein, Saul, 72-421
Wire plane, particle detection, 71-344
Witten, Victor H., 72-160
Witter, Richard L., 72-255
Wittig, Jörge, 71-351
Wolfgang, Richard L., 69-162
Wollman, Elie, 71-393
Wood, John A., 72-36
Wood, W. Barry, Jr., 72-424
Woods Hole (Mass.)
  Oceanographic Institution, 72-122
Worden, Alfred M., 72-367
Works Progress Administration
  (WPA), 71-399
Worley, Joseph F., 71-276
Worthington, Lawrence (Val),
  71-402
Wren, Sir Christopher, 72-12
Wright, D. J. M., 72-256
Wu, Chien-Shiung, 71-210

X

Xenon hexafluoride, structural
  considerations, 72-290
X-Ogen, radio signal of, 72-73
X rays: acne treatment, 72-160;
  angiography, 72-178; astronomy,
  high energy, 72-269; cancer etiology,
  69-85; cosmic, 72-269, 69-264; Crab
  Nebula pulsar, 72-270, 71-264, 69-49;
  diffraction methods, 72-289; galactic
  emissions, 72-270, 71-41, 46; insulin
  examination, 71-283; stellar
  emissions, 71-262

Y

Yamashiro, Donald, 72-277
Yellowtail fish, 72-110
Yoshikawa, Shoichi, 71-348
Young, John W., 71-362
Yukawa, Hideki, 71-221

Z

Zenkevich, Lev A., 72-424
Zimmerman, James, 71-351
Zinc, superplastic, 71-280
Zinjanthropus specimen, 71-254
Zond (probe) series, 72-370, 71-370
Zones of upwelling, marine life
  concentrations, 72-118
Zoology: 72-376, 71-374, 69-373;
  anticancer ingredient, kaunaoa worm,
  72-377; bats, fog avoidance, 72-380;
  Books of Science, 72-280, 71-275,
  69-272; fertilization, test-tube, 72-379,
  69-373; Gnathostomulida, new
  phylum, 71-377; head transplant, in
  axolotl, 69-374; hermit crabs, and polyp
  hitchhikers, 72-377; hibernation
  studies, squirrels, 69-375;
  immunological rejection mechanism,
  69-375; intelligence, brain weight and,
  69-376; light, role in animal cycles,
  71-375; limu-make-o-Hana, 72-376;
  Loch Ness monster, 71-378; Mariner 6,
  71-374; memory experiments, 71-374;
  mouse plague, Australian, 71-326;
  palytoxin, 72-376; pineal gland, 71-376;
  regeneration, 72-377, 69-375; Siamese
  twin fetuses, 69-373, il., 373; smell,
  identification of mother, 72-379; tiger,
  white, 71-378; whooping cranes,
  71-377. See also Brain; Dolphin;
  Primate behavior
Zooplankton, 72-303

Acknowledgments

The publishers of *Science Year* gratefully acknowledge the courtesy of the following artists, photographers, publishers, institutions, agencies, and corporations for the illustrations in this volume. Credits should be read from left to right, top to bottom, on their respective pages. All entries marked with an asterisk (*) denote illustrations created exclusively for *Science Year.* All maps were created by the *World Book* Cartographic Staff.

Cover

Frank Armitage

Essays

10   Royal Library, Windsor Castle, London. Reproduced by Gracious Permission of Her Majesty Queen Elizabeth II
12   Civic Museum, Piacenza, Italy, photo by B. Pellegrini
13   Bibliothèque Nationale, Paris, (*Science Year* photo by Hubert Josse*); Bodleian Library, Oxford, Eng.
14   Royal Library, Windsor Castle, London. Reproduced by Gracious Permission of Her Majesty Queen Elizabeth II
15   Ars Medica Collection, Philadelphia Museum of Art. Purchased: The Smith Kline and French Laboratories Collection
16   From *"Sir Charles Bell's Manuscript of Drawings of the Arteries,"* courtesy National Library of Medicine, Bethesda, Maryland Historical Library, Pennsylvania Hospital, Philadelphia. (*Science Year* photo by Charles Mills*)
17   Ars Medica Collection, Philadelphia Museum of Art: Given by Dr. S. B. Sturgis; A. J. Nystrom & Co.; From *"Handatlas Der Anatomie Des Menschen,"* published 1920 by S. Hirzel, Leipzig, Germany
18-19   Property of the Department of Art as Applied to Medicine, The Johns Hopkins Medical School, Baltimore, Md. (Walters Collection #16, #699, #15, and #974)
20   Property of the Department of Art as Applied to Medicine, The Johns Hopkins Medical School, Baltimore, Md.; Georgetown University, photo courtesy of Abbott Laboratories
21   Property of the Wilmer Ophthalmological Institute, The Johns Hopkins Medical School, Baltimore, Md.; Property of the Department of Art as Applied to Medicine, The Johns Hopkins Medical School (Walter E. Dandy Collection); Georgetown University, photo courtesy of Appleton-Century-Crofts; Georgetown University, photo courtesy of Appleton-Century-Crofts
22   Courtesy of Professor Becker, Professor Lepier and Urban & Schwarzenberg; © Paul Peck
23   © 1969 CIBA Pharmaceutical Company, Division of CIBA Corporation. Reproduced with permission. From *The CIBA Collection of Medical Illustrations* by Frank H. Netter, M.D.
24   Biagio John Melloni; © 1970 RASSEGNA MEDICA LEPETIT, Milan, Italy, *Rassegna Medica E. Culturale.* Plates by Giorgio Gondoni, drawings by Elisa Paternagni; Courtesy Allergan Pharmaceuticals
26-27   Yoichi Okamoto*

Special Reports

36-37   NASA; Herb Herrick*
39   NASA; Bendix Corporation
41   NASA

42   Floyd C. Clark, California Institute of Technology
43   NASA
44   Herb Herrick*; NASA
45   Herb Herrick*
46   Herb Herrick*; Robert N. Clayton, University of Chicago
47   Herb Herrick*
48   Herb Herrick*; NASA
49   Herb Herrick*
50   NASA
53   E. F. Hoppe, Alfa Studio*
55-57   Product Illustration, Inc.*
58   Joseph Weber; E. F. Hoppe, Alfa Studio*; E. F. Hoppe, Alfa Studio*; E. F. Hoppe, Alfa Studio*; E. F. Hoppe, Alfa Studio*
60-61   Product Illustration, Inc.*
62   Joel Cole*; Joseph Weber; Joel Cole*; Joel Cole*; Joel Cole*
63   Staff photo
64   Product Illustration, Inc.*
65   Bendix Corporation
66-68   Joel Cole*
69-70   Bob Keyes*
71   Bob Keyes*; © California Institute of Technology and Carnegie Institution of Washington. From Hale Observatories.
72-73   Bob Keyes*
74-75   Joel Cole*
76-77   Bob Keyes*
78   © California Institute of Technology and Carnegie Institution of Washington. From Hale Observatories.
80-90   Natural Resource Ecology Laboratory, Colorado State University
91   Dr. Charles Proctor, University of Georgia
92   Natural Resource Ecology Laboratory, Colorado State University
94-95   NASA
96-97   Product Illustration, Inc.*
98   Russ Kinne, Photo Researchers; Joseph Rychetnik, Van Cleve Photography
99   Deep Sea Drilling Project, Scripps Institution of Oceanography; Allan W. H. Bé, Lamont Geological Observatory
101   Goddard Space Flight Center
102   Michael Friedel, Black Star
103   David Muench, Van Cleve Photography; Ted Spiegel, Rapho Guillumette
104   H. Near, Photo Researchers; Thomas Hollyman, Photo Researchers
106-107   Pine Hill Catfish Farm, Aliceville, Ala.
108   Lauren R. Donaldson
110   Pine Hill Catfish Farm, Aliceville, Ala.; Monsanto Company
111   Monsanto Company
112-113   Stanley Meltzoff*; Product Illustration, Inc.*
114   Dr. B. Andréu
115   John Bardach; Masaru Fujiya
116-117   Product Illustration, Inc.*; Stanley Meltzoff*
118   Product Illustration, Inc.*
120-128   Deep Sea Drilling Project, Scripps Institution of Oceanography

129-130 Product Illustration, Inc.*
131-133 Deep Sea Drilling Project, Scripps Institution of Oceanography
136-137 Joel Cole*
140 Joel Cole*; Reprinted from "A Multi-Fontal Alphabet for Dyslexic Children" by Bernard Sklar and John Hanley, *Journal of Learning Disabilities* (in press)
144 Joel Cole*; Dr. D. N. Spinelli, Stanford University; Joel Cole*
145-153 Joel Cole*
154 Dr. Marion B. Sulzberger
155 Product Illustration, Inc.*
156-157 Dr. C. M. Papa, Johnson & Johnson Research Laboratories
157 Dr. Constantine Orphanos, University of Cologne, Germany
158 Edward F. Glifort*
159 Edward F. Glifort, University of Pennsylvania School of Medicine
160 Edward F. Glifort*
162 Edward F. Glifort, University of Pennsylvania School of Medicine
163-171 Lou Bory and Associates*
173 Department of Dermatology, Wayne State University School of Medicine, Detroit; Department of Dermatology, Wayne State University School of Medicine, Detroit; Dr. Marion B. Sulzberger; Edward F. Glifort, University of Pennsylvania School of Medicine; Department of Dermatology, Wayne State University School of Medicine, Detroit
175 From *A Manual of Dermatology* by Donald M. Pillsbury. Published by W. B. Saunders Company, 1971
176 E. F. Hoppe, Alfa Studio*
178 Product Illustration, Inc.*
179 Joel Cole*; Joel Cole*; Dept. of Cardiovascular Thoracic Surgery, Rush-Presbyterian- St. Luke's Medical Center, Chicago
180-188 E. F. Hoppe, Alfa Studio*
189 Cleveland Clinic
190-191 Frank Armitage*
192 Robert Issacs
194-195 Frank Armitage*
197 Henry Harris, University of Oxford, Oxford, England
198-199 Frank Armitage*
200 Henry Harris, University of Oxford, Oxford, England
201-202 Product Illustration, Inc.*
204-205 Tony Frelo, National Accelerator Laboratory
206 Tim Fielding, National Accelerator Laboratory
207-209 Tony Frelo, National Accelerator Laboratory; Joel Cole*
210 Joel Cole*
212-217 Ken Dallison*
219 Product Illustration, Inc.*
220-221 Ken Dallison*
223 Product Illustration, Inc.*
226 Sonja Bullaty and Angelo Lomeo, Rapho Guillumette
230 Product Illustration, Inc.*
231 Product Illustration, Inc.*; C. W. Ferguson
233 Anne F. Hubbard, Photo Researchers; © Peter Clayton; Stephen G. Armytage, Photo Researchers
234-235 Sonja Bullaty and Angelo Lomeo, Rapho Guillumette; Anthony Saris*
236-237 Jim Hubbard, Photo Researchers; Anthony Saris*
238 British Broadcasting Corporation; Peter Bartlett
241 Fred Stein, *Chicago Daily News*; WBBM-TV; Robert Kotalik, *Chicago Daily News*
243-244 WBBM-TV
246-247 Fred Stein, *Chicago Daily News*
250 Don Burk, *Chicago Sun-Times*; M. Leon Lopez, *Chicago Daily News*
251 M. Leon Lopez, *Chicago Daily News*

Science File

254 Funk Brothers, Bloomington, Illinois
257 Drawing by James Stevenson, © 1971 The New Yorker Magazine, Inc.
258 United Press Int.
259 Athens Archaeological Museum, Greece
260 © 1971 Alexander Marshack
261 University of Michigan; Dr. Charles K. Davenport, San Bernardino County Museum Association
262 Smithsonian Institution
263 NASA
264 United Press Int.
266 Hale Observatories
268 Australian National University
269 American Science & Engineering, Inc.
270 G. Swarup, Tata Institute of Fundamental Research, Bombay, India
272 Product Illustration, Inc.*
276 Drawing by Chas. Addams, © 1971 The New Yorker Magazine, Inc.
277 Product Illustration, Inc.*
281 Pictorial Parade; *The New York Times*
282 Dr. D. H. Northcote, Cambridge University
283 Drawing by Booth, © 1970 The New Yorker Magazine, Inc.
287 National Bureau of Standards
289-291 Product Illustration, Inc.*
293-294 Bell Telephone Laboratories
296 Product Illustration, Inc.*
299 Governor Medical Center, Providence, R. I.
300 Neil F. Hadley, Arizona State University
301 Drawing by Lorenz, © 1970 The New Yorker Magazine, Inc.; Drawing by Shirvanian, © 1970 The New Yorker Magazine, Inc.; Drawing by Richter, © 1971 The New Yorker Magazine, Inc.; © *Punch*, London; © *Punch*, London
304 Wide World
307 *Philadelphia Bulletin*
309 Bell Telephone Laboratories
310 Donald Reilly, *Look*
312 NASA
314 M. L. O'Riordan, V. H. Robinson, K. E. Buckton, H. J. Evans, Western General Hospital, Edinburgh, Scotland
317 NASA
319 National Earthquake Information Center, NOAA
322 © 1970 by the New York Times Co. Reprinted by permission.
323 Ed Fisher, © 1970 Saturday Review, Inc.
324 © 1970 by The New York Times Co. Reprinted by permission.
325 Warner-Lambert Research Institute
326 Massachusetts Institute of Technology
327 Dr. Marion I. Barnhart, Wayne State University School of Medicine, Detroit
328 Dr. Gene Saccomanno, St. Mary's Hospital, Grand Junction, Colo.
329 Dr. Reuben Hoppenstein
330 Ted Russell
331 General Motors Research Laboratories
332 Lederle Laboratories
333 Don Getsug
334-336 National Oceanic and Atmospheric Administration
337 Stuart M. Ridley, University of York, Helsington, York, England
338 Drawing by W. Miller, © 1971 The New Yorker Magazine, Inc.
339 Dr. Lee D. Simon, Institute for Cancer Research
341 Re-Entry & Environmental Systems Division, General Electric Company
342 National Oceanic and Atmospheric Administration
343 Product Illustration, Inc.*

Typography

Display—Univers
Hayes-Lochner, Inc., Chicago
LST Typography, Chicago
Monsen Typographers, Inc., Chicago
Text—Baskerville monotype (modified)
Hayes-Lochner, Inc., Chicago
LST Typography, Chicago
Monsen Typographers, Inc., Chicago

Offset Positives

Colorcraft Litho, Chicago
Process Color Plate Company, Chicago
Schawk Graphics Inc., Chicago

Printing

R. R. Donnelley & Sons Company,
Crawfordsville, Ind.

Binding

R. R. Donnelley & Sons Company,
Crawfordsville, Ind.

Paper

Text
White Field Web Offset (basis 60 pound)
Westvaco, Luke, Md.

Cover Material

White Offset Blubak
Holliston Mills, Inc., Kingsport, Tenn.